ATLANTIC AND INDIAN OCEANS: NEW OCEANOGRAPHIC RESEARCH

ATLANTIC AND INDIAN OCEANS: NEW OCEANOGRAPHIC RESEARCH

EDWARD S. ASKEW
AND
JAMES P. BROMLEY
EDITORS

Nova Science Publishers, Inc.
New York

Library of Congress Cataloging-in-Publication Data
Available upon request.

ISBN 978-1-60692-475-4

Published by Nova Science Publishers, Inc. ✢New York

CONTENTS

PREFACE

This book describes in detail two kinds of models that simulate the dispersion of radionuclides in the sea. Biases have a strong impact on the predictable skill and the structures of the model predicted patterns in the tropical Atlantic Ocean. These biases and the related effects are presented as well as the atmospheric teleconnections between the tropical Atlantic and the African continent. Special focus is given to recent developments in the nitrogen cycle in the open ocean and the new challenges that have arisen from these advancements. This book also provides a review of the history, geography, and climate of the Atlantic and Indian Oceans, as well as the role they play in natural disasters. A review of the tsunami hazard along the coastal area of the Gulf of Bengal is explored and the seismicity of the Indian Ocean area. The Indian Ocean Dipole (IOD) and its affects, such as monsoon strength over the Indian subcontinent, climatic extremes and socio-economic hardship throughout the tropical Indian Ocean region are presented. This book considers a selection of statistical time series models to analyze and predict the IOD. Finally, using a data-assimilative model of the North Indian assimilating altimetry, the seasonal and inter-annual variability of the meridional heat fluxes in the Indian Ocean are explored.

Chapter 1 – In this chapter, we examined the biases and the most predictable patterns over the tropical Atlantic Ocean by using hindcasts and a long-term integration of the National Centers for Environmental Prediction (NCEP) Climate Forecast System (CFS). CFS is a state-of-the-art coupled system, which is now being used at NCEP for operational seasonal climate predictions. It is shown that CFS reproduces the major features of the leading variability modes in the tropical Atlantic Ocean and the associated physical processes. However, there are significant biases, particularly over the southeastern tropical Atlantic Ocean. These biases have a strong impact on the predictable skill and the structures of the model predicted patterns in the tropical Atlantic Ocean.

Chapter 2 – Numerical models that simulate the dispersion of radionuclides in the sea have been widely developed in the last years. Two different kinds of models may be identified: rapid response models used to support the decision-making process after an emergency situation and environmental models aimed at studying the marine system dynamics using radionuclides as tracers of environmental processes. Examples of both types of models are described in this chapter. First, a rapid response model for the Strait of Gibraltar. Secondly, a model developed to study dispersion processes and dynamics of the Albor'an Sea is described. Both models solve the hydrodynamic equations to obtain tidal and residual (mean) currents in the domain, which are then used to solve the dispersion equations. The Albor'an Sea model is a two-layer model including sediment transport and interactions of dissolved radionuclides with the sed- iments. A detailed model description is given together with results and applications. The rapid-response model for the Strait of Gibraltar is a particle-tracking lagrangian model suitable for simulating the dispersion of a radioactive spill occurring at the sea surface. Again the model is described in detail and examples of results are given.

Chapter 3 – Marine invertebrates, algae and microorganisms produce a considerable number of structurally unique and biologically active compounds, also called secondary metabolites. The ecological roles of these natural products, although not completely understood, range from chemical defense against predation to intra-specific cues for larval settlement. Since the late 1960's marine natural product bioprospecting has demonstrated that most of these molecules may be drug candidates.

In addition to their pharmaceutical applications, secondary metabolites may also be employed as chemotaxonomic markers useful for elucidating classification problems and phylogenetic relationships. Drug discovery and chemotaxonomy are two main tracks in marine natural products chemistry which have attracted the attention of numerous biologists and chemists throughout the world. However, despite this intense global interest, the natural products chemistry of marine flora and fauna of the western Indian Ocean islands is still largely ignored.

This paper will provide therefore an outline of the contribution made by the Chemistry Laboratory of Natural Substances and Food Sciences (LCSNSA) at Reunion Island University to the isolation and identification of natural bioactive molecules as well as chemotaxonomic markers from marine organisms of Mayotte, a French Overseas Territory of the Comoros archipelago located in the western Indian Ocean.

Chapter 4 – A review of the tsunami hazard along the coastal area of the Gulf of Bengal, Indian Ocean, is provided. We consider here earthquakes as tsunami-triggering geophysical sources. For that purpose, the seismicity of the Indian Ocean area is described, i.e. the tectonics of the area, the earthquake recurrence, as well as historical tsunami events. In that context, recent events like the 26th December 2004 megathrust earthquake and associated tsunami are discussed. Finally, we describe the main physical processes that are responsible for wave amplification/attenuation and for the coastal vulnerability that may vary significantly along the Gulf of Bengal. Beside earthquake magnitude and modes of co-seismic rupture, we show in particular the importance of local effects for wave amplification (shape of the coast and local bathymetry), and of specific features, like underwater canyons, that are likely to inhibit wave amplification at the coast.

Chapter 5 – Drawing on research conducted in coastal communities in Thailand following the 26th December Indian Ocean tsunami, this chapter discusses how mainstream, pre-existing community characteristics derived from dealing with the challenges posed by daily life predict community members' resilience and their capcity to experience posttraumatic growth in the aftermath of the tsunami impact. The chapter also discusses the implications of these findings for developing cross-cultural understanding of natural hazard resilience. Specifically, it explores the procedures that need to be adopted to assess the cross-cultural validity of resilience and growth (collective efficacy, response efficacy and posttraumatic growth) measures developed in Western, individualistic cultures with members of collectivistic communities. Having established the cross-cultural validity of the measures, the authors proceed to identify the predictors of each outcome. The chapter concludes with a discussion of the need to develop a better understanding of indigenous and culture-specific resilience.

Chapter 6 – The Indian Ocean Dipole (IOD) (Saji et al., 1999) is an ocean-atmosphere phenomenon in the Indian Ocean. It is normally characterized by a positive phase which sees greater-than-average sea-surface temperatures (SST) leading to greater precipitation in the western equatorial Indian Ocean, and a negative phase which sees a corresponding cooling of waters in the south eastern Indian Ocean which tends to cause droughts in adjacent land areas of Indonesia and Australia. The IOD also affects the strength of monsoons over the Indian subcontinent. The variability of the IOD causes climatic extremes and socio-economic hardship throughout the tropical Indian Ocean region (Birkett et al., 1999). The ability to predict the strength of the IOD two to three seasons ahead will help communities prepare, if need be for adverse climatic events. This study considers a selection of statistical time series models to analyze and predict the IOD. These models are based only on the IOD index time series and do not require the input of any other climate time series variables. We will show that one of these time series models produces credible forecasts that are comparable to those produced by a forecasting system developed by Luo et al, 2007, and which incorporates several time series variables.

Chapter 7 – Investigations in the Indian Ocean with respect to sea routes, harbor locations and favorable winds became increasingly important when Vasco da Gama rounded the Cape of Good Hope and discovered the sea route to India at the end of the 15th century. However, it took more than two centuries before the Danish Arabia Felix Expedition with the Danish Navy ship Grønland investigated the Indian Ocean between 1761 and 1767. Since then, several worldwide expeditions were performed which visited also the Indian Ocean. The first voyage into this area with a focus on biology was the British John Murray Expedition in the years 1933 to 1934.

After World War II, the numbers of expeditions to the Indian Ocean increased. Based on fundamental scientific interest, the International Indian Ocean Expeditions (IIOE) took place between 1959 and 1965. Twenty-three countries were involved in 180 research cruises on 40 different research vessels in this venture. After this period only occasional cruises were obtained by several countries until, between 1990 and 1998, participants from nine countries investigated the northeastern part of the Indian Ocean, the Arabian Sea, on 80 cruises with 18 ships in the course of the international Joint Global Ocean Flux Study (JGOFS).

Chapter 8 – The seasonal cycle of the cross-equatorial heat transport is of great importance to South Asian monsoons since it is a component of the heat budget and hence determines the SST of the North Indian Ocean. In this chapter, we use a data-assimilative model of the North Indian Ocean assimilating altimetry and MCSST data and driven by ECMWF and QuikSCAT wind stresses, to explore the seasonal and inter-annual variability of the meridional heat fluxes during the years 1993 to 2005. In addition to a strong seasonal variability, the cross-equatorial heat transport also exhibits considerable inter-annual variability. There are also strong, short period intra-seasonal fluctuations on time scales of 20-50 days. The meridional heat flux values are 20-30% higher when the hindcast is driven by observed QuikSCAT wind stresses than when it is driven by ECMWF winds.

Chapter 9 – Equatorial currents at 90° E from a numerical hindcast of the North Indian Ocean are analyzed for the years 2001 to 2004 along with ADCP currents from a mooring deployed there by the Japan Agency for Marine-Earth Science and Technology (JAMSTEC). The analysis indicates that the model captures a substantial portion of the low frequency variability in zonal, and to a lesser extent, meridional currents. Currents from the 1/4° model are more energetic than the 1/2° model, and QuikSCAT-derived wind stresses produce more energetic and more realistic currents than ECMWF-derived ones. The dominant variability is annual and semi-annual. Wavelet analysis highlights the significant events during this period. The 13-year hindcast of the years 1993-2005, driven by ECMWF winds, clearly highlights the anomalous 1997-98 period.

Short Commentary – For the last 30 years, oceanographic research in the nitrogen cycle has stood firmly on two important concepts: new and regenerated production (Dugdale and Goering 1967). The former defines the part of primary production fueled by allochtonous nitrogen (mainly nitrate coming from vertical advection), while the latter represents the uptake of regenerated forms of N (usually ammonium). These concepts opened the way for estimating the strength of the "biological pump", i.e., the capacity of the biological compartment of the ocean to take carbon dioxide from the atmosphere and to export it to the deep ocean. Through the formulation of the *f* ratio (Eppley and Peterson 1979), the fraction of primary production sustained by allochtonous nitrate could be assimilated to the export of carbon to the deep ocean. In recent years, the assumption necessary for the use of *f*—that is, negligible nitrification (microbial oxidation of ammonium into nitrate) within the euphotic layer—has been challenged by increasing evidence of surface nitrogen regeneration in the open ocean. A few years ago, the largest data set with a seasonal frequency gathered in the NE Atlantic Ocean showed that in situ nitrification can account for up to 30% of nitrate assimilation within the euphotic zone (Fernández et al. 2005b). The consequences of this unaccounted nitrate supply into the euphotic zone (biasing our estimation of export production) are only beginning to be quantified at the global scale (Yool et al. 2007). In this chapter, I will focus on these recent developments in the nitrogen cycle and discuss the methodological improvements that allowed us to get to the present point in our understanding of nitrogen regeneration. Also, I will point out the new challenges that have arisen from these advancements, which should outline future research.

Expert Commentary – Atmospheric teleconnections between the tropical Atlantic and the African continent, extending to the Indian Ocean basin, modulate the interannual variability of precipitation in highly vulnerable and densely populated regions like the Sahel and the Indian peninsula. Additionally, increasing evidence supports the notion that in the Atlantic basin low-latitude, wind-driven convection cells contribute to the transport of heat and salt potentially interacting with the Atlantic Meridional Overturning Circulation (AMOC). Nevertheless, while the progress in our understanding of high latitude oceanography has been remarkable, the nature and causes of the variability of tropical Atlantic complex dynamics are poorly understood. Alone, observations and interpretations of Atlantic tropical variability have been controversial. Debate about interpretations of both trends and natural cycles within the observational data set is expected because the instrumental record is not long enough and too sparse. Furthermore, tropical Atlantic sea surface temperatures (SSTs) have been identified as a major source of errors in coupled general circulation models. It is therefore extremely difficult to discern the interannual variability signal, which is potentially linked both to the Atlantic Ocean internal dynamics and to the El Nino Southern Oscillation (ENSO), from longer period oscillations such as the Atlantic Multidecadal Oscillation (AMO) and secular trends perhaps related to anthropogenic activity. Here we explore recent research opportunities and challenges that the scientific community is facing in the Tropical Atlantic and we suggest possible avenues of collaboration between atmo- spheric scientists and physical and paleo-oceanographers, linking modeling efforts and oceanographic observational campaigns with the goal of promoting a coordinated and multi-disciplinary research effort in this region.

In: Atlantic and Indian Oceans
Editors: E. S. Askew and J. P. Bromley, pp. 1-15
ISBN 978-1-60692-475-4

Chapter 1

BIASES AND THE MOST PREDICTABLE PATTERNS IN THE NCEP CFS OVER THE TROPICAL ATLANTIC OCEAN

Zeng-Zhen Hu*[*1], Bohua Huang[1, 2] and Kathy Pegion[1]
[1]Center for Ocean-Land-Atmosphere Studies, Calverton, Maryland, USA
[2]Department of Climate Dynamics, George Mason University, Fairfax, Virginia, USA

ABSTRACT

In this chapter, we examined the biases and the most predictable patterns over the tropical Atlantic Ocean by using hindcasts and a long-term integration of the National Centers for Environmental Prediction (NCEP) Climate Forecast System (CFS). CFS is a state-of-the-art coupled system, which is now being used at NCEP for operational seasonal climate predictions. It is shown that CFS reproduces the major features of the leading variability modes in the tropical Atlantic Ocean and the associated physical processes. However, there are significant biases, particularly over the southeastern tropical Atlantic Ocean. These biases have a strong impact on the predictable skill and the structures of the model predicted patterns in the tropical Atlantic Ocean.

Compared with observations, the CFS produces too few low clouds over the southeastern Atlantic and too many high clouds over the whole tropical Atlantic. The center of the low clouds is shifted westward in the model away from the cold tongue region towards the central ocean. The underestimation of low clouds in the southeastern Atlantic may be one of the potential sources of the warm biases of sea surface temperature (SST) in this region. The low clouds are linked to a stable or inversion layer between 850 hPa and 925 hPa in observations, but are associated with a stable layer between 700 hPa and 850 hPa in the CFS, which suggests that low clouds are generated in a higher layer in the model than in the real world. On average, the lower troposphere is less stable in the CFS than in the observations over the tropical Atlantic region, which is more favorable for the development of deep convection and generation of high clouds.

On average, for predictions with initial conditions (IC) of all months, the predictability of SST in the tropical Atlantic is higher in the west than in the east. Seasonally, the skill is higher for predictions with IC in summer or autumn and lower for

* Corresponding author: Dr. Zeng-Zhen Hu; E-mail: hu@cola.iges.org; Tel: 301-902-1270; Fax: 301-595-9793

those with IC in spring. The CFS poorly predicts the meridional gradient across the equator. The superiority of the CFS predictions to the persistence forecasts depends on IC month, region, and lead time. The CFS prediction is generally better than the corresponding persistence forecast when lead time is longer than 3 months. The most predictable pattern of SST in March has the same sign in almost the whole tropical Atlantic. The corresponding pattern in March is dominated by the same sign for geopotential height at 200 hPa in most of the domain and by significant opposite variation for precipitation between the northwestern tropical North Atlantic and the regions from tropical South America to the southwestern tropical North Atlantic. These predictable signals mainly result from the influence of the El Niño-Southern Oscillation (ENSO). For the CFS, the predictive skill in the tropical Atlantic Ocean is largely determined by its ability to predict ENSO. This is due to the strong connection between ENSO and the most predictable patterns in the tropical Atlantic Ocean in the model. The higher predictive skill of tropical North Atlantic SST is consistent with the ability of the CFS to predict ENSO on inter-seasonal time scales, particularly for the ICs in warm months from March to October. In the southeastern ocean, the systematic warm bias is a crucial factor leading to the low skill in the region.

1. CFS Model and Data

As a state-of-the-art coupled system, the National Centers for Environmental Prediction (NCEP) Climate Forecast System (CFS) is now being used at NCEP for operational seasonal climate predictions. Other studies have demonstrated that the CFS reproduces the major features of the El Niño-Southern Oscillation (ENSO) (Wang et al. 2006; Saha et al. 2006; Zhang et al. 2007) and the Asian and Australian monsoons (Liang et al. 2008). In this paper, we will examine its simulation of the climatological state in the tropical Atlantic Ocean and its predictive skill in this region. A brief description of the coupled system as well as the datasets we have used for this examination is given below.

The atmospheric component of the CFS has horizontal resolution of T62 and 64 vertical sigma levels. The oceanic component is configured from the version 3 of the Modular Ocean Model of Geophysical Fluid Dynamics Laboratory (Pacanowski and Griffies 1998). The ocean model has 40 levels vertically, with 27 of them in the upper 400 meters. The domain of the ocean model extends from 74°S to 64°N with a horizontal grid of 1°x1° poleward of 30°S and 30°N, and with gradually increased meridional resolution to $1/3^{\circ}$ between 10°S and 10°N. The atmospheric and oceanic components exchange surface fluxes on a daily interval without flux adjustment.

The NCEP CFS hindcasts analyzed in this work were initialized in all calendar months from 1981 to 2003. In a given start month, 15 nine month predictions were produced. The oceanic initial conditions (ICs) were taken from the NCEP Global Ocean Data Assimilation (GODAS) (Behringer et al. 1998, Behringer and Xue 2004) analyses on the 11th and 21st of lead-month 0 and 1st of lead-month 1 (Saha et al. 2006). The atmospheric ICs were from the NCEP reanalysis II (Kanamitsu et al., 2002). For each of the three chosen oceanic ICs, five atmospheric ICs from 2 days before to 2 days after at daily intervals were chosen to form 15 ocean-atmosphere ICs. The details about the ICs were given in Saha et al. (2006) and Huang et al. (2007). The monthly mean of the 15 individual member predictions and their ensemble mean are available from the NCEP website at http://nomad6.ncep.noaa.gov/cfs/monthly.

In addition, a 52-year coupled integration using the CFS was conducted at the Center for Ocean-Land-Atmosphere Studies (COLA). The ocean-atmospheric IC is from January 1, 1985. The monthly data of the last 42 years, after a 10-year spin-up, are used in this analysis.

For comparison with the model simulations, we also use the monthly mean data derived from the NCEP/NCAR reanalyses (Kalnay et al. 1996) and from the extended reconstruction of the sea surface temperature (SST) analysis (ER-v2) on a 2°x2^{o} grid (Smith and Reynolds 2003). The data periods used in this investigation are from 1948 to 2003 for the NCEP/NCAR reanalysis, and from 1950 to 2003 for the ER-v2 SST data.

Observation-based monthly mean cloud coverage and net short-wave radiative fluxes at the surface from July 1983 to December 2004 are also used. The cloud coverage data were produced by the International Satellite Cloud Climatology Project (ISCCP), which infer the global distribution of clouds and their properties from diurnal to interannual time scales using a series of satellite radiance measurements (Rossow and Dueñas 2004). By their definitions, the low clouds in the CFS are different from that in the ISCCP. The former refers to the clouds formed above a given boundary layer (BL) within 10% of the surface pressure but below 650 hPa, while the later is simply formed below 680hPa. In order to make the low clouds more comparable between the CFS and ISCCP, in the following, the sum of low cloud and BL cloud from the CFS output is redefined as the CFS low clouds. It is clear that this definition overestimates the actual low clouds in the CFS due to the overlap of low and BL clouds. Furthermore, in the ISCCP, the low-level clouds are counted and considered only in regions not covered by higher clouds. Therefore, for the same cloud distribution, the low-level cloud fraction will be greater in the CFS definition than in the ISCCP. However, the differences in the definitions between the CFS and ISCCP do not affect the conclusion, since the CFS low cloud amount is persistently low compared with that in the ISCCP, even with the inflated definition in the CFS. The monthly net short-wave radiative fluxes at the surface were derived from a comprehensive radiation data set calculated by Zhang et al. (2004) using a complete radiative transfer model with input from surface and atmospheric observations and ISCCP cloud properties. In this work, only the percentage of sky covered by low, middle, and high clouds and the net short-wave radiation at the surface on a 2.5^{o}x2.5^{o}grid are used to compare with the corresponding variables in the CFS. For convenience, all the observation based data, such as the ISCCP cloud and radiation data, the reanalyses, and ER-v2 SST, are referred to as observations in this work.

Based on the results from the CFS hindcasts and the long-term integration, we have examined the model mean climate, variability (Hu et al. 2008a), leading modes and their physics (Hu et al. 2008b), as well as the predictive skill and the most predictable patterns (Hu and Huang 2007) in the tropical Atlantic Ocean. The main results are high-lighted in the following.

2. CFS Mean Climate and Variability

By comparing the CFS simulations with corresponding observations, it is shown that the CFS captures the seasonal mean climate, including the zonal gradients of SST, as well as the wind stress pattern in the equatorial Atlantic Ocean (Figure 1), even though the CFS produces warm mean biases and underestimates the variability over the southeastern ocean (right column of Figure 2).

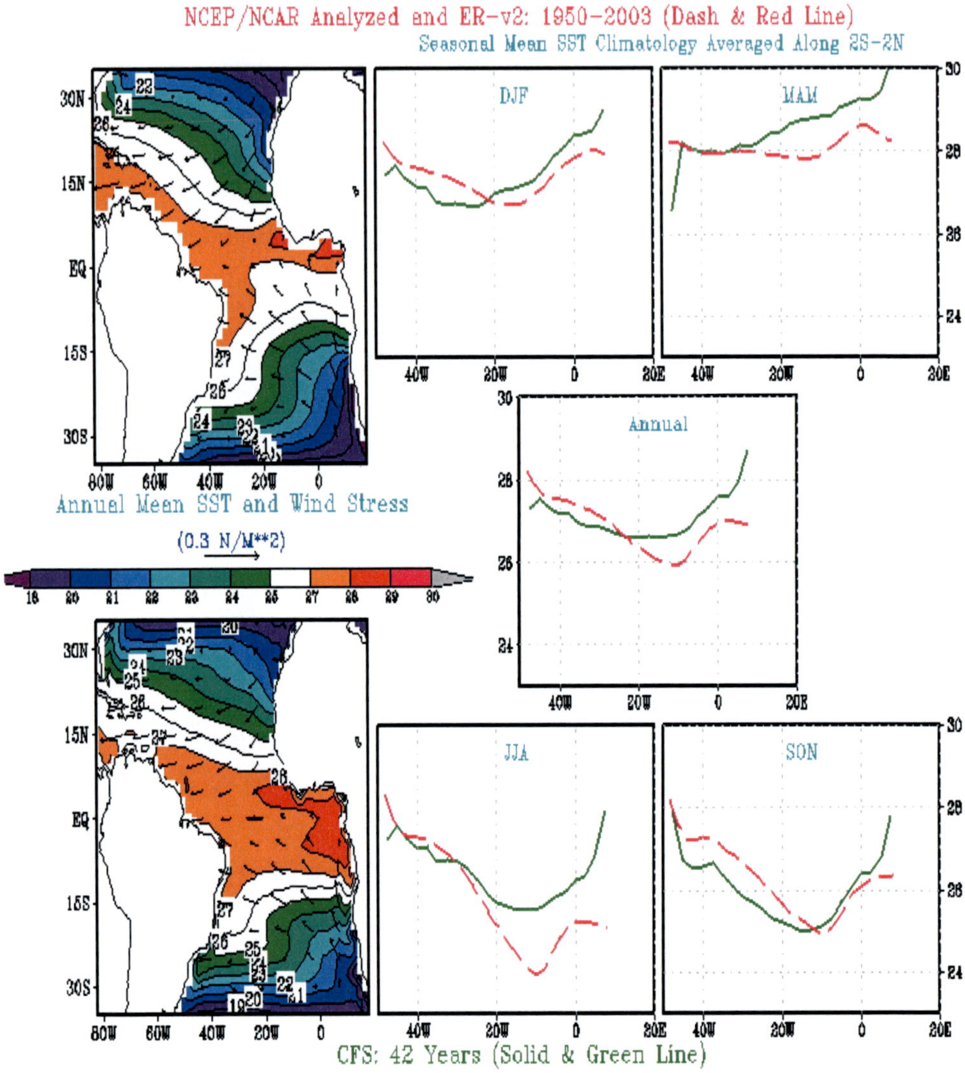

Figure 1. Annual mean climatology of SST and wind stress in observations (top pane of left column) and in the CFS (bottom panel of left column), and seasonal and annual mean SST along the equator (central and right columns). In the central and right columns, the dished and red curves are the observations and the solid and green ones are the CFS. The contour interval is 1oC in the left column.

The biases in the tropical Atlantic in the long-term integration are consistent with the result derived from the hindcasts (Huang et al., 2007). Huang et al. (2007) also studied the seasonal evolution of the biases in the hindcasts and found the errors of the CFS hindcasts in the southeastern tropical Atlantic grow much faster for the prediction starting in boreal summer and fall than those starting in late winter and spring. In addition, the CFS produces too few low-level clouds (left column of Figure 2) and too many high-level clouds (not shown) over the southeastern part of the tropical Atlantic. The center of the low clouds in the model is shifted westward, away from the cold tongue region, to the central ocean (left column of Figure 3). The underestimation of low-level clouds in the southeastern Atlantic could be a cause of the warm SST biases in this region.

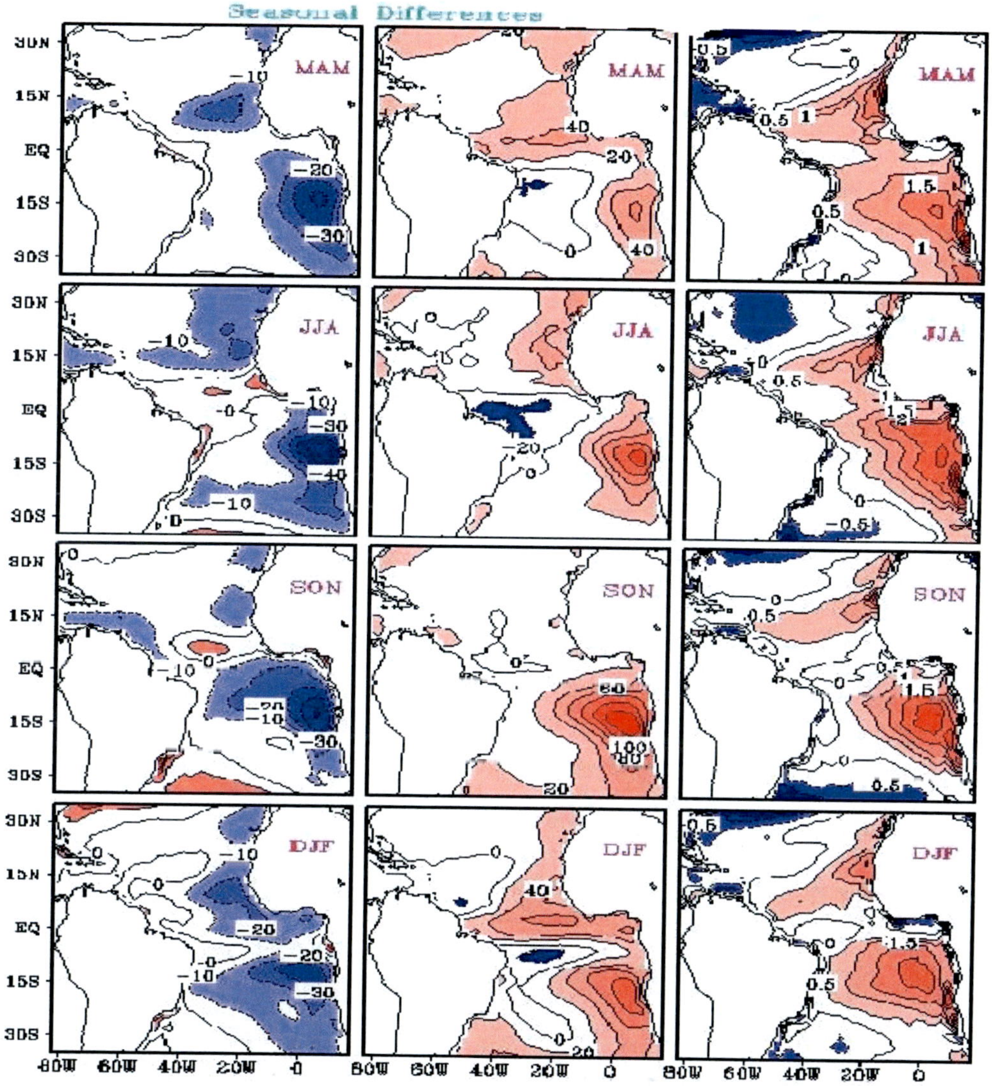

Figure 2. Seasonal mean climatology differences of low clouds (left column), downward short-wave radiation (central column), and SST (right column) between the CFS simulation and corresponding analyses. The contour interval is 10% for the low clouds, 20 W/m2 for the short-wave radiation, and 0.5oC for the SST. (Reproduced by permission of American Geophysical Union from Fig. 2 of Hu et al. 2008a).

The association between low clouds and the atmosphere stability in the lower troposphere is different between the observations and CFS. The low-level clouds are linked to a stable or inversion layer between 850 hPa and 925 hPa in the observations, but found to be associated with a stable layer between 700 hPa and 850 hPa in the CFS (central and right columns of

Figure 3), which suggests that low clouds are generated in a higher layer in the CFS than in the real world.

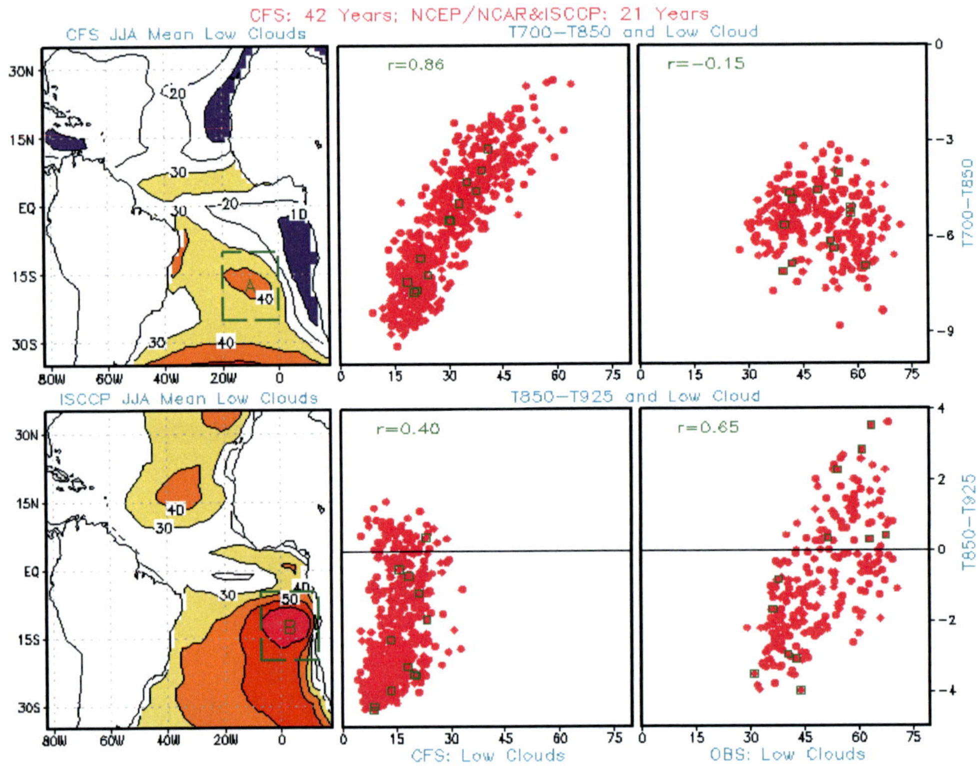

Figure 3. Left column: climate mean low clouds in June, July, and August in the CFS (top panel) and ISCCP (bottom panel). The contour interval is 10%. Central and right columns: scatter plot of low-level clouds (%) vs. T700-T850 (oC, top panels), and vs. T850-T925 (oC, bottom panels) averaged in the region "A" (25oS-15oS, 20oW-0; see the green square in the left top panel) in the CFS (central column) and in the region "B" (20oS-5oS, 7oW-13oE; see the green square in the left bottom panel) in the observations (right column). The circles represent individual monthly means and the small green squares are the monthly climate mean. (Modified by permission of American Geophysical Union from Fig. 5 of Hu et al. 2008a).

This is confirmed by further examination, which shows that for the global tropical average (30°S-30°N), maximums of the low clouds occur in a layer between 750 hPa and 650 hPa in the CFS; however, the mean top pressure of low clouds from ISCCP is in the range of 750 hPa to 900 hPa (Figure 4).

In addition, the maximums in the cloud cover over the equator around 200 hPa in the CFS are associated with deep convection (see bottom panel of Figure 4).

The scenario can be explained as follows. Assuming that the initial temperature stratification is correct, the inappropriate cloud scheme or other physical processes or parameterizations do not generate enough low clouds in the model, which then weakens the initial vertical temperature inversion at the top of the marine atmospheric boundary layer. In turn, the weakened stratification further prevents low clouds from developing. Therefore, as the model spins up from an observed initial condition, it gradually approaches the state shown in Figures 2 and 3, with an incorrect stratification and biased cloud fields.

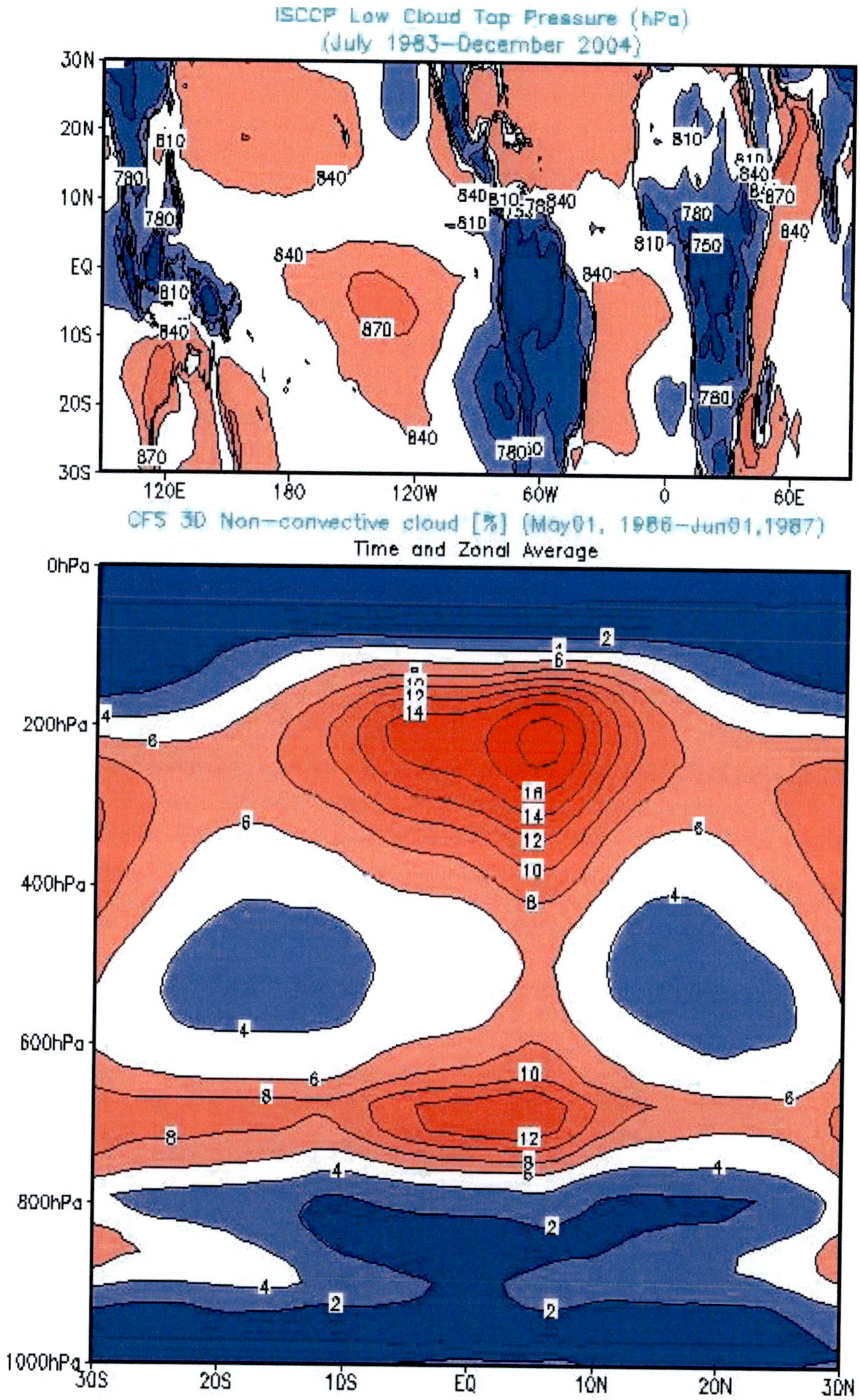

Figure 4. Top panel: ISCCP mean low cloud top pressure (hPa). The contour interval is 30 hPa. Bottom panel: X-Z cross section of the CFS simulated time averaged zonal mean cloud cover. The contour interval is 2%.

The scenario is confirmed when low cloud amount in the NCEP/NCAR reanalyses is compared with that in the CFS and ISCCP (not shown). Low cloud amount over the southeastern Atlantic in the NCEP/NCAR reanalyses is much closer to the ISCCP than to that in the CFS.

This may be due to the fact that temperature stratification in the lower troposphere in the NCEP/NCAR reanalyses is more analogous to observations than that in the CFS.

Furthermore, on average, the lower troposphere is less stable in the CFS than in the observation over the tropical Atlantic, a condition more favorable for the development of deeper convection and generation of high clouds. Moreover, in the observation, the low-level cloud variability in the South Atlantic is mainly associated with anomalies in the southeastern Atlantic and along the African coast, but in the CFS simulations the anomalies are moved to the subtropical South Atlantic.

Overall, the interactive processes among the low-level clouds, vertical inversion layer, and SST are important. The inadequate low-level cloud (left column of Figure 2) and excessive short-wave radiation (central column of Figure 2) are associated with the warm SST biases in the southeastern Atlantic (right column of Figure 2). In return, the warm biases do not favor an inversion layer and low cloud formation (Figure 3). Details about this are given in Hu et al. (2008a).

3. Leading Modes and Physics in CFS and Observations

Correctly simulating the mean annual cycle and the leading patterns of the tropical Atlantic Ocean variability is still a challenge for current state-of-the-art general circulation models (CGCMs). The mean annual cycle, interannual variability, and leading patterns of the tropical Atlantic Ocean simulated in a long-term integration of the CFS have been examined (Hu et al. 2008b). Besides the warm biases in the tropical southeastern Atlantic (see the previous section), it is found that seasonal transition from a warm to a cold phase along the equator is delayed by one month in the CFS compared with the observations. This might be related to the model failure in simulating the cross-equatorial meridional wind associated with the African monsoon.

The CFS realistically simulates both the spatial structure and spectral distributions of the three major leading patterns of the SST anomalies in the tropical Atlantic Ocean: the south tropical Atlantic pattern (STA), the north tropical Atlantic pattern (NTA), and the southern subtropical Atlantic pattern (SSA) (Figure 5). The CFS also simulates the seasonal dependence of these patterns and partially reproduces their association with the ENSO. Furthermore, the dynamical and thermodynamical processes associated with these patterns are similar in the simulation and the observations. The air-sea interaction processes associated with the STA pattern are well simulated in the CFS. The primary feature of the anomalous circulation in the Northern Hemisphere (NH) associated with the NTA pattern resembles that in the Southern Hemisphere (SH) linked with the SSA pattern, implying a similarity of the mechanisms in the evolution of these patterns and their connection with the tropical and extratropical anomalies in their respective hemispheres. The anomalies associated with both the SSA and NTA patterns are dominated by atmospheric fluctuations of equivalent-barotropic structure in the extratropics including zonally symmetric and asymmetric components.

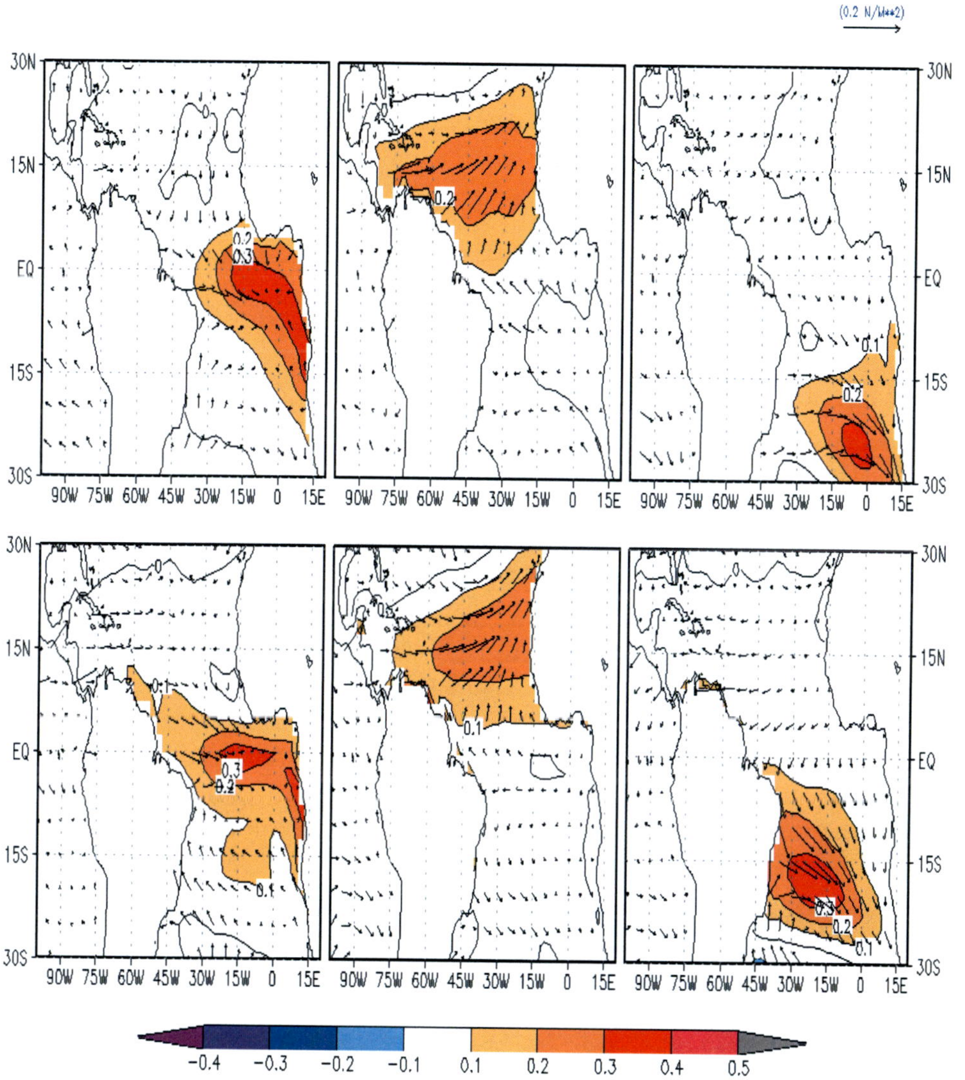

Figure 5. Spatial patterns (shading and contour) of rotated empirical orthogonal functions (REOFs) of ER-v2 (top panels) and the CFS simulated (bottom panels) seasonal mean SST in the tropical Atlantic (30oS-30oN, 100oW-20oE). The vectors represent the simultaneous regression of seasonal mean wind stress onto the corresponding time series of the REOF. The contour interval is 0.1oC. The percentage of the variance explained by each pattern are 16%, 15%, 10% (top panels from left to right), 10%, 11%, and 12% (bottom panels from left to right). (Modified with kind permission of Springer+Business Media from Fig. 7 of Hu et al. 2008b).

The zonally symmetric variability is associated with the annular modes, the Arctic Oscillation in the NH and the Antarctic Oscillation in the SH. The zonally asymmetric part of the anomalies in the Atlantic is teleconnected with the anomalies over the tropical Pacific.

The misplaced teleconnection center over the southern subtropical ocean may be one of the reasons for the deformation of the SSA pattern in the CFS.

4. Predictive Skill and the Most Predictable Patterns in CFS Hindcasts

These biases in the mean climate and the leading variability modes in the tropical Atlantic Ocean can affect the model's ability to correctly predict the anomalies in the basin. The spatial distribution of the biases alters the mean state and changes the path of wave propagation, which may result in an incorrect signal of ENSO in the tropical Atlantic Ocean. The predictive skill and most predictable patterns in the tropical Atlantic Ocean, as well as their connection with ENSO and the influence of the biases have been studied using the CFS hindcasts (Hu and Huang 2007). The skill is measured by SST anomaly correlation between the predictions and the corresponding analyses. On average, for predictions with IC of all months, the predictability of SST is higher in the west than in the east (Figure 6). The highest skill is near the tropical Brazilian coast and in Caribbean Sea, and the lowest skill occurs along the eastern coast. Seasonally, the skill is higher for predictions with IC in summer or autumn and lower for those with IC in spring. The CFS poorly predicts the meridional gradient across the equatorial Atlantic Ocean. The superiority of the CFS predictions to the persistence forecasts depends on IC month, region, and lead-time. The CFS prediction is generally better than the corresponding persistence forecast when lead-time is longer than 3 months.

The predictable patterns are identified by applying an empirical orthogonal function (EOF) analysis with maximized signal-to-noise ratio (MSN EOF hereafter) to the predicted time series of 1981-2003 with given lead-time of the hindcasts. The most predictable pattern of SST in March has the same sign in almost the whole tropical Atlantic (Figure 7). The corresponding pattern of geopotential height at 200 hPa (H200) is dominated by the same sign in most of the domain (Figure 7), while the corresponding precipitation pattern showed a distinct opposite variation between the northwestern tropical North Atlantic and the regions from tropical South America to the southwestern tropical North Atlantic (not shown). Moreover, the time series of the most predictable pattern of SST, as well as the H200 and precipitation, are highly correlated with the SST in the eastern tropical Pacific (Figure 7), implying that the predictable signals in the tropical Atlantic Ocean mainly result from the ENSO influence. The significant values in the most predictable precipitation pattern in the regions from tropical South America to the southwestern tropical North Atlantic in March are associated with the regional excessive divergence (convergence) at low (high) level in the CFS hindcasts. Due to the strong connection between ENSO and the most predictable patterns in the tropical Atlantic in the model, the predictive skill of the CFS in the tropical Atlantic is largely determined by its ability to predict ENSO. The relative high predictive skill of the tropical North Atlantic SST is consistent with the CFS ability to predict ENSO on interseasonal time scales, particularly for the ICs in warm months from March to October. In the southeastern ocean, the systematic warm bias is considered a crucial factor for the low skill in that region.

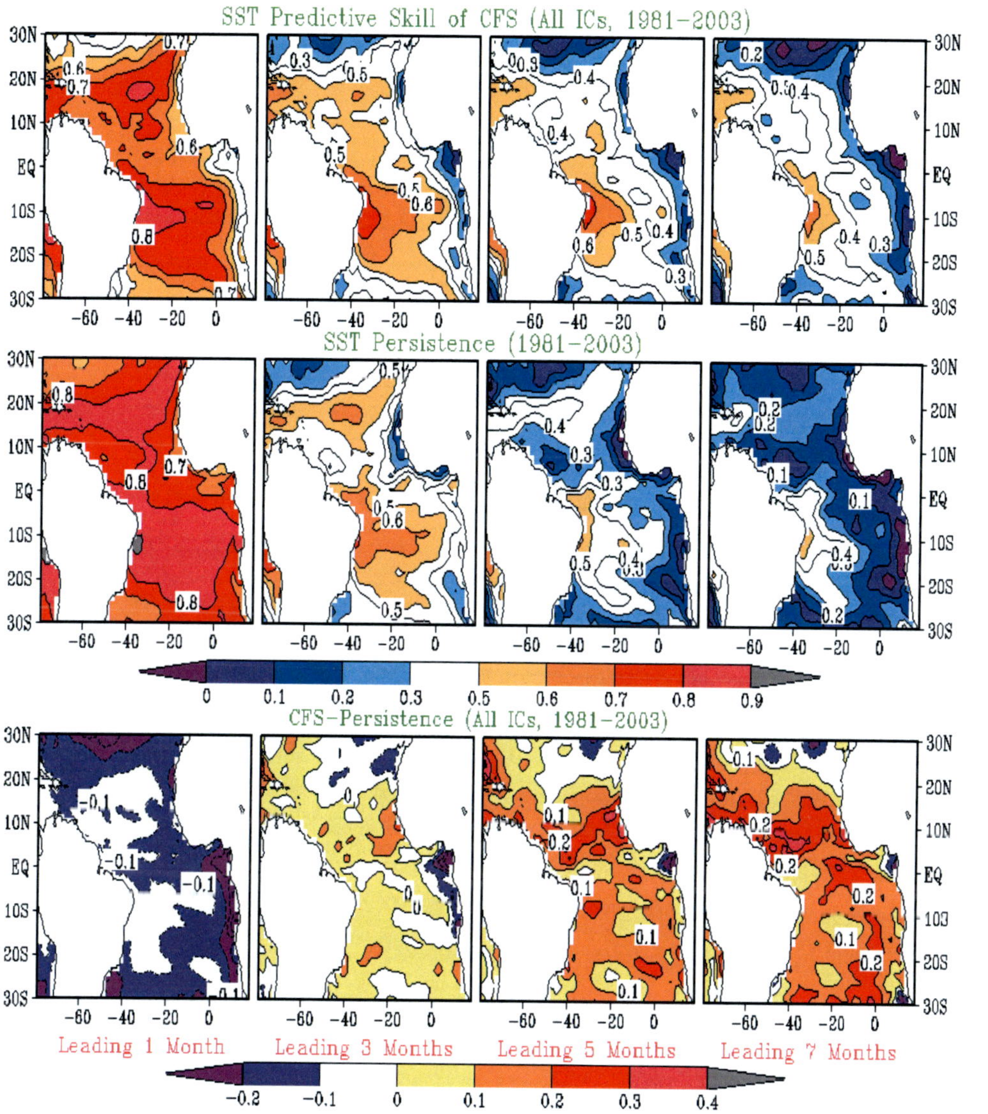

Figure 6. Correlations between OIv2 analyzed and CFS predicted SST anomalies (top row), persistence correlations of OIv2 analyzed SST anomalies (central row), and the differences (bottom row) at different lead months for all ICs in 1981-2003. The contour interval is 0.1 and the shading represents values larger than 0.5 or smaller than 0.3 in the top and central rows, and positive values in the bottom row. (Adapted by permission of American Meteorological Society from Fig. 1 of Hu et al. 2007).

The warm biases in the southeastern Atlantic are a potential reason to explain why the most predictable pattern of SST in the CFS is almost homogeneous in the tropical Atlantic Ocean (See Figure 7). The warm biases in the CFS (Figure 2) may also lead to excessive atmospheric convection in the model. According to Chiang and Sobel (2002), the existence of atmospheric convection is crucial for temperature variations to propagate from the troposphere to the surface. This is confirmed by the zonal and vertical cross section averaged between 8°-12°S for temperature regression onto the time series of MSN EOF1 of precipitation (Figure 8).

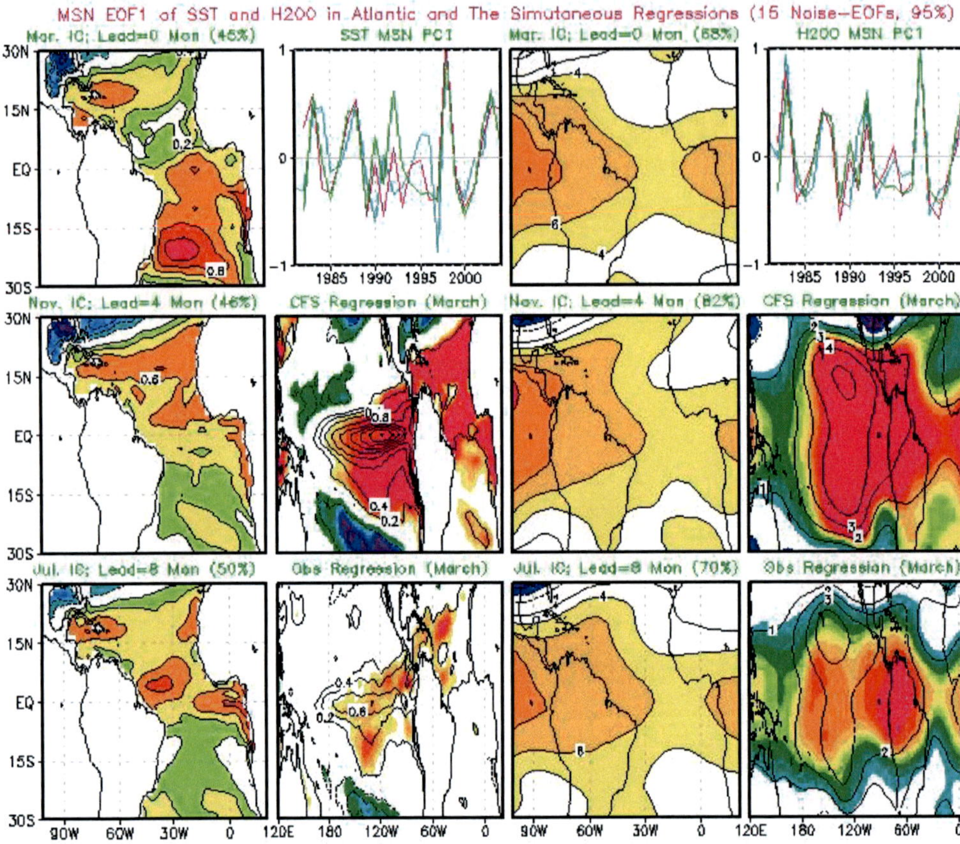

Figure 7. Left column: spatial patterns of MSN EOF1 of March SST, lead 0 month and March IC (top panel), lead 4 months and November IC (middle panel), and lead 8 months and July IC (bottom panel). Second column: the time series corresponding to the left column (top panel), the CFS predicted (middle panel) and observed (bottom) SST regression onto the time series with lead 4 months and November IC. Third and fourth columns: similar to the left and second columns, respectively, but for geopotential height at 200 hPa (H200). The contour interval is 0.2oC for the MSN EOF patterns of SST and 2 m for the MSN EOF patterns of H200, the zero lines are omitted. The contour interval of the SST (H200) regression is 0.2 oC (1m) per standard deviation of the corresponding time series. The real magnitude of the SST and H200 anomalies can be restored by multiplying the values in the spatial patterns with the corresponding time series. The percentage of the explained variance for the ensemble mean anomalies is indicated in each panel. (Adapted by permission of American Meteorological Society from Figures 5-8 of Hu et al. 2007).

It is shown that the MSN EOF pattern is associated with temperature variations at all levels over the tropical southeastern Atlantic in the CFS. In contrast, in the reanalyses, the relationship over the tropical southeastern Atlantic is constricted to the upper troposphere and the regressions become negative for the lowest levels in the southeastern tropical Atlantic. Therefore, the theory of Chiang and Sobel (2002) can explain why the most predictable pattern of SST in the CFS is almost homogeneous in the tropical Atlantic Ocean (See Figure 7). This result suggests that the systematic warm bias in the southeastern ocean is a crucial factor leading to the low skill and unrealistic most predictable pattern in this region.

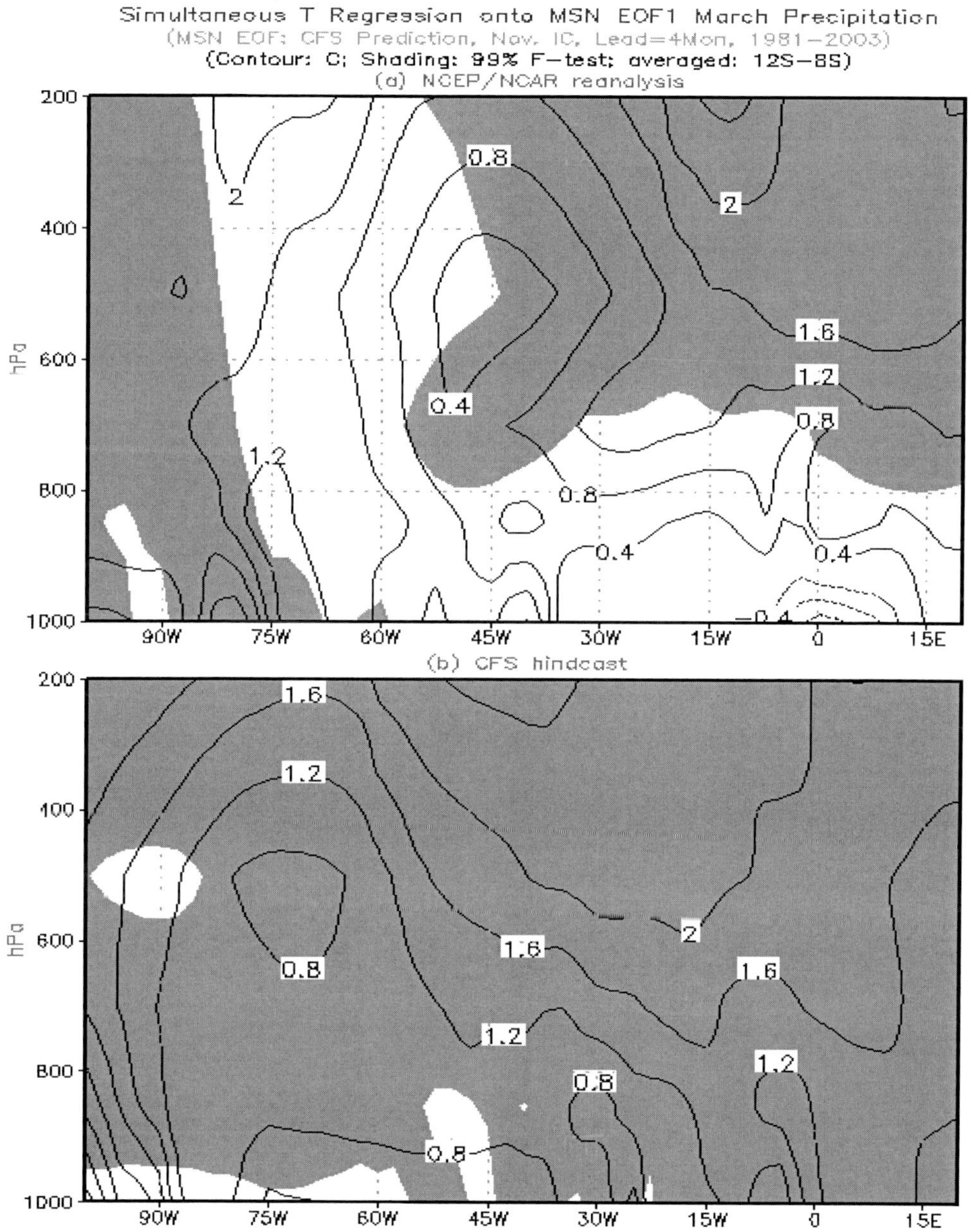

Figure 8. Zonal and vertical cross section averaged between 12-8oS of simultaneous regression of temperature in the NCEP/NCAR reanalyses (a) and in the CFS predictions (b) onto the time series of precipitation MSN EOF1 for lead 4 months and November IC. The contour interval is 0.5oC per standard deviation of the MSN EOF1 time series, the zero lines are omitted, and the shading is for significant regression at the 95% level. (Adapted by permission of American Meteorological Society from Fig. 12 of Hu et al. 2007).

SUMMARY AND DISCUSSION

Correctly simulating the mean annual cycle and the leading patterns of tropical climate variability and predicting the anomalies are still a challenge for current state-of-the-art CGCMs. This chapter demonstrated this challenge in the tropical Atlantic Ocean by using large volumes of hindcasts and long-term simulations of the NCEP CFS as examples. In the investigations, it is shown that the seasonal cycle and zonal gradient along the equator, as well as some major features of leading variability modes are qualitatively reproduced by the model. However, the model has significant biases in the tropical Atlantic Ocean, particularly in the southeastern domain. The warm biases in the ocean surface are strongly related to the systematic errors of low clouds. The underestimated low clouds are connected with the vertical temperature gradients in the lower troposphere and air-sea interaction. It seems that there is an interaction process between the low-cloud biases, stability of the lower troposphere, and SST.

The predictive skill and the most predictable patterns are affected by the model biases. The warm biases in the southeastern Atlantic may also lead to excessive atmosphere convection in the model, which may cause the ENSO signal to un-realistically propagate to the lower atmosphere over the tropical South Atlantic Ocean. This may be a reason for the low predictive skill and unrealistic most predictable pattern in this region.

ACKNOWLEGEMENT

We appreciated the suggestion and comments of Drs. S. Saha, S. Yang, and K. Campana. This work is supported by the NOAA CLIVAR Program (NA07OAR4310310 and NA04OAR4310115).

REFERENCES

Behringer, D., M. Ji, and A. Leetmaa, 1998: An improved coupled model for ENSO prediction and implications for ocean initialization. Part I: The ocean data assimilation system. *Mon.. Wea. Rev.*, 126, 1013-1021.

Behringer, D., and Y. Xue, 2004: Evaluation of the global ocean data assimilation system at NCEP: The Pacific Ocean. *Eighth Symposium on Integrated Observing and Assimilation Systems for Atmosphere, Ocean, Land Surface. AMS 84th Annual Meeting*, Seattle, Washington, January 11-15, 2004.

Chiang, J. C. H., and A. H. Sobel, 2002: Tropical tropospheric temperature variations caused by ENSO and their influence on the remote tropical climate. *J. Climate,* 15, 2616-2631.

Hu, Z.-Z., B. Huang, and K. Pegion, 2008a: Low cloud errors over the southeastern Atlantic in the NCEP CFS and their association with lower-tropospheric stability and air-sea interaction. J. Geophys. *Res. (Atmosphere)* (in press).

Hu, Z.-Z., B. Huang, and K. Pegion, 2008b: Leading patterns of the tropical Atlantic variability in a coupled general circulation model. *Clim. Dyn.,* 30 (7-8), 703-726, DOI 10.1007/s00382-007-0318-x.

Hu, Z.-Z., and B. Huang, 2007: The predictive skill and the most predictable pattern in the tropical Atlantic: The effect of ENSO. *Mon. Wea. Rev.,* 135 (5), 1786-1806.

Huang, B., Z.-Z. Hu, and B. Jha, 2007: Evolution of model systematic errors in the tropical Atlantic basin from the NCEP coupled hindcasts. *Clim. Dyn*., 28 (7/8), 661-682, DOI 10.1007/s00382-006-0223-8.

Kalnay, E., M. Kanamitsu, R. Kistler, W. Collins, D. Deaven, L. Gandin, M. Iredell, S. Saha, G. White, J. Woollen, Y. Zhu, M. Chelliah, W. Ebisuzaki, W. Higgins, J. Janowiak, K. C. Mo, C. Ropelewski, J. Wang, A. Leetmaa, R. Reynolds, R. Jenne, and D. Joseph, 1996: The NCEP/NCAR 40-year reanalysis project. *Bull. Amer. Meteor. Soc.,* 77, 437-471.

Kanamitsu, M., W. Ebisuzaki, J. Woollen, S.-K. Yang, J. J. Hnilo, M. Fiorino, and G. L. Potter, 2002: NCEP-DOE AMIP-II Reanalysis (R-2). *Bull. Amer. Meteor. Soc.,* 83, 1631-1643.

Liang, J., S. Yang, Z.-Z. Hu, B. Huang, A. Kumar, and Z. Zhang, 2008: Predictable patterns of the Asian and Indo-Pacific summer climate in NCEP CFS. *Clim. Dyn.,* (revised).

Pacanowski, R. C., and S. M. Griffies, 1998: MOM 3.0 manual, NOAA/Geophysical Fluid Dynamics Laboratory, Princeton, New Jersey, USA 08542, 668 pp.

Rossow W. B., and E. N. Dueñas, 2004: The International Satellite Cloud Climatology Project (ISCCP) web site: An online resource for research. *Bull. Amer. Meteor. Soc.,* 85, 167-172.

Saha, S, S. Nadiga, C. Thiaw, J. Wang, W. Wang, Q. Zhang, H. M. van den Dool, H.-L. Pan, S. Moorthi, D. Behringer, D. Stokes, M. Pena, S. Lord, G. White, W. Ebisuzaki, P. Peng, and P. Xie, 2006: The NCEP climate forecast system. *J. Climate*, 19, 3483-3517.

Smith, T. M., and R. W. Reynolds, 2003: Extended reconstruction of global sea surface temperatures based on COADS data (1854-1997). *J. Climate,* 16, 1495-1510.

Wang, W., S. Saha, H.-L. Pan, S. Nadiga, and G. White, 2005: Simulation of ENSO in the new NCEP coupled forecast system model. *Mon. Wea. Rev.,* 133, 1574-1593.

Zhang, Y., W. B. Rossow, and A. A. Lacis, 2004: Calculation of radiative fluxes from the surface to top of atmosphere based ISCCP and other global data sets: refinements of the radiative transfer model and the input data. *J. Geophys. Res. (Atmos.),* 109, doi: 10.1029/2003JD004457.

Zhang, Q., A. Kumar, Y. Xue, W. Wang, and F. F. Jin, 2007: Analysis of the ENSO cycle in the NCEP coupled forecast model. *J. Climate*, 20, 1265–1284.

In: Atlantic and Indian Oceans
Editors: E. S. Askew and J. P. Bromley, pp. 17-76
ISBN 978-1-60692-475-4

Chapter 2

NUMERICAL MODELLING OF MARINE POLLUTANTS: RADIOACTIVITY DISPERSION

Raúl Periáñez
Dpt Física Aplicada 1, EUITA
Universidad de Sevilla
Crta Utrera km 1, 41013-Sevilla, Spain

Abstract

Numerical models that simulate the dispersion of radionuclides in the sea have been widely developed in the last years. Two different kinds of models may be identified: rapid response models used to support the decision-making process after an emergency situation and environmental models aimed at studying the marine system dynamics using radionuclides as tracers of environmental processes. Examples of both types of models are described in this chapter. First, a rapid response model for the Strait of Gibraltar. Secondly, a model developed to study dispersion processes and dynamics of the Alborán Sea is described. Both models solve the hydrodynamic equations to obtain tidal and residual (mean) currents in the domain, which are then used to solve the dispersion equations. The Alborán Sea model is a two-layer model including sediment transport and interactions of dissolved radionuclides with the sediments. A detailed model description is given together with results and applications. The rapid-response model for the Strait of Gibraltar is a particle-tracking lagrangian model suitable for simulating the dispersion of a radioactive spill occurring at the sea surface. Again the model is described in detail and examples of results are given.

1. Introduction

A model can be defined as a mathematical tool that simulates the behaviour of a given system. It is a huge simplification of an usually very complex reality. The modeller must select the most relevant processes to be included in the model, and derive or select the appropriate mathematical equations that describe such processes. Such equations are, usually, differential equations that have to be solved numerically in most of the cases. Thus, a computational code has to be developed to carry out the integration of the equations.

A model can predict the behaviour of the modelled system in response to a given external forcing. Thus, they can be used to provide inputs to decision-making processes. Also, a model can be used to test hypotheses concerning the environmental behaviour of a system when direct observations cannot be carried out or, for instance, are too expensive.

A model for radioactivity dispersion in the environment is a good example concerning the comments given above. The model describes the behaviour of a given radionuclide in a given environmental system. The model will include the most relevant processes governing the behaviour of the radionuclide, probably both physical and biogeochemical. These models can be used to assess the radioactive impact arising from deliberate or routine releases of radioactivity into the environment in a quantitative manner. They may provide concentrations of radionuclides in several environmental compartments, also including biota and the foodchain. Thus, they can be used in the prediction of doses for radiological assessments. As discussed in [1], the objectives of any modelling study for radionuclide dispersion can be classified as:

- Description and characterisation of the environment.
- Prediction of the state of the environment as a result of an action. For instance, in the case of an accident where there is only limited data available in the early stages, a model can be used to predict concentrations of radionuclides and whether they will exceed defined limits.
- Monitoring. Models are used in a regulatory situation, to demonstrate compliance with standards set to protect man, biota and the environment.

A model should also give an insight into the processes responsible for the observed behaviour, not only a set of radionuclide concentrations at certain locations, compartments or instants of time.

A particular case would be a model for simulating radioactivity dispersion in the marine environment. Frequently, they are used as tools to forecast the consequences of accidental releases and to support decision-making concerning remedial counter-measures. Model forecasts can help to detect and observe areas of high contamination, but may also serve for further dose assessments. They also help to monitor areas that are affected by permanent releases from reprocessing plants or other nuclear facilities.

But it is not only radioecology that benefits from numerical modelling of dispersion. Oceanographers are also interested in radionuclide dispersion modelling results: anthropogenic radioactivity can be a valuable oceanic tracer of currents and ocean circulation. Releases from nuclear fuel reprocessing plants are specially useful for this purpose since the input function is usually well known. Also, isotopic ratios for certain elements are constant and source specific. Thus, water is labeled and its transit through the ocean can be followed.

The process of building a model has several stages. First, environmental processes to be simulated must be selected, as well as model structure (temporal and spatial resolution etc). This is followed by the creation of a set of differential equations that captures the selected processes and model structure. This mathematical model must be translated into a computational code. Model calibration follows. It consists of selecting the appropriate parameter values that allow the reproduction of measurements. In some cases, it has to be done through simple trial and error. The model must be verified through the comparison of model output with a set of observations different to that used for model calibration. Finally,

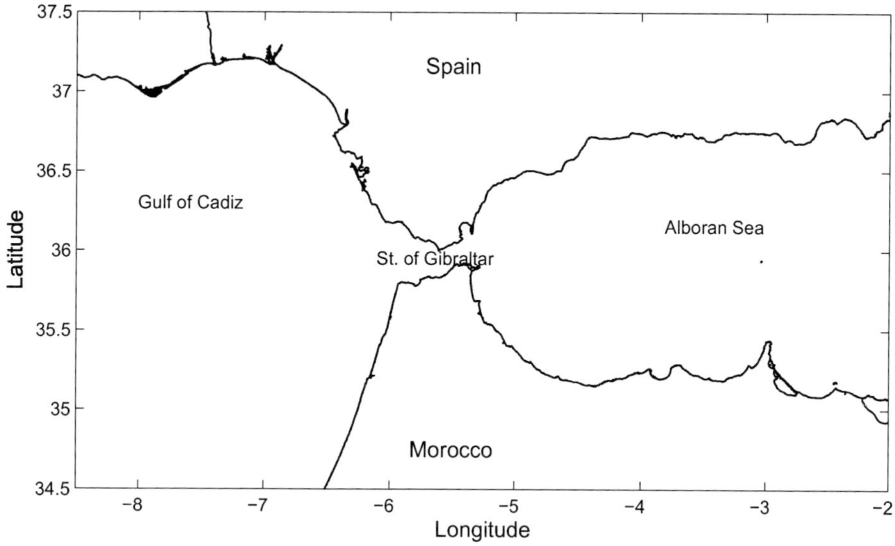

Figure 1. General localization of the study areas.

sensitivity analysis has to be carried out to study the model response to changes in input parameters.

Two examples of models developed with different purposes are presented in this chapter. First, a rapid-response lagrangian model for the Strait of Gibraltar which can be used to support the decision-making process after an accidental or deliberate radioactivity release in the Strait. Next, a model developed to study radionuclide dynamics in the Alborán Sea is presented. This is an example of a model developed with the purpose of describing an environment and giving an insight into the processes that are effective there. Before going into the model presentations, a description of the area of interest, the Strait of Gibraltar-Alborán Sea, is given in the next section. It must be taken into account that water currents are required to calculate the dispersion of any pollutant released to the sea, thus the hydrodynamics of the area of interest must be solved, and a precise knowledge of its physical oceanography is necessary to achieve that.

2. The Strait of Gibraltar-Alborán Sea Region

The Strait of Gibraltar connects the Atlantic Ocean and the Mediterranean Sea. This part of the western Mediterranean is known as the Alborán Sea. The Strait is a narrow channel with length about 40 km and minimum width of some 14 km. A map showing the general localization of the studied areas, the Strait of Gibraltar and the Alborán Sea, between Spain and Morocco, may be seen in Fig. 1. The Gulf of Cádiz is in the Atlantic Ocean and the Alborán Sea is part of the Mediterranean. The Strait of Gibraltar is the only connection between both.

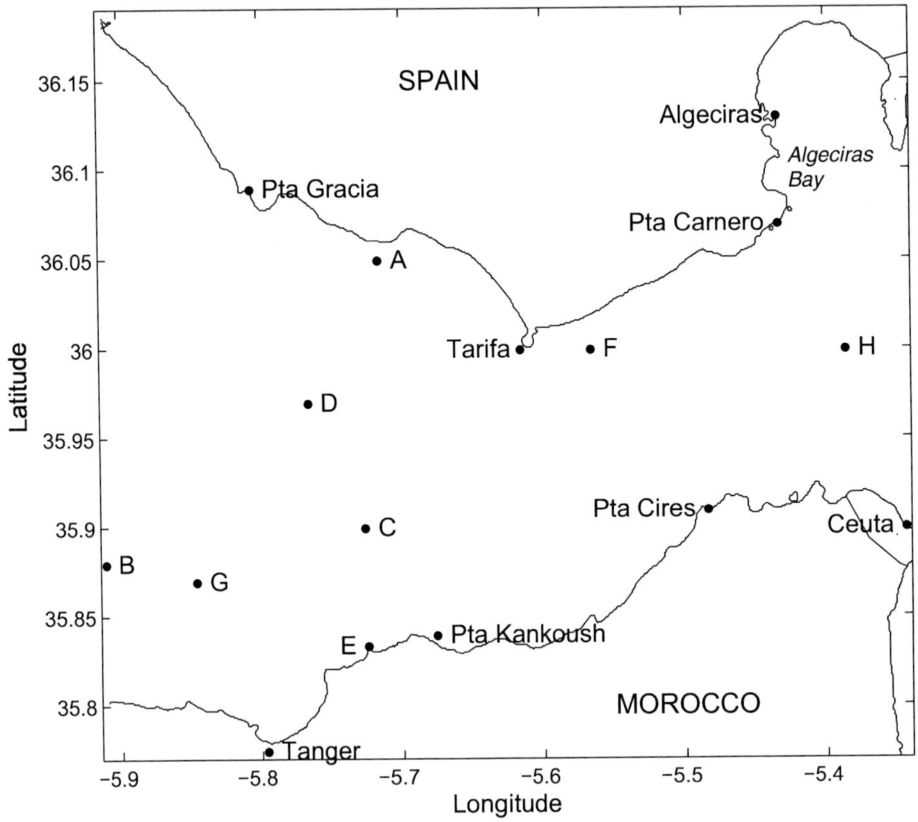

Figure 2. Map of the Strait of Gibraltar showing some important towns and locations where tide amplitudes and currents have been compared with observations.

The area of the Strait (Fig. 2) has a high ecological value, being essential in marine and aerial migratory processes. As the only connection between the Mediterranean Sea and the Atlantic Ocean, marine turtles and mammals (dolphin, porpoise, sperm whale, killer whale) travel through the Strait. By decree 57/2003 (4th March, 2003) of the Junta de Andalucía, the Natural Park of Algeciras-Tarifa Coast was created. It consists of 191.3 km^2, 92.47 of which correspond to the marine part of the Park. A review about geomorphology, flora and fauna of the Park may be consulted elsewhere [2], but some data are summarized here. Over 1900 species of marine flora and fauna living in the Park have been described. Some of the species with a major interest (due to their endemic character and/or rareness) are sponges like Axinella estacioi and jellyfishes (like Merona ibera, Cervera atlantica and Scleranthelia microsclera). There are other 23 species under strict conservation, like Patella ferruginea (the largest limpet of European coasts), Lithophaga lithophaga, Pinna nobilis, Centrostephanus longispinus, Astroides calicularis etc. Also, the region has a high tourist interest, since many kilometers of sand beaches attract thousands of tourists each year. Finally, there are also some important towns. A release of contamination in the area as a consequence of an accident (or a deliberate release) can lead to relevant ecological and economic impacts.

Shipping activities in the area of interest are very intense, again due to the fact that it is the only connection between the Atlantic and the Mediterranean. There is a traffic over 70000 merchant vessels per year, 30 % of them declaring hazardous cargos. Traffic of oil tankers is about 5000 vessels per year. Transit of nuclear submarines and of vessels transporting radioactive waste must also be considered. Only in the Strait of Gibraltar there are over 12000 vessels (mostly passenger ferries) crossing between the north and south coasts. Fishing activities have to be added. Algeciras is the most important port in Spain (and number 25 in the world), with 61.7 Mt of cargo handled in 2004 [3].

It is usual to have adverse meteorological conditions in the Strait, with more than 54 % of days of moderate to poor visibility and 13 % of days with persistent fog conditions. Winds must be added, with frequent east and west gales. East winds (*levantes*) blow an average of 165 days per year, predominantly from April to October, with an average speed of the order of 50 km/h. Maximum speed reaches 125 km/h. Gusts of winds can remain up to 7-10 days. West winds (*ponientes*) blow an average of 60 days per year, from November to March predominantly. Minimum and maximum speeds are 30 and 90 km/h. West winds are not as persistent as *levantes*, lasting for some 12-36 hours.

The particular conditions in the area (intense traffic and adverse meteorology) make navigation difficult. Indeed, 81 accidents have occurred in the Strait in the last years [4], with 14 collisions and 16 groundings. For instance, in 1990 there was a collision between the oil tanker Hesperus and the chemical tanker Sea Spirit. More recently, a ferry and a gas tanker collided off Ceuta.

The water circulation in the Strait of Gibraltar is characterized by a surface inflow of Atlantic water and a deep outflow of more dense Mediterranean water (Figure 3). Exchanged flows are of the order of 1 Sv [5] (1 Sv$=10^6$ m^3/s) with a net inflow into the Mediterranean Sea of about 0.05 Sv. This net inflow compensates the excess of evaporation over precipitation and river supply in the Mediterranean. The exchanged flows show some variability. In particular, Tsimplis and Bryden [5] gave an estimation of 0.66 ± 0.47 Sv for the Atlantic inflow, where the large error is due to seasonal variability, and 0.57 ± 0.26 Sv for the outflow. Nevertheless, these values are about 10 % lower than the lowest previous estimates. A review on transport estimations through the Strait of Gibraltar may be seen in Bryden et al. [6]: from the very first carried out in 1877 and the widely cited value of 1.2 Sv for both outflow and inflow [7], to their own estimations of 0.72 and 0.68 Sv for inflow and outflow respectively.

The water exchanges through the Strait of Gibraltar have an influence in the biogeochemistry of the entire Mediterranean Sea. Substances with higher concentrations in deep water (nutrient profile) present a net loss through the Strait, experiencing a dilution in the Mediterranean. Thus, the inverse estuarine circulation of the whole basin determines a negative budget for nutrients at the Strait of Gibraltar. On the other hand, substances with higher concentrations in the surface waters of the Gulf of Cádiz will be concentrated in an evaporative basin as the Mediterranean Sea. A considerable effort has been made in recent years to determine the budget of elements in the western Mediterranean and their fluxes through the Strait of Gibraltar [8, 9, 10, 11, 12].

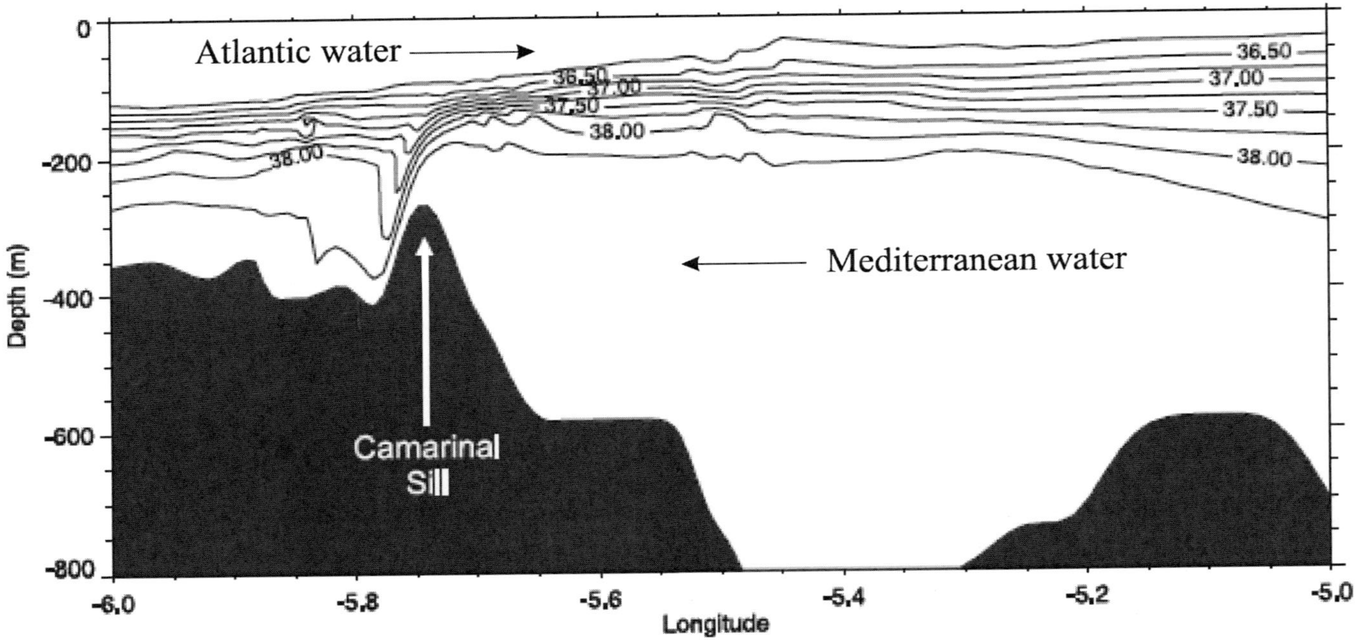

Figure 3. Salinity profile along the axis of the Strait of Gibraltar. A strong pycnocline separating Atlantic and Mediterranean waters, flowing in opposite directions, is apparent.

The water masses that may identified in the Alborán Sea are: the Surface Atlantic Water (SAW) which has its origin in the North Atlantic Central Water (NACW) that has been modified by air-sea interactions, thus strictly speaking is not a water mass [13]; the Surface Mediterranean Water (SMW) which occupies the surface layer in areas not reached by the SAW; the Levantine Intermediate Water (LIW) that flows from the Mediterranean to the Atlantic at depths between some 200 and 600 m and, finally, the Western Mediterranean Deep Water (WMDW), below the LIW and also flowing towards the Atlantic.

However, this description may be simplified to a two-layer system with water flowing in opposite directions [14]. Thus, from an operative point of view, SAW and MSW constitute the upper layer, and LIW and WMDW constitute the bottom layer, flowing to the west. Indeed, a steady westward flow of LIW and WMDW has been recorded in several experiments [14, 15]. This approach has already been used to study the water exchanges between the Atlantic and the Mediterranean by means of numerical models [16, 17, 18].

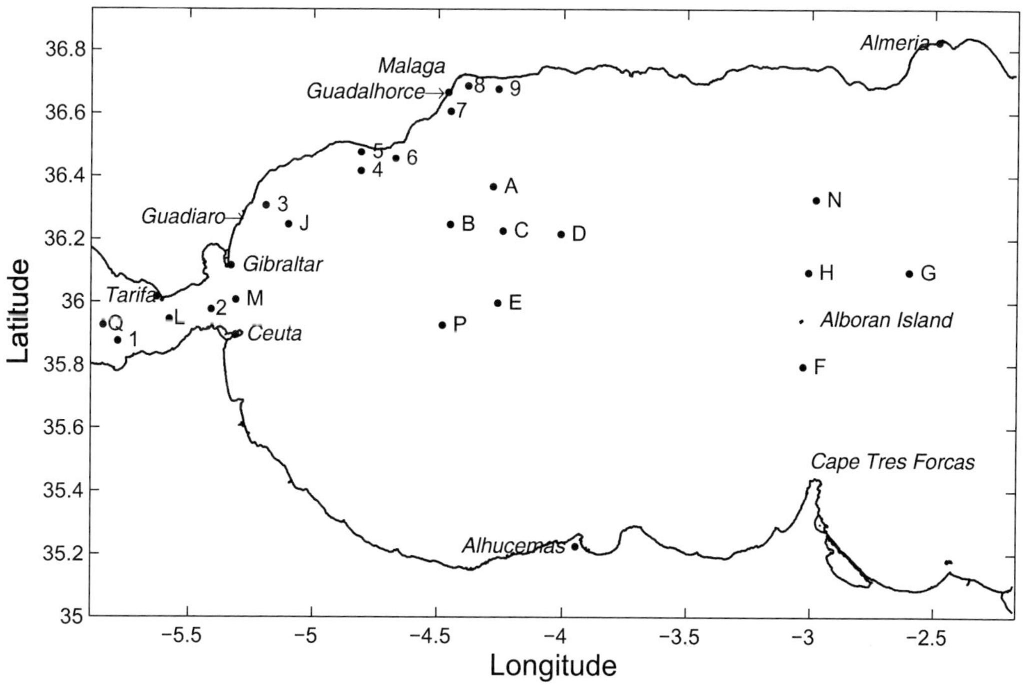

Figure 4. Map of the Alborán Sea showing position of sampling points and geographical sites mentioned in the text. The two main rivers discharging in the area, Guadiaro and Guadalhorce, are also shown. Numbers indicate points where computed current ellipses have been compared with observations and letters are the same but for sedimentation rates and radionuclide concentrations.

The surface circulation in the Alborán Sea is as follows. The inflowing SAW penetrates the Strait of Gibraltar (see Fig. 4 for geographic names) and reaches mean velocities of the order of 0.6 m/s. This water forms a jet that enters the Alborán Sea in a east-northeast direction. The jet flows along the Spanish coast and curves to the south at about $-3.5°$ longitude. Then part of it flows to the west, incorporating to an anticyclonic gyre, while the remainder flows to the African coast between Cape Tres Forcas and Alborán Isle [19]. The anticyclonic gyre is known as the Western Alborán Gyre (WAG). Another gyre, more elusive, fills the eastern basin of the Alborán Sea: it is the Eastern Alborán Gyre (EAG), that is out of the domain of our model. A detailed description may be seen for instance in [20, 21] and references cited in these papers, which include the classical works carried out since the 70s. The WAG is an almost permanent feature, although presents variations in intensity and even there are periods in which disappears [20]. The spatial scale of the WAG is about 100 km in diameter and some 200 m in depth.

Bryden and Stommel [22] have found that the outflow of water from the Mediterranean to the Atlantic occurs along the Moroccan continental slope, and that this circulation appears to be a nearly permanent feature of the southern side of the Alborán Sea.

With respect to the tides in the area, an important feature of the tidal flow in the Strait is that it can be considered, as a first approach, as barotropic. Indeed, 93 % of the variance of current velocity in the semidiurnal band has a barotropic character in the Strait [23]. Tsimplis and Bryden [5] have pointed out that tidal currents are barotropic and larger than the mean inflow or outflow. The semidiurnal tide dominates ADCP (acoustic doppler current profiler) records in the Strait, obscuring the expected two-layer character of the mean flow. The tidal signal is so strong that it reverses the currents near the bottom for a part of each tidal cycle. As a consequence, 2D depth-averaged models have already been applied to simulate surface tides in the Strait [24]. Tsimplis et al. [25] have even used a 2D barotropic model for simulating tides in the whole Mediterranean Sea.

In the case of the main tidal constituent, M_2, tide amplitude at the Atlantic entrance of the Strait is about 0.8 m. This amplitude decreases towards the east. Thus, at the Mediterranean entrance is only about 0.3 m. The amplitude of the tide is reduced more in the Alborán Sea, reaching 0.09 m at the east limit of our domain (see map in Figure 4). The associated currents decrease in a similar way: from tidal currents of the order of 1 m/s in the Strait to currents of a few cm/s in the Alborán Sea basin. Similar behaviour is observed for the S_2 constituent.

The topography of the Strait of Gibraltar-Alborán Sea may be seen in Figure 5. It has been reported [26] that essentially all the western Alborán basin and southern part of the eastern one (which indeed is the whole domain of the model) are covered with muds, with less than 7% of coarse material ($> 63\ \mu$m). The composition of these muds is controlled by river inputs, marine carbonate production in shallow areas, biogenic production in surface water and water mass circulation [27]. Lithogenic particles are mostly introduced to the Alborán Sea by fluvial and eolian mechanisms and, to a small extent, by coastal erosion (less than 5% of the total [28]). Eolian supply is strongly controlled by the proximity of the Sahara Desert. Thus, some episodes of atmospheric dust deposition, mostly of Saharan origin, have been recorded [28]. The fluvial input of particles has been extensively investigated by Liquete et al. [29]. These authors have found that supply from the north coast is at least one order of magnitude larger than from the southern one. They provide detailed data on sediment supply by the main rivers discharging into the Alborán Sea.

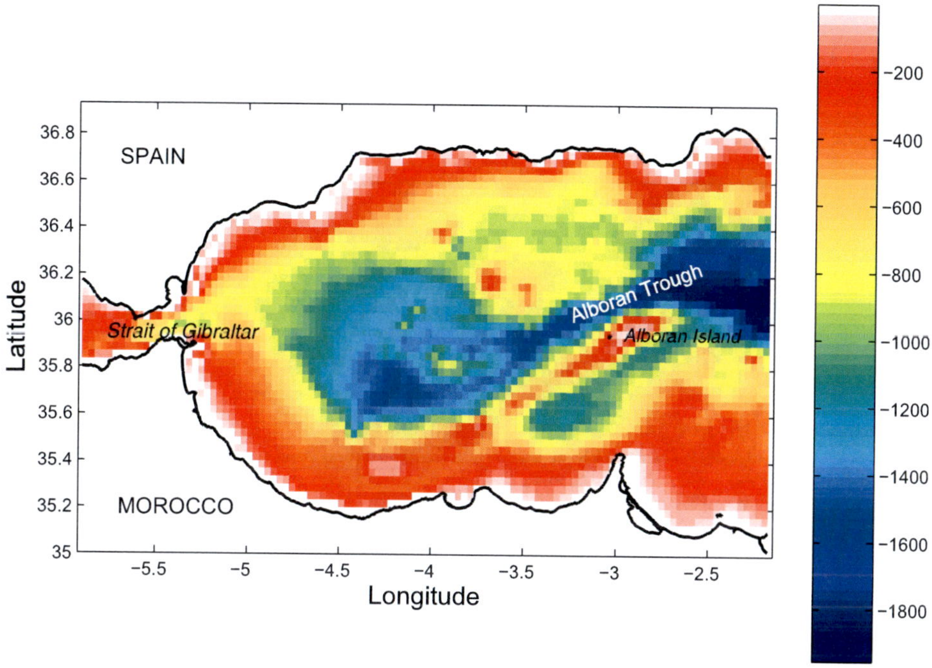

Figure 5. Topography of the Strait of Gibraltar-Alborán Sea. Water depths are given in m.

3. Rapid Response Radioactivity Dispersion Model for the Strait of Gibraltar

A 2D depth-averaged barotropic model, run in advance, is used to obtain currents in the region since this is a suitable approach, as commented before. Tidal and average circulation are stored in files that will be read by the dispersion code to compute water current at any time and position. Dispersion is solved using a particle tracking method. Thus, the radioactive spill is simulated by a number of particles, each of them equivalent to a number of units (for instance Bq), whose paths are followed in time. Radioactive decay is included in the model using Monte Carlo techniques. Radionuclide concentrations may be obtained at the desired time from the density of particles per water volume unit.

3.1. Hydrodynamics

The depth-averaged hydrodynamic equations may be written as [30]:

$$\frac{\partial z}{\partial t}+\frac{\partial}{\partial x}(Hu)+\frac{\partial}{\partial y}(Hv)=0 \tag{1}$$

$$\frac{\partial u}{\partial t}+u\frac{\partial u}{\partial x}+v\frac{\partial u}{\partial y}+g\frac{\partial z}{\partial x}-\Omega v+\frac{\tau_u}{\rho H}=A\left(\frac{\partial^2 u}{\partial x^2}+\frac{\partial^2 u}{\partial y^2}\right) \tag{2}$$

$$\frac{\partial v}{\partial t} + u\frac{\partial v}{\partial x} + v\frac{\partial v}{\partial y} + g\frac{\partial z}{\partial y} + \Omega u + \frac{\tau_v}{\rho H} = A\left(\frac{\partial^2 v}{\partial x^2} + \frac{\partial^2 v}{\partial y^2}\right) \quad (3)$$

where u and v are the depth averaged water velocities along the x and y axis, D is the depth of water below the mean sea level, z is the displacement of the water surface above the mean sea level measured upwards, $H = D + z$ is the total water depth, Ω is the Coriolis parameter ($\Omega = 2w \sin\beta$, where w is the Earth rotational angular velocity and β is latitude), g is acceleration due to gravity, ρ is water density and A is the horizontal eddy viscosity. τ_u and τ_v are friction stresses that have been written in terms of a quadratic law:

$$\begin{aligned} \tau_u &= k\rho u\sqrt{u^2+v^2} \\ \tau_v &= k\rho v\sqrt{u^2+v^2} \end{aligned} \quad (4)$$

where k is the bed friction coefficient.

The solution of these equations provides the water currents at each point in the model domain and for each time step. Currents are treated through standard tidal analysis [31] and tidal constants are stored in files that will be read by the dispersion code to calculate the advective transport of particles. The model includes the two main tidal constituents, M_2 and S_2. Thus, the hydrodynamic equations are solved for each constituent and tidal analysis is also carried out for each constituent separately. A residual transport cannot be produced by the pure harmonic currents given by the tidal analysis, thus tidal residuals have been calculated as well as explained below.

Some open boundary conditions must be provided to solve the hydrodynamic equations. Surface elevations are prescribed as periodic functions of time along the open boundaries of the computational domain (Figure 2). These elevations are obtained from observations [32, 33]. A radiation condition is applied to the water velocity component that is normal to the open boundary:

$$\frac{\partial \phi}{\partial t} = c\frac{\partial \phi}{\partial n} \quad (5)$$

where ϕ is the current component normal to the boundary, in the direction n, and c is a phase speed calculated as in [34]. Water flux across a land boundary is set to zero as usual.

The hydrodynamic equations are solved using an explicit finite difference scheme on a grid with resolution $\Delta x = \Delta y = 1000$ m. The numerical scheme has already been described in detail [30]. Time step, limited by the CFL condition [30], is $\Delta t = 5$ s. Once a stable periodic solution is achieved, tidal analysis is carried out to determine tidal constants that are used by the particle-tracking code. Tidal residual transports are also calculated. This is done for the M_2 and S_2 tides separately. Tidal residuals for each constituent are calculated from the equation:

$$\vec{q_r} = \frac{< H\vec{q_t} >}{< H >} \quad (6)$$

which corresponds to the Eulerian residual transport velocity [35]. In this equation $<>$ is the time averaging operator, $\vec{q_r}$ is the tidal residual and $\vec{q_t}$ is the instantaneous tidal current. In addition, the mean current due to residual flow is calculated from an additional run of the hydrodynamic model. In this run constant values of the sea surface elevation along the two open boundaries of the domain are specified [36]. The total mean current is obtained adding the tidal residuals of the M_2 and S_2 constituents plus the mean flow. Current $\vec{q}$ at any time

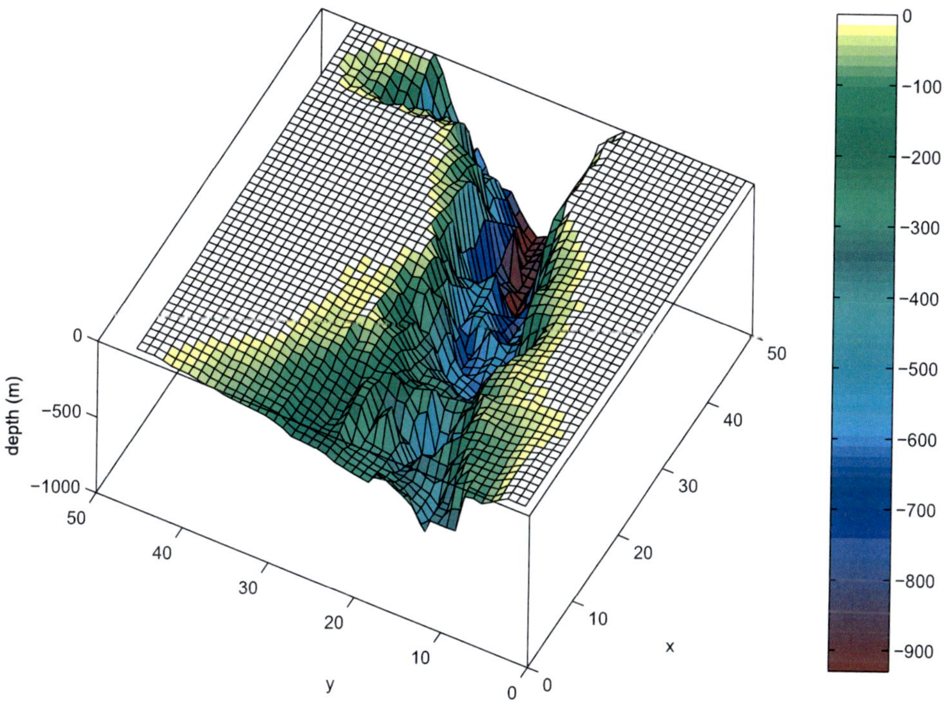

Figure 6. Computational domain showing the detailed topography of the Strait. Each unit in x and y axis is grid cell number (thus equal to 1000 m). North is in the y direction. Units in the color scale are m.

and position in the Strait is obtained adding this total mean current and the instantaneous tidal currents at the corresponding point. This is the current value used in the dispersion calculations.

The computational grid extends from 35.77°N to 36.19°N and from 5.91°W to 5.34°W. Water depths were introduced from a bathymetric chart (Figure 6).

3.2. Lagrangian Dispersion Model

The dispersion of radionuclides is calculated using a particle-tracking method. Essentially, the radionuclide discharge is simulated by a number of discrete particles, as commented above. The path followed by each particle is computed, turbulent diffusion being modelled as a three-dimensional random walk process. Advection is calculated through the addition of the tidal and the residual current. Radioactive decay is also simulated using a Monte Carlo method. The density of particles per water volume unit is finally computed to obtain radionuclide concentrations over the domain at the desired time.

Advection is computed solving for each particle the equation:

$$\frac{d\vec{r}}{dt} = \vec{q} \tag{7}$$

where $\vec{r}$ is the position vector of the particle and $\vec{q}$ is the current vector solved in components u and v.

The particle-tracking model is three-dimensional, but the hydrodynamic calculations provide depth-averaged currents. In the main body of water above the logarithmic layer, the flow gradually increases in a manner which may be represented as [31]:

$$u_{z'} = u_s \left(\frac{D - z'}{D} \right)^{1/m} \tag{8}$$

where $u_{z'}$ is the current speed at a level z' below the sea surface and u_s is the surface flow. From observations, it has been deduced that m ranges between 5 and 7. The surface current can be deduced from the depth averaged one [31]:

$$u_s = \frac{m+1}{m} \overline{u} \tag{9}$$

where $\overline{u}$ is the depth averaged current. Thus, components u and v of the current at any depth can be obtained from their depth-averaged values (provided by the hydrodynamic model) applying equations 8 and 9. This current profile has already been used in particle-tracking dispersion models [37].

Wind is typically included in particle-tracking models assuming that the surface wind-induced current is 3 % of the wind speed measured 10 m above the sea surface [31, 38]. This current decreases logarithmically to zero at a depth z_1. This depth is assumed to be 20 m [39]. The wind induced current at any depth z' below the surface is written as [31]:

$$u_{z'} = \begin{cases} u_0 - \frac{u^*}{\kappa} \ln\left(\frac{z'}{z_0}\right) & \text{if} \quad z' < z_1 \\ 0 & \text{if} \quad z' \geq z_1 \end{cases} \tag{10}$$

where u_0 is the surface wind-induced current, $\kappa = 0.4$ is the von Karman constant, u^* is a friction velocity and z_0 is the sea surface roughness length, which has values between 0.5 and 1.5 mm. It has been obtained [31] that the friction velocity can be estimated as

$$u^* = 0.0012W \tag{11}$$

for a wide range of conditions, where W is wind speed 10 m above the sea surface. From these equations, the wind effect on the advection of particles can be calculated. Of course, the current profile is solved in the u and v components. It may be noted that wind is not included in the hydrodynamic calculations, but only in dispersion. This is the standard approach in rapid response models, which require current fields computed in advance so that computation time is not drastically increased.

Three dimensional diffusion is simulated using a random walk (Monte Carlo) method. It has been shown [38] that it is a simulator of Fickian diffusion provided that the maximum size of the horizontal step given by the particle, D_h, is:

$$D_h = \sqrt{12 K_h \Delta t} \tag{12}$$

in the direction $\theta = 2\pi RAN$, where RAN is a random number between 0 and 1. This equation gives the maximum size of the step. In practice, it is multiplied by RAN to obtain the real size at a given time and for a given particle. Similarly, the maximum size of the vertical step is:

$$D_v = \sqrt{2K_v\Delta t} \tag{13}$$

given either towards the sea surface or the sea bottom. K_h and K_v are the horizontal and vertical diffusion coefficients respectively.

Radioactive decay can be treated using a stochastic method if it is assumed that the probability p of removal of a particle at each time step is:

$$p = 1 - e^{-\lambda\Delta t} \tag{14}$$

where λ is the radioactive decay constant. In practice, a random number is generated for each particle at each time step. If $RAN \leq p$ then the particle is removed from the computation. Obviously, in the case of a stable element $\lambda = 0$.

Both instantaneous and continuous releases can be simulated. From the total amount of pollutant discharged, concentration maps can be obtained by counting the density of particles per water volume unit. Date and time of the discharge (and duration in the case of continuous releases) must be specified since the fate of the release will depend on the tidal state when it took place. Thus, the appropriate phase of each tidal constituent at $t = 0$ must be specified. The values used in this model correspond to the origin of time being January 1, 2003 at 0:15 hours Greenwich time.

The adsorption of pollutants by suspended and bottom sediments can also be simulated with a particle-tracking model [40]. However, these processes are neglected in the present study since suspended matter concentrations are very low in the Strait, typically 0.1-0.5 mg/L [11]. Also, average depth is 350 m (reaching 900 m in the eastern part) and, as a consequence, interactions of pollutants with bed sediments can be neglected as well.

While there is no stability criterion equivalent to the CFL condition in the particle tracking calculations, it is wise to ensure that each particle does not move through a distance that exceeds the grid spacing during each time step. This was satisfied using a time step of 100 s.

All the information required to run the model is summarized in Table 1. The release point is introduced as geographic coordinates. Wind data is included into a file externally edited. The mean current in the Strait is also affected by other factors as for instance atmospheric pressure differences between the Atlantic and the western Mediterranean, and thus presents some variability. Thus, a factor that acts as a modulator of the residual current amplitude must be introduced. If 1 is used, the residual current for the mean water inflow through the Strait is used in the calculations. These mean currents may be amplified or reduced by specifying values for the modulator larger or smaller than 1 respectively.

It is worth commenting that it is difficult to provide a value for this modulator: let us imagine that an accident occurs just now. How do we run the model? In other words: Which is the water inflow through the Strait just now? Presently it is not possible to have an answer. Thus, it is recommended to carry out calculations under the most probable conditions in a first guess (using the mean current, with a modulator equal to 1). Additional simulations may then be carried out using other current modulators to increase and reduce water velocities. This method will, at least, allow to estimate if there is any chance that a given sensible point (a coastal town for instance) is affected by contamination. Given the

Table 1. Information required by the model to be introduced by user. [a]In case of a continuous release, release rate is assumed to be constant along the duration of the accident.

Release point coordinates
Select instantaneous/continuous release option
Wind data file
Release date (day,month,year)
Release time, UTC, (hours,minutes)
Residual current modulator
Simulation time (days)
Magnitude of the release in the corresponding units [a]
Radioactive decay constant

short running times of the model (12 seconds per day of simulation on a Pentium 4 PC in the case of an instantaneous release), this is not a problem. This procedure has already been suggested for a radioactive spill model recently developed for the Alborán Sea [41].

Twelve snapshots at constant intervals during the simulation may be plotted to show the evolution of the radioactive patch over time. Photographs may be seen one by one or as a movie. Another graphic consists of a map of the final radionuclide concentration over the Strait computed from the density of particles per water volume unit. If the option was selected, the time evolution of the number of particles at a given selected point may be finally seen.

3.3. Results and Discussion

3.3.1. Hydrodynamics

After a calibration process, the bed friction coefficient was fixed as $k = 0.070$ and the horizontal eddy viscosity as $A = 10$ m^2/s. In general, good agreement between model results and observations are obtained with these values. Computed tidal charts for the M_2 and S_2 tides are presented in Figure 7. They are in agreement with previously computed ones [24] and observations [32]. They show an amplitude reduction in a factor 2 approximately along the Strait in both tides, and essentially constant amplitudes across the Strait. Cotidal lines are essentially oriented in a northwest-southeast direction. Good quantitative agreement is also obtained. Observed and computed amplitudes and phases of both tides at several locations in the Strait, shown in Figure 2, are given in Table 2.

A comparison between computed and barotropic current amplitudes and phases deduced from measurements [23] in the Strait can be seen in Table 3. The agreement in currents is not so good as in the case of tidal elevations, specially for the M_2 tide. However, the difficulty in appropriately defining the barotropic current has already been commented [24] when these authors compared barotropic currents predicted by their model with those derived from observations in the Strait.

The computed mean currents in the Strait are presented in Figure 8. It can be seen

Table 2. Observed and computed amplitudes (A, cm) and phases (g, deg) of tidal elevations at some locations indicated in Figure 2.

	M_2				S_2			
station	A_{obs}	g_{obs}	A_{comp}	g_{comp}	A_{obs}	g_{obs}	A_{comp}	g_{comp}
Pta Gracia	64.9	49	71.9	59	22.3	74	24.9	82
D	60.1	52	57.5	53	22.5	74	21.4	79
C	54.0	62	54.1	54	21.1	83	20.0	82
A	52.3	48	57.5	54	18.5	73	20.9	79
E	57.1	67	59.8	60	20.6	92	21.4	87
B	78.5	56	78.5	64	29.0	82	27.0	89
Pta Kankoush	51.8	69	51.3	54	20.1	90	19.1	83
Tarifa	41.5	57	42.9	46	14.2	85	16.6	73
F	44.4	48	39.8	45	16.1	74	15.4	72
Pta Cires	36.4	47	37.8	52	14.1	74	14.0	79
Algeciras	31.0	48	29.1	47	11.1	74	11.0	72
Pta Carnero	31.1	48	29.1	46	11.5	71	11.0	71
Ceuta	29.7	50	29.6	52	11.4	76	10.9	74

Table 3. Observed and computed amplitudes (q, m/s) and phases (g, deg) of M_2 and S_2 barotropic tidal velocities at three locations in Figure 2.

	M_2				S_2			
station	q_{obs}	g_{obs}	q_{comp}	g_{comp}	q_{obs}	g_{obs}	q_{comp}	g_{comp}
C	0.91	147	0.94	126	0.31	171	0.37	177
H	0.25	160	0.31	115	0.12	178	0.13	154
G	0.65	157	0.62	129	0.23	182	0.25	172

that maximum currents are obtained in the central part of the Strait, and they are directed along its axis. The magnitude of the mean currents increases as going to the east. The u component (along-strait) of the mean current magnitude along the axis of the Strait is presented in Figure 9 together with earlier calculations [36]. It may be seen that both models give very similar residual currents along the Strait.

3.3.2. Dispersion Model

Ideally, results from the dispersion model should be compared with observations after a real accident in the Strait. However, there are no data until now to carry out this work. Thus, we could only simulate some hypothetical accidents simply to show that results are logical and consistent.

Values for the diffusion coefficients have to be provided. The horizontal diffusion coef-

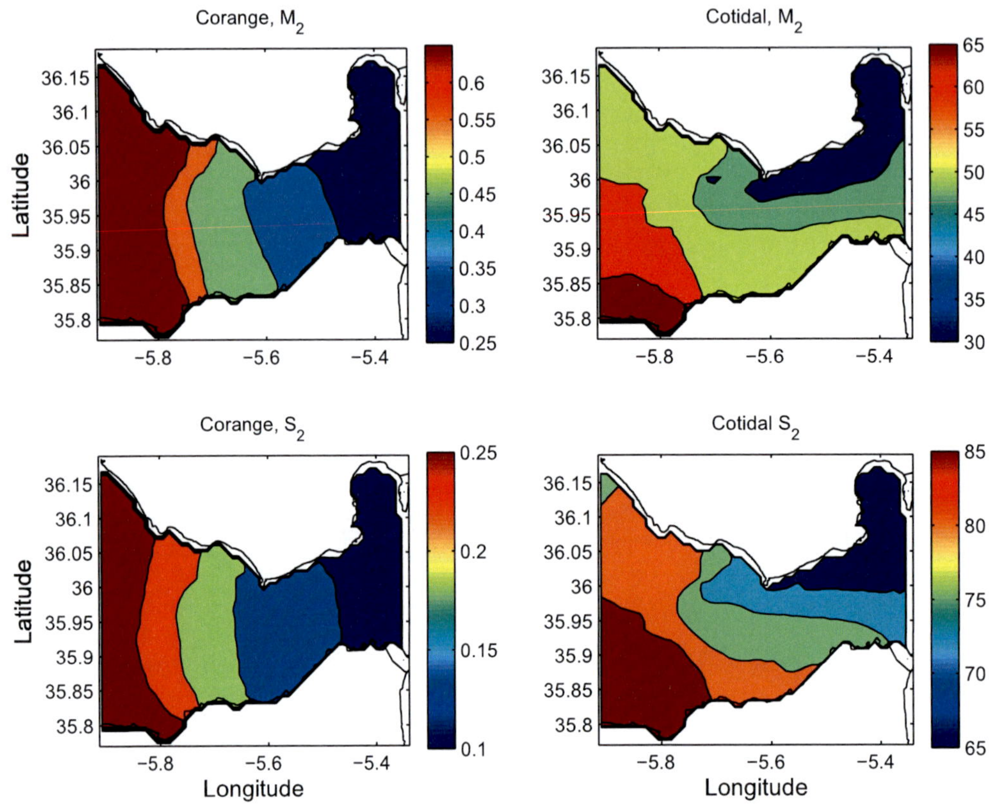

Figure 7. Computed corange and cotidal charts for M_2 (up) and S_2 (down) tides. Amplitudes are given in m and phases in degrees.

ficient depends on the horizontal grid spacing. Following Dick and Schonfeld [42]:

$$K_h = 0.2055 \times 10^{-3} \Delta x^{1.15} \tag{15}$$

The present grid resolution gives $K_h = 0.58$ m^2/s. For the vertical diffusion coefficient a typical value of 0.001 m^2/s is used [43, 44, 45].

A first numerical experiment consisted of simulating an instantaneous release of a long live radionuclide ($\lambda \simeq 0$) at coordinates -5.86° longitude, 35.95° latitude, that corresponds to grid cell (5,20). Date and time of the release are January 1, 2006 0:00 hours UTC (there is not any reason for selecting this instant of time, it is just an example). The total amount released was supposed to be 1.0×10^{12} units (Bq for instance). No wind is assumed. The position of the contamination patch at 8, 16, 20 and 24 hours after the release can be seen in Figure 10. Although it cannot be appreciated in the figure, the patch oscillates forward and backward because of tides. Nevertheless, there is a net transport directed to the Mediterranean caused by the residual current. The mean transit velocity along the Strait results 0.50 m/s (43 km in one day), in excellent agreement with the typical inflow velocity [46]. Winds may vary this transport velocity, thus the time-scale of interest ranges from several hours to several days [46] and the spatial and temporal resolution of the model

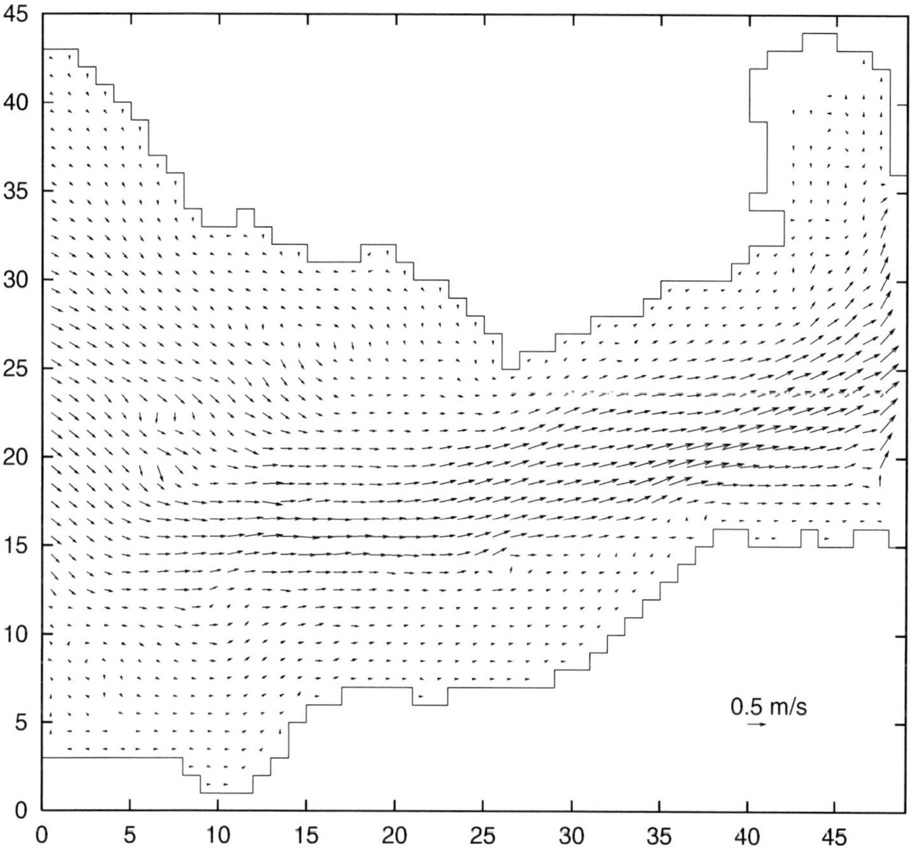

Figure 8. Computed residual currents in the Strait. Each unit in x and y axis is grid cell number (thus equal to 1000 m).

seems appropriate for solving dispersion processes in the Strait of Gibraltar.

The same experiment has been repeated but considering a constant 15 m/s wind from the east (*levante* wind is the most frequent in the Strait). Results are shown in Figure 11, that may compared with Figure 10. It can be clearly observed that east winds tend to retain contaminants in the Strait for a longer time. Also, shear dispersion is enhanced and, as a consequence, the size of the patch increases in the wind direction. The result is that after one day, the contamination patch extends over some 20 km. As another example of the results that may be obtained from the model, the map of radionuclide concentrations over the Strait for this experiment, from the density of particles per water volume unit, is presented in Figure 12.

Winds from the west, in the same direction as the mean current, produce a faster contaminant flush off from the Strait.

Generally speaking, a radioactive patch released in the central region of the Strait will be flushed out the model domain after a few days (depending on wind conditions). However, these initial days are the most relevant from a decision-making point of view, since radionuclide concentrations are higher and, as a consequence, risk (for instance radioactive

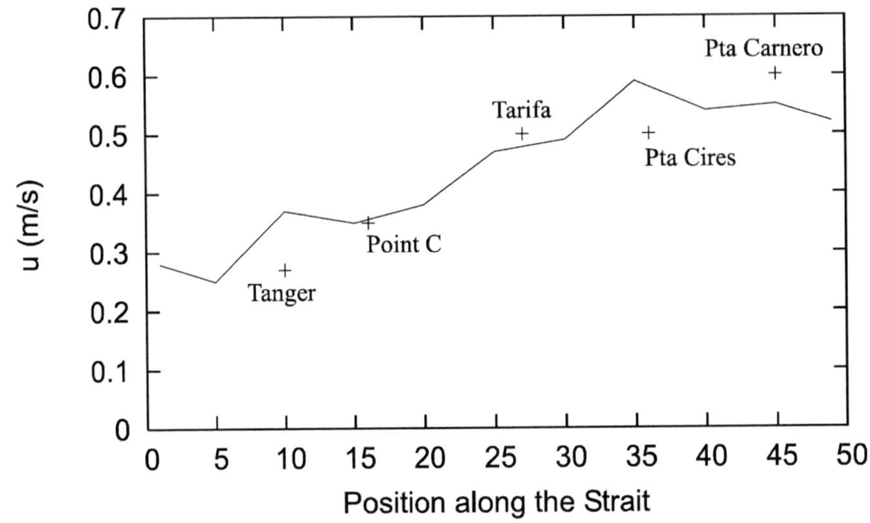

Figure 9. Computed u component of mean current along axis of the Strait (line). Points correspond to earlier calculations [36]. Names (see Figure 2) help to identify position.

dose) is also higher. Nevertheless, if the accident occurs close to the shoreline, the time required to flush out contaminants increases. This may be seen with the example presented in Figure 13, where the instantaneous release occurs close to the Spanish shore: -5.78 ° longitude, 36.04° latitude. Residual currents are weaker along the shores of the Strait (Figure 8), thus contaminants stay in the area for a longer time (in this case the patch is still in the Strait after 48 hours). Forward and backward oscillations of the patch are now more evident, although the full sequence of images is not shown. Also, the patch increases its size in the direction of the current because of the strong horizontal current shear. Finally, it may be seen that the Spanish coast is affected by contamination. As an example, the time evolution of the radionuclide concentration south of Tarifa (coordinates -5.61 ° longitude, 35.99° latitude) is shown in Figure 14. It may be seen that contamination arrives this point 32.3 hours after the accident, and that the maximum concentration (755 units/m^3) occurs 35.6 hours after the release. The effect of tidal oscillations may be appreciated as well in Figure 14.

The situation is similar if the accident occurs close to the coast of Morocco: contaminants are retained for a longer time in the Strait, the patch increases its size in the longitudinal direction and the coast is affected by contamination.

More examples of model results, in particular applications in the case of continuous releases may be seen in [47].

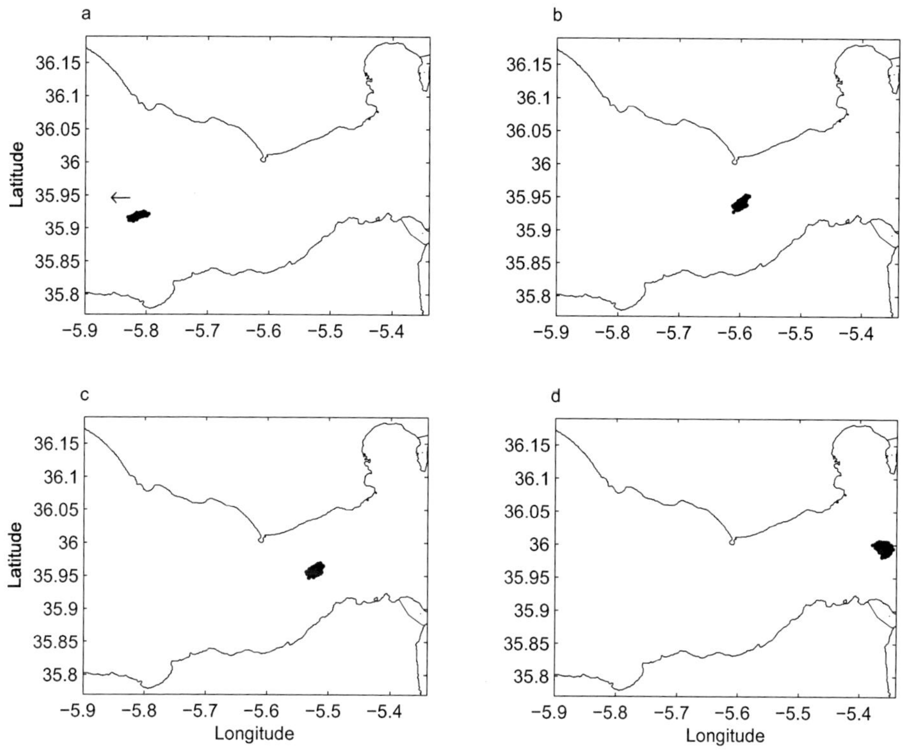

Figure 10. Position of particles 8 (a), 16 (b), 20 (c) and 24 (d) hours after an instantaneous release at the arrow position.

4. A Modelling Study on Radionuclide Dynamics in the Alborán Sea

The objective of this work consists of studying, by means of numerical models, radionuclide dispersion processes in the Alborán Sea. Some excellent works have already been published, which describe water circulation models for the area. Due to the particular dynamics of the Alborán Sea they are reduced gravity or two-layer models [17, 18, 48]. Nevertheless, sediment transport and sedimentation models have not been described for the Alborán Sea to the moment, although a high effort has been made in recent years to measure sedimentation rates and particle fluxes in the Alborán Sea [27, 28] because of the interest of the area, as commented in Section 2.. Moreover, only a particle-tracking rapid-response dispersion model for dissolved radionuclides has been described for the region [41]. Water circulation was obtained from a reduced-gravity model, thus only the dispersion of radionuclides released in the surface could be simulated. Consequently, interactions of radionuclides with sediments were neglected in such study. Now, these processes are taken into account.

Two different hydrodynamic models are applied to describe water circulation in the present work, which is required to solve the advection terms in the dispersion equations as commented before. A depth-averaged model is used to calculate tides and a two-layer model is used to obtain the gestrophic flow. As will be shown, these are reasonable ap-

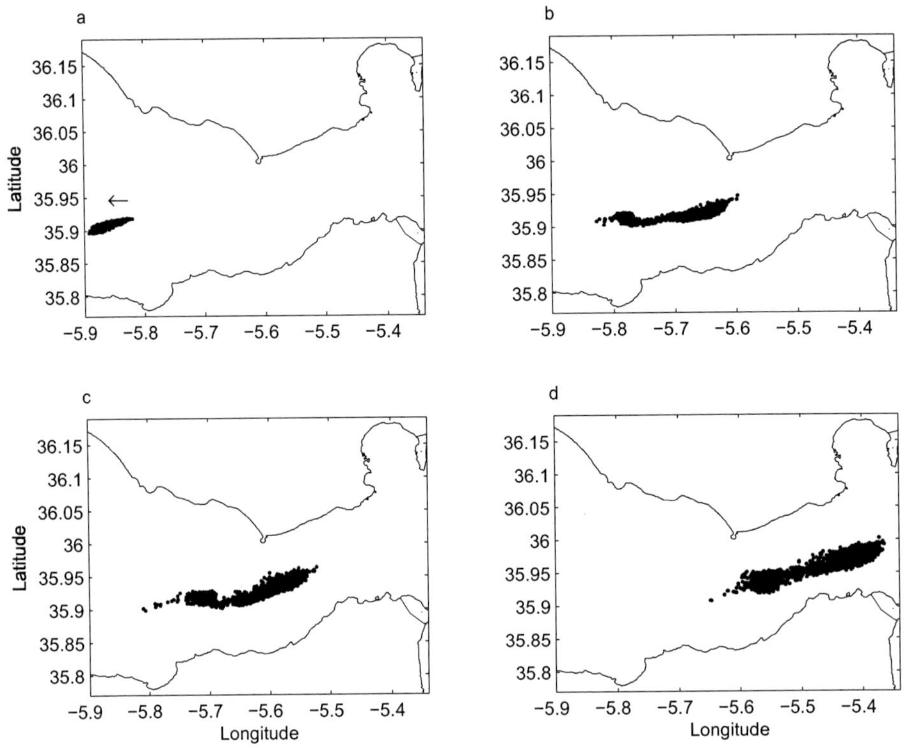

Figure 11. Position of particles 8 (a), 16 (b), 20 (c) and 24 (d) hours after an instantaneous release at the arrow position. Wind from the east, speed 15 m/s.

proaches. Of course, although a full 3D model could be used, we preferred the simplest approach that may be reasonably applied.

The sediment transport model consists of an advection/diffusion equation to which some terms are added: settling, deposition, erosion, diffusion through the pycnocline and external sources of particles. The sediment model is run off-line. Thus, tidal analysis is applied to the output of the depth-averaged model to allow a fast calculation of tidal currents at any time and position in the sea (the two main constituents M_2 and S_2 are included). The residual flow is obtained (from the two-layer model) for a mean water exchange through the Strait of Gibraltar. It is stored in a file that will be read by the sediment model together with files containing tidal constants provided by the tidal analysis. The sediment transport model is integrated until a steady sediment distribution is obtained. Sedimentation rates (SR) are calculated as the balance between deposition and erosion terms. The computed sediment distributions in both water layers and sedimentation rates over the domain are stored in files as well, that are then used by the radionuclide dispersion model to calculate transport and radionuclide exchanges between the water column and the sediments.

Nevertheless, the problem is very complicated since, although tides are periodic, the residual flow presents seasonal variability and is also affected by meteorological conditions as pressure differences between the Atlantic and Mediterranean. Winds have to be considered as well. It was decided to run the two-layer model for average inflow and outflow

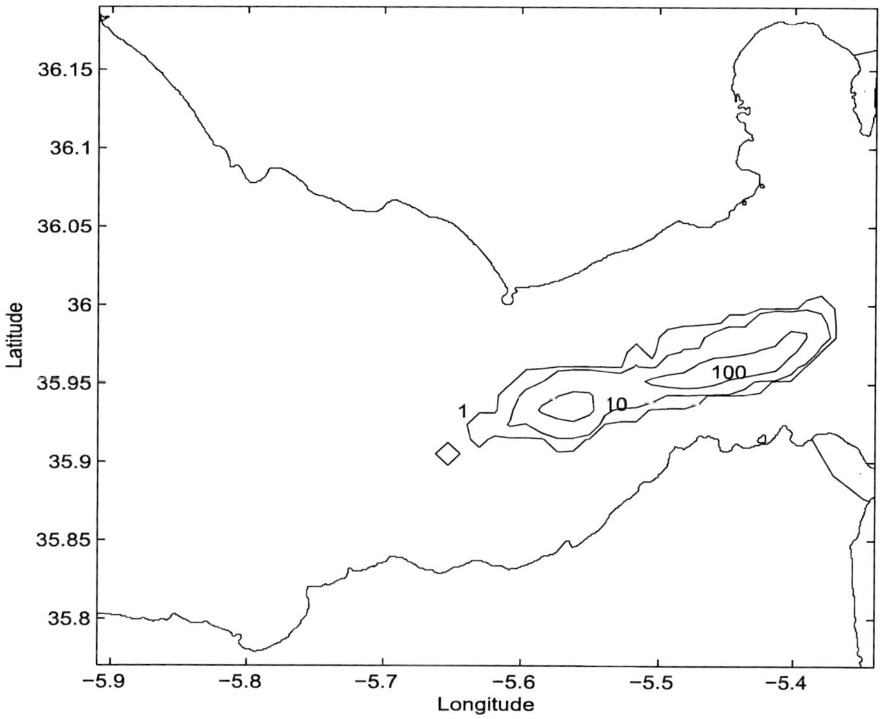

Figure 12. Computed contaminant concentrations (units/m^3) for experiment in Figure 11, 24 hours after release.

through the Strait of Gibraltar and without wind, thus a constant residual flow is obtained. This may be a too simple approach, but can give a realistic view of the main transport processes in the sea, given the generally good agreement between measured and computed sedimentation rates and contaminant distributions that will be shown below. Nevertheless, the sensitivity of suspended sediment and pollutant transport in the Alborán Sea to changing winds is an interesting problem that has to be addressed in the next future since dispersion processes related to meteorological conditions may become apparent. Moreover, erosion events may be induced by wind waves in the shallower areas. Nevertheless, wave-induced erosion is not included in the model since calculations are carried out under calm conditions. These approaches have been successfully used before when studying the dispersion of suspended particles and radionuclides by the Rhone River plume, northwest Mediterranean Sea [49, 50]. Following these approaches, the mean particle supplies from the rivers are used [29] and deposition events of Sahara dust are not considered since cannot be quantified. Coastal erosion has finally been neglected in comparison with river supply [28]. Only suspended matter is considered in this study, thus bedload transport is not included in the model. The sources of radionuclides will be discussed below.

In spite of all these simplifications, the model retains the essential aspects of water circulation in the Alborán Sea and gives a correct overview of the sedimentation process and suspended particle and pollutant distributions, as will be shown below.

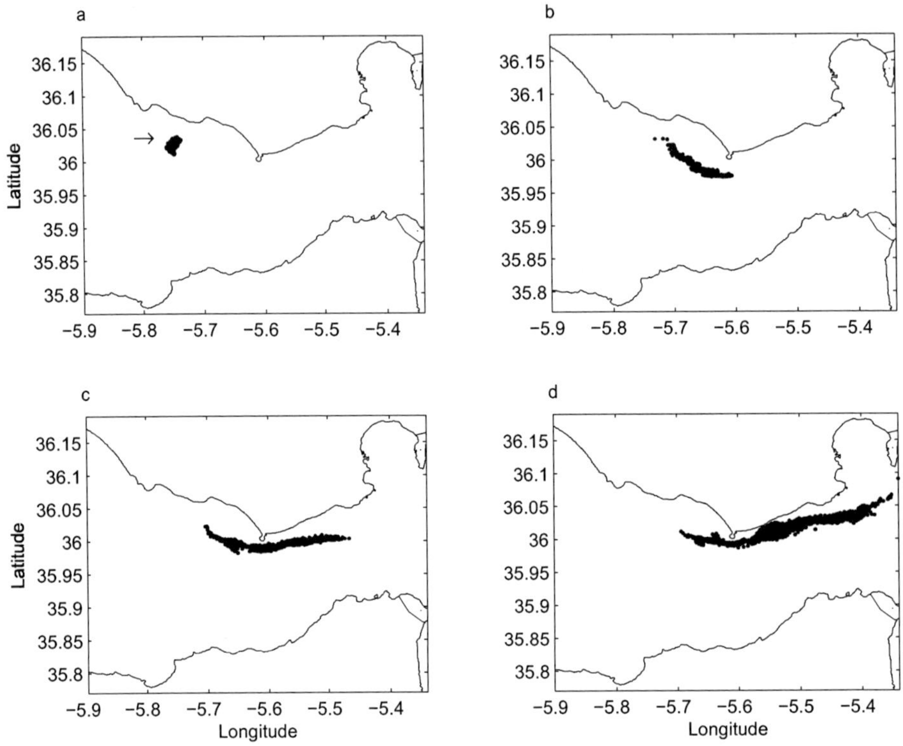

Figure 13. Position of particles 8 (a), 24 (b), 36 (c) and 48 (d) hours after an instantaneous release at the arrow position.

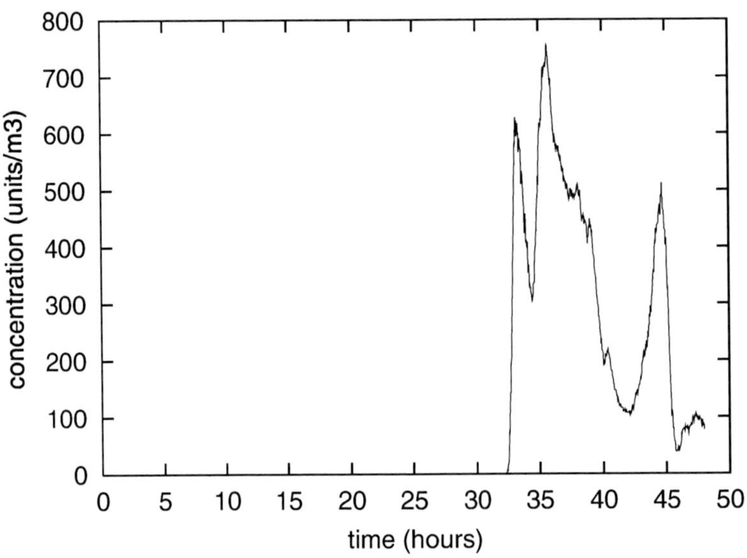

Figure 14. Time evolution of pollutant concentration at position -5.61 ° longitude, 35.99° latitude (south of Tarifa) for experiment shown in Figure 13.

4.1. Hydrodynamics

The hydrodynamics of the area is described, as commented before, by means of two models: a barotropic model is used to calculate tides and a two-layer model is applied to obtain the mean circulation. Tides have been explicitly calculated since tidal currents, specially in the shallower areas in the vicinity of the Strait of Gibraltar and during spring tides, will increase bed stress and thus enhance erosion of the bed sediment. These erosion episodes would be missed if only the residual flow is considered.

The 2D barotropic model is the same as in the Strait of Gibraltar rapid response model (Section 3.1.), thus its description will not be repeated here.

For the mean flow, two water layers with different densities are flowing in opposite directions in the Alborán Sea. The equations describing this flow are the following (in the formulation of Izquierdo et al. [16]):

$$\frac{\partial(H_1\vec{u_1})}{\partial t}+\nabla(H_1\vec{u_1}\cdot\vec{u_1})+\Omega\vec{k}\times H_1\vec{u_1}+gH_1\nabla\zeta_1=-\vec{\tau_1}/\rho_1+AH_1\nabla^2\vec{u_1} \tag{16}$$

$$\frac{\partial H_1}{\partial t}+\nabla\cdot H_1\vec{u_1}=0 \tag{17}$$

$$\frac{\partial(H_2\vec{u_2})}{\partial t}+\nabla(H_2\vec{u_2}\cdot\vec{u_2})+\Omega\vec{k}\times H_2\vec{u_2}+g\frac{\rho_1}{\rho_2}H_2\nabla\zeta_1+g'H_2\nabla\zeta_2=(\vec{\tau_1}-\vec{\tau_2})/\rho_2+AH_2\nabla^2\vec{u_2} \tag{18}$$

$$\frac{\partial H_2}{\partial t}+\nabla\cdot H_2\vec{u_2}=0 \tag{19}$$

with indexes 1 and 2 for the upper and lower layers respectively. In these equations H_i is the thickness of the water layer, ρ_i is water density in each layer, A is a horizontal friction coefficient and g' is the reduced gravity:

$$g'=g\frac{\rho_2-\rho_1}{\rho_2} \tag{20}$$

ζ_1 is the elevation of the sea surface with respect to the mean level and ζ_2 is the depth of the interface between layers. Finally $\vec{\tau_1}$ and $\vec{\tau_2}$ are friction stresses between water layers and between the lower layer and the seabed respectively. They are formulated in terms of a quadratic law as usual:

$$\begin{aligned}\vec{\tau_1}&=c_1\rho_1|\vec{u_1}-\vec{u_2}|(\vec{u_1}-\vec{u_2})\\ \vec{\tau_2}&=c_2\rho_2|\vec{u_2}|\vec{u_2}\end{aligned} \tag{21}$$

where c_1 and c_2 are the interfacial and bottom friction coefficients. No significant differences in model results were observed between a $f-$plane and a $\beta-$plane approach, thus the first was used.

As commented before, current at a given time and position is obtained through the addition of the instantaneous tidal current plus the residual flow. Tidal currents in the bottom layer are obtained from the depth-averaged tidal current provided by the barotropic model using the standard current profile already described in Section 3.1..

4.2. Sediment Transport

The transport of sediments is described by an advection-diffusion equation to which some terms are added. These are external sources of particles, terms describing particle deposition on the seabed and erosion from the bed to the water column, vertical settling and diffusive transport through the interface separating the two water layers. Equations are written in the following form:

$$\frac{\partial (Hm)_k}{\partial t} + \frac{\partial (uHm)_k}{\partial x} + \frac{\partial (vHm)_k}{\partial y} = \frac{\partial}{\partial x}\left(H_k K_h \frac{\partial m_k}{\partial x}\right) + \frac{\partial}{\partial y}\left(H_k K_h \frac{\partial m_k}{\partial y}\right) \quad (22)$$

$$-w_{s,1} m_1 (\delta_{k,1} - \delta_{k,2}) + \frac{\partial}{\partial z}(K_v m_k) + (ER - DEP)\delta_{k,2} + S\delta_{k,1}$$

where $k = 1, 2$ for the upper and lower layers respectively, m is the suspended matter concentration, H is the water layer thickness, u and v are layer-averaged currents, K_h and K_v are the diffusion coefficients, respectively, for horizontal mixing and mixing through the pycnocline, S is the external particle source and ER and DEP are the erosion and deposition terms respectively. The settling velocity of particles $w_{s,k}$ is different for each layer due to the different water density and delta functions are introduced to consider that erosion and deposition processes only affect to the bottom water layer and that the external sources, due to river supplies, are introduced into the upper layer.

The deposition term is usually written in the following form [51, 52]:

$$DEP = w_{s,2} m_2 \left(1 - \frac{|\vec{\tau}|}{\tau_{cd}}\right) \quad (23)$$

where $\vec{\tau}$ is the total bottom friction stress and τ_{cd} is a critical deposition stress above which no deposition occurs since particles are kept in suspension by turbulence. The settling velocity of particles can be obtained from Stokes's law:

$$w_{s,k} = \frac{\rho - \rho_k}{\rho_k} \frac{2gR^2}{9\nu} \quad (24)$$

where ρ and R are suspended particle density and radius respectively, ρ_k is density of water in layer k and ν is the kinematic viscosity of water.

The erosion rate is generally written in terms of the erodability constant [51, 53]:

$$ER = Ef\left(\frac{|\vec{\tau}|}{\tau_{ce}} - 1\right) \quad (25)$$

where E is the erodability constant, f gives the fraction of small particles (diameter less than 62.5 μm) in the bed sediment (which are particles that may be resuspended) and τ_{ce} is a critical erosion stress below which no erosion occurs. Finally, it is also possible to calculate net sedimentation rates as the balance between the deposition and erosion terms:

$$\mathrm{SR} = DEP - ER \quad (26)$$

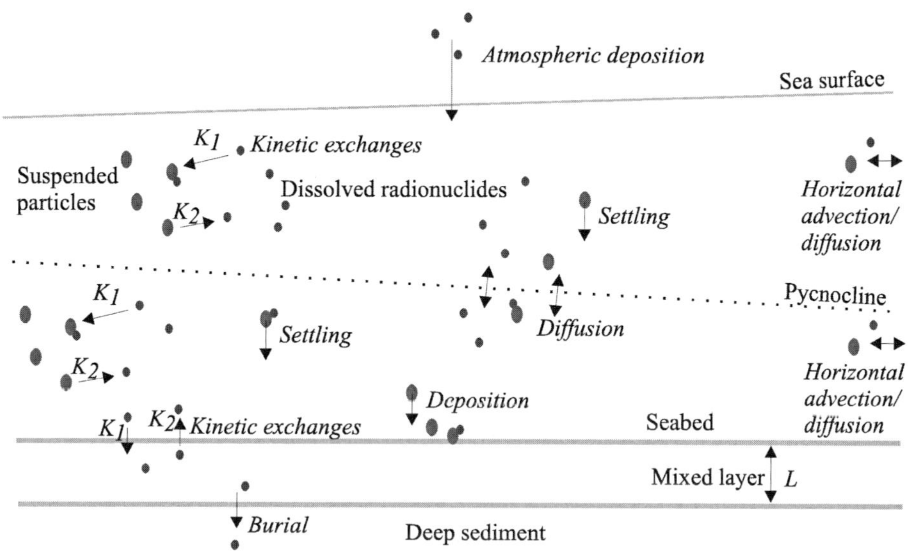

Figure 15. Processes affecting the dispersion of non conservative radionuclides in a two-layer sea. Black circles are suspended particles and the red ones are radionuclides.

4.3. Radionuclide Dispersion in a 2-layered Sea

Non conservative radionuclides are those which do not remain dissolved in the water column, but have a certain affinity to be fixed to particles. If the radionuclide is introduced in the surface water layer, it will be fixed to settling suspended particles and their deposition on the sea bed will contaminate the bottom sediment. Of course there are also advection/diffusion processes in both water layers, diffusion of dissolved radionuclides and of particles through the pycnocline and direct adsorption of radionuclides on the seabed. The exchanges of radionuclides between the dissolved and solid phases may be described in terms of kinetic transfer coefficients. Thus, assuming that adsorption/release reactions are governed by a single reversible reaction, a coefficient k_1 governs the transfer from the liquid to the solid phase and a coefficient k_2 governs the inverse process. Also, the migration of radionuclides to the deep sediment must be included since simulations over several years are carried out. Thus, radionuclides deposited on the sediment surface will be buried by particle deposition and will migrate below the mixed sediment layer that directly interacts with the dissolved phase (see for instance the discussion in Monte et al. [54]). This effect may be easily treated as a decay process with constant λ_{burial} given by:

$$\lambda_{burial} = \frac{\mathrm{SR}}{\rho_s L} \tag{27}$$

where L is the sediment mixing depth (the distance to which the dissolved phase penetrates the sediment) and ρ_s is the sediment bulk density (dry mass divided by wet volume). A summary of all the processes involved may be seen in Figure 15.

It is known that adsorption depends on the surface of particles per water volume unit at each point and time. This quantity has been denoted as the exchange surface [58, 59, 55,

56, 57]. Thus, the kinetic coefficient $k_{1,k}$ is written as:

$$k_{1,k} = \chi(S_{m,k} + S_s \delta_{k,2}) = k_{1,k}^m + k_1^s \tag{28}$$

where S_m and S_s are the exchange surfaces for suspended matter and bottom sediments respectively (dimensions $[L]^{-1}$) and χ is a parameter with the dimensions of a velocity denoted as the exchange velocity (see references cited above). The index $k = 1, 2$ represents each water layer. Assuming spherical particles, the exchange surfaces are written as:

$$S_{m,k} = \frac{3m_k}{\rho R} \tag{29}$$

and

$$S_s = \frac{3Lf\phi(1-p)}{RH_2} \tag{30}$$

where R and ρ are particle radius and density respectively, f is the fraction of small particles in the sediment, p is sediment porosity and ϕ is a correction factor that takes into account that part of the sediment particle surface may be hidden by other sediment particles. This formulation has been successfully used in all modelling works cited above. Real particles are not spheres, but with this approach it is possible to obtain an analytical expression for the exchange surface [60]. The kinetic coefficient k_2 is considered to be constant.

The equation that gives the time evolution of the radionuclide concentration in the dissolved phase, C_d, is:

$$\frac{\partial (HC_d)_k}{\partial t} + \frac{\partial (uHC_d)_k}{\partial x} + \frac{\partial (vHC_d)_k}{\partial y} = \tag{31}$$

$$\frac{\partial}{\partial x}\left(H_k K_h \frac{\partial C_{d,k}}{\partial x}\right) + \frac{\partial}{\partial y}\left(H_k K_h \frac{\partial C_{d,k}}{\partial y}\right) + \frac{\partial}{\partial z}(K_v C_{d,k}) -$$

$$k_{1,k}^m H_k C_{d,k} + k_2 H_k m_k C_{s,k} - k_1^s H_2 C_{d,2} + \delta_{k,2} k_2 A_s L \rho_s f \phi - \lambda H_k C_{d,k}$$

where C_s and A_s are radionuclide concentrations in suspended matter and the bed sediment mixed layer respectively and λ is the radioactive decay constant.

The equation for the time evolution of radionuclide concentration in suspended matter is

$$\frac{\partial (HC_s m)_k}{\partial t} + \frac{\partial (uHC_s m)_k}{\partial x} + \frac{\partial (vHC_s m)_k}{\partial y} = \tag{32}$$

$$\frac{\partial}{\partial x}\left(H_k K_h \frac{\partial (C_s m)_k}{\partial x}\right) + \frac{\partial}{\partial y}\left(H_k K_h \frac{\partial (C_s m)_k}{\partial y}\right) +$$

$$\frac{\partial}{\partial z}(K_v C_{s,k} m_k) - w_{s,1} m_1 C_{s,1}(\delta_{k,1} - \delta_{k,2}) + k_{1,k}^m C_{d,k} -$$

$$k_2 m_k C_{s,k} - \delta_{k,2} \mathrm{SR} C_{s,k} - \lambda H_k m_k C_{s,k}$$

and for the bed sediment radionuclide concentration we have

$$\frac{\partial A_s}{\partial t} = k_1^s \frac{H_2 C_{d,2}}{\rho_s L f} - k_2 A_s \phi + \mathrm{SR}\frac{C_{s,2}}{\rho_s L f} - (\lambda + \lambda_{burial}) A_s \tag{33}$$

in the mixed sediment depth. The total radionuclide content, A_p, in the sediment below the mixed depth is given by the following equation:

$$\frac{\partial A_p}{\partial t} = \lambda_{burial}\rho_s L f A_s - \lambda A_p \tag{34}$$

The radionuclide concentration in the sediment mixed depth is given by fA_s. This is what would be obtained from a sediment sample where a coarse sediment fraction $(1 - f)$ exists with negligible radionuclide concentration.

4.4. Computational Scheme

The 2D barotropic equations are solved using the same explicit finite difference scheme as in the Strait of Gibraltar model [30]. Some boundary conditions must be provided to solve them: surface elevations are specified, from observations, along open boundaries of the computational domain. Water flux across a land boundary is set to zero as usual.

The model is integrated until steady periodic oscillations are obtained, then tidal analysis is carried out to calculate and store tidal constants. This procedure is followed for each tidal constituent separately.

The equations of the two-layer model are solved using the same scheme as in the barotropic model. Water flows are specified along open boundaries and equations are integrated until a steady circulation is obtained. Mean surface inflow and deep outflow used to integrate the model are, respectively, 1.25 and 1.20 Sv. The inflow magnitude corresponds to the value previously used by Preller [17] and the outflow is smaller by 0.05 Sv, as discussed before. Flows are gradually increased from zero to their maximum values over a finite time to minimize the excitation of high frequency waves [17]. Densities of the upper and lower layers are, respectively, 1027 and 1029 kg/m^3 [16]. Friction coefficients were obtained from a calibration procedure, thus friction between layers was selected as 0.0001 and friction with the seabed as 0.0025. The horizontal eddy viscosities were fixed as $A = 10$ m^2/s in the barotropic model and $A = 50$ m^2/s in the two-layer model to improve stability.

The computed steady circulation is then stored in files that will be read by the sediment and pollutant transport models. The total water velocity at any point in the domain is obtained adding the residual velocity obtained from the 2-layer model (constant at each point) and the tidal velocities (time dependent).

Explicit finite differences are also used to solve the advection/diffusion equation. In particular, the second order accuracy MSOU (monotonic second order upstream) scheme [61, 30] has been applied for the advective terms. A second order scheme is also applied for diffusion [62, 30]. The sediment transport model is again integrated until a steady situation is achieved. In practice, simulations over 4 months are carried out. The source of sediment is river supply [29]: the 7 main rivers discharging in the Alborán Sea have been considered. The suspended particle concentration in the inflowing Atlantic waters is fixed as 0.5 g/m^3 [11] and a zero gradient condition is used along the other open boundaries. This kind of boundary conditions has already been used [63].

The present grid resolution (see below) gives $K_h = 2.0$ m^2/s for the horizontal diffusion coefficient through the application of equation 15. Diffusion through the pycnocline was fixed as $K_v = 1.0 \times 10^{-5}$ m^2/s.

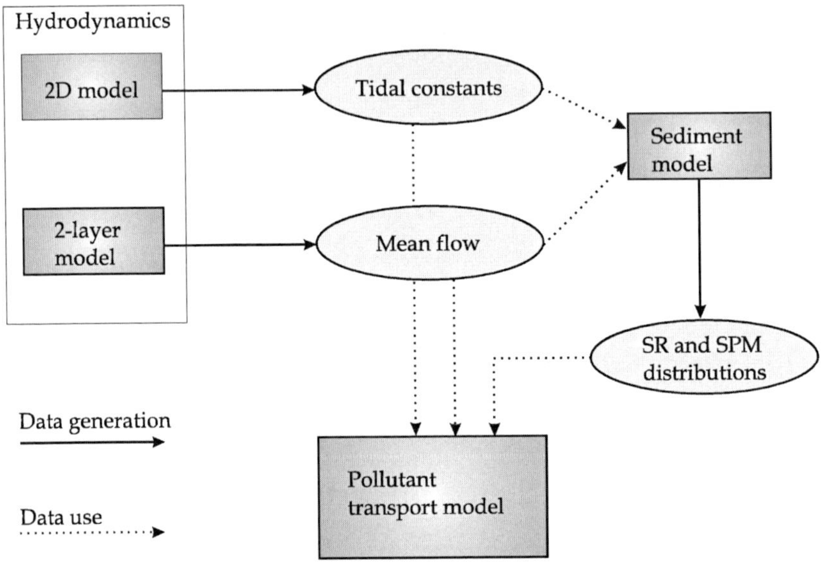

Figure 16. Modelling procedure. Yellow ellipses are data generated by the corresponding model (grey rectangles).

The computed steady distribution of suspended particles and sedimentation rates are again stored in files that are read by the pollutant dispersion model.

The computational domain extends from 35°00′ N to 36°56′ N and from 5°54′ W to 2°10′ W. Resolution of the grid is 2 minutes. Water depths were downloaded from the NOAA (US National Ocean and Atmosphere Administration) Geodas database (Fig. 5), which is available on-line. Time steps used for integration of the models are the following: $\Delta t = 10$ s for both hydrodynamic models and $\Delta t = 300$ s for the transport models. With these selections all stability limitations are satisfied.

The following stability condition is introduced by the terms describing the transfers of pollutants between the liquid and the solid phases [30]:

$$\Delta t < \frac{1}{k_{max}} \tag{35}$$

where k_{max} is the maximum kinetic rate involved in the equations. It means that the pollutant mass transferred from one phase to another in a time step must be smaller than the content in the origin phase. This condition is more restrictive for a highly reactive radionuclide as ^{239,240}Pu and ^{210}Pb than for more conservative ones as, for instance, ^{137}Cs. The same time step as in the sediment transport model was used when the pollutant dispersion model was applied to ^{137}Cs, but the model was not stable in the case of ^{239,240}Pu. Thus, a time-splitting procedure was used to treat the water-sediment interactions. This is essentially the same as in hydrodynamic models, when time-splitting is used to solve the external and internal modes. The time step used to solve the water-sediment interaction process is reduced to 10 s in the case of plutonium. A simple scheme of the general modelling procedure may be seen in Fig. 16.

Table 4. Observed (subindex *obs*) and computed (subindex *comp*) amplitudes (A, cm) and phases (g, deg) of tidal elevations at several locations indicated in Figure 4.

	M_2				S_2			
station	A_{obs}	g_{obs}	A_{comp}	g_{comp}	A_{obs}	g_{obs}	A_{comp}	g_{comp}
Tarifa	42	57	41	45	14	85	16	79
Ceuta	30	50	32	52	11	76	12	86
Málaga	17	59	18	42	7	72	8	77
Alhucemas	18	58	18	56	7	80	7	89
Almería	9	51	9	49	4	78	4	77
Gibraltar	30	46	29	41	11	72	12	77

4.5. Results and Discussion

4.5.1. Hydrodynamics

Computed values of the amplitude and phase of the water surface elevation due to the M_2 and S_2 tides have been compared with observed values [25, 64] at different locations in the Strait of Gibraltar and the Alborán Sea. Results are given in Table 4. Locations may be seen in the map in Figure 4. It can be observed that generally there is a good agreement between the measured and computed tidal constants for both constituents. A tidal chart for the M_2 tide is shown in Figure 17. The same color joins points with the same tide amplitude (corange chart). The amplitude decreases quickly in the Strait of Gibraltar, from some 0.75 m in the Atlantic entrance to 0.30 in the section of Gibraltar. There is a further, although slower, amplitude reduction in the Alborán Sea, reaching only some 0.10 m at the eastern boundary. It can be seen that corange lines are oriented in a south-north direction, while cotidal lines (not shown) are essentially in a northeast direction, in agreement with observations and earlier computations [25].

The spatial distribution of the M_2 current amplitude is presented in Figure 18 as an example, and a comparison of computed current ellipse parameters with those deduced from measurements [65] is presented in Table 5. In general, there is a good agreement between both set of data. Returning to Figure 18, the current amplitude in most of the Alborán Sea is some 0.03 m/s. Only in some specific areas larger currents are obtained. This is the case of the south-north section going from Cape Tres Forcas to Alborán Isle and the Spanish coast. Currents of the order of 0.06 to 0.09 m/s are computed close to the Cape, around Alborán Isle and close to the Spanish coast. This is in good agreement with earlier computations [25]. Current amplitude increases as approaching Gibraltar Strait, where, as expected, maximum values are obtained. Indeed, currents of the order of 0.8 m/s are computed in some locations. For clarity reasons, however, the scale maximum in Figure 18 is limited to 0.2 m/s. In the case of the S_2 tide, results show a similar decrease in the tide amplitude, from some 0.25 m at the Atlantic entrance of the Strait to some 0.04 m at the eastern open boundary of the domain. Current amplitudes decrease from maximum values of the order of 0.25 m/s in the Strait of Gibraltar to currents weaker than 1 cm/s in the Alborán Sea.

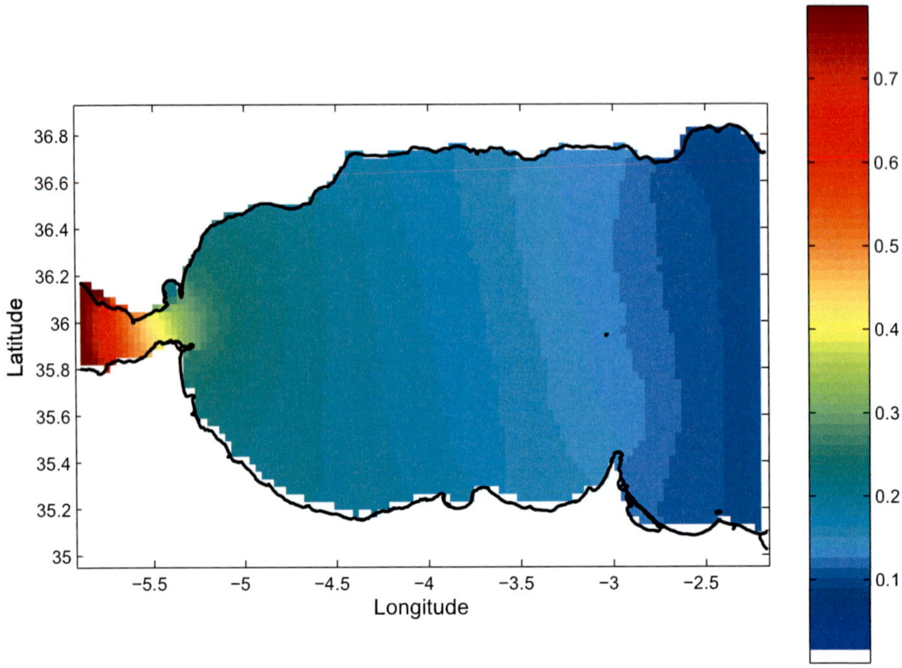

Figure 17. Corange chart for the M_2 tide. Amplitude of the tide is given in m.

Table 5. Computed and measured current ellipse parameters for the M_2 tide at points indicated with numbers in Figure 4. M and m are the magnitudes of the major and minor semiaxes respectively and dir is the direction of the major semiaxis measured anticlockwise from east.

point	measured values			computed values		
	M (cm/s)	dir (deg)	m (cm/s)	M (cm/s)	dir (deg)	m (cm/s)
1	65	15	3	68	2	2
2	21	20	6	34	22	1
3	9	54	2	14	57	1
4	6	4	0.5	5	7	0.1
5	3	22	1	3	-5	0.03
6	7	22	2	6	-8	0.4
7	4	45	2	2	67	1
8	2	40	0.2	4	2	1
9	1	-5	0.5	2	17	0.1

The calculated mean flows for the upper and lower layers, for the mean water exchanges through the Strait of Gibraltar, are presented in Figure 19. In the upper layer, the jet of

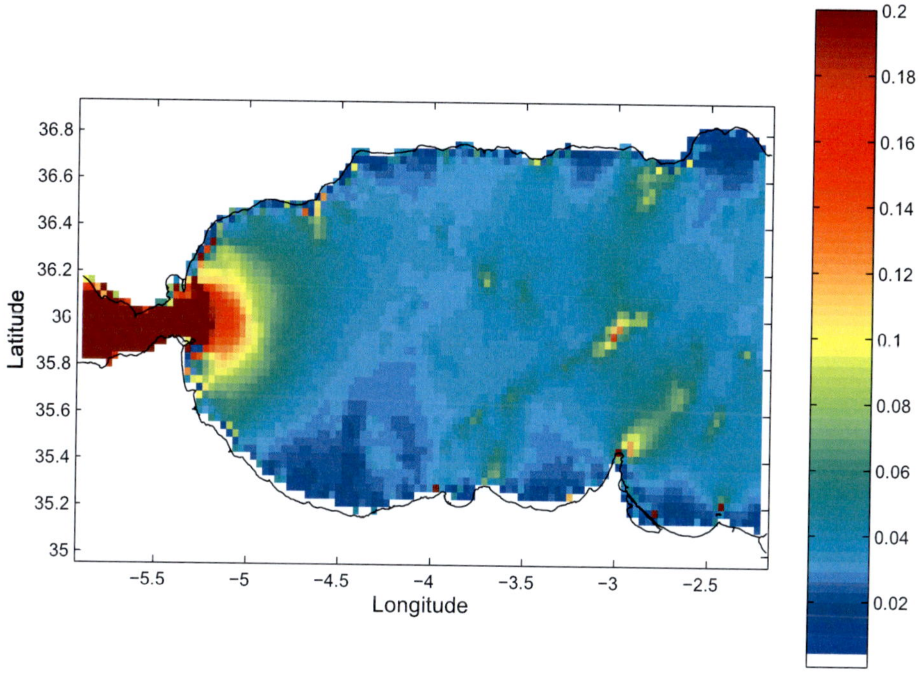

Figure 18. Calculated current amplitude (m/s) for the M_2 tide.

Atlantic water entering through the Strait of Gibraltar flows towards the east along the Spanish coast and partially curves to the south before reaching Alborán Island. Part of this flow continues to the east between Cape Tres Forcas and Alborán Island and the remaining rotates towards the west. A gyre of anticyclonic circulation is thus completed. Surface water velocity in the Strait reaches 0.53 m/s, in agreement with the current speeds of the order of 0.6 m/s reported from measurements [19] and models [36]. In the jet, along the Spanish coast, maximum surface velocities are of the order of 0.4 m/s. The measurements of Perkins et al. [19] in this area range between 0.1 and 0.53 m/s and the model of Werner et al. [48] produces maximum currents in the north of the gyre of 0.25 m/s. Vargas-Yáñez et al. [20] have measured eastward residual velocities in the passage between Cape tres Forcas and Alborán Island (point F in Fig. 4). They are 11 ± 5, 5 ± 3 and 1 ± 3 cm/s at depths of 74, 117 and 178 m respectively (all depths correspond to the surface layer). Computed current at the same point for the surface layer is 3.8 cm/s.

A westward circulation is obtained for the bottom layer. In the western Strait of Gibraltar computed velocities are of the order of 0.20 m/s, in good agreement with the 0.15 m/s obtained by Béranguer et al. [66]. Maximum outflow current calculated in the Strait is 0.36 m/s; Sannino et al. [67] obtained a figure of 0.35 m/s. In the Alborán Sea, water velocities are reduced to some 0.1 m/s in the southern area and less in the northern part of the basin. This circulation pattern for the deep layer is in good agreement with earlier calculations [17], that also show that most of the deep water flows along the south coast with speed below 0.15 m/s, and with measurements [14], which give velocities in the southern shelf, near

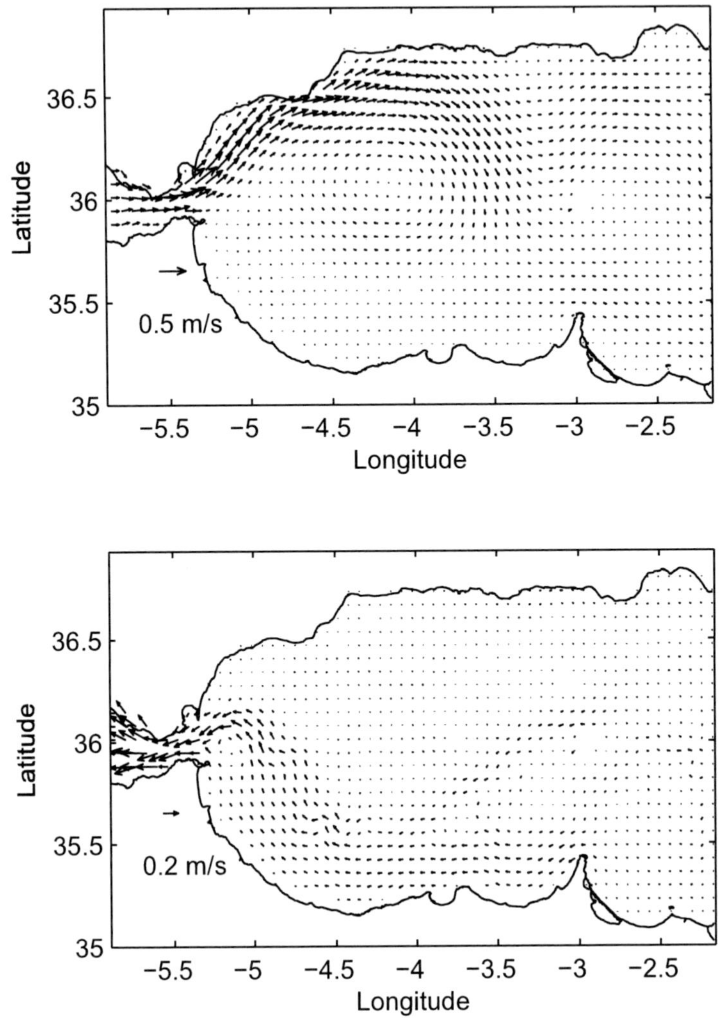

Figure 19. Computed residual currents in the upper (up) and lower (down) water layers. Only one of each four computed vectors is shown for clarity.

the Strait, of 0.1 m/s. As commented before, Bryden and Stommel [22] have also found, from CTD transects, that the westward deep flow occurs along the Moroccan continental slope. These authors have measured a mean (using a 341-day record) outflow velocity of 4.6 ± 0.6 cm/s in a point located at about $-4.6°$ longitude and over the 500 m isobath in the Moroccan continental slope. They do not give the exact position of the current meter mooring, but the model predicts (for such longitude and a depth of 535 m) a westward current equal to 3.6 cm/s. In the northern part of the sea, Fabres et al. [28] have measured westward currents of 1.93 and and 0.95 cm/s in stations C and E (see Figure 4) respectively. The corresponding calculated currents are 1.0 and 0.60 cm/s.

As expected for geostrophic flow, the interface between both water layers in the Strait of Gibraltar is deeper in the south than in the north. The computed slope along a south-north section at the Atlantic entrance of the Strait is 1.78×10^{-3}, in good agreement with Sannino et al. [67] who found a value of 1.7×10^{-3}.

Some nonlinear interaction between tides and the mean flow may be expected, as happens in the model of Nikiema et al. [68] for the Amazon River plume, for instance. However, despite such effects, it seems that, generally speaking, the present models give a representation of the Alborán Sea tides and circulation that is realistic enough to implement on it the sediment and pollutant transport models.

4.5.2. Sedimentation

Some parameters have to be provided for the sediment transport calculations. Indeed, one of the main difficulties in sediment transport modelling is the method of obtaining the erosion and deposition critical stresses and the erodability constant. These parameters are site-specific, but usually are not available since they are not easily determined from field data [69]. Thus, they have been taken from the common values found in literature.

Since less than 7% of the bed sediments consists of coarse material [26], it has been taken an uniform value of $f = 0.95$. The critical deposition stress typically ranges between 0.04 and 0.1 N/m^2 [69], while the critical erosion stress ranges 0.1-1.5 N/m^2 [69, 70]. In the present application intermediate values of 0.07 and 1.0 N/m^2 have been taken for the critical deposition and erosion stresses respectively. The erodability constant is fixed as $E = 1.6 \times 10^{-3}$ kg/m^{2}s. This parameter typically varies between 2×10^{-4} and 3×10^{-3} kg/m^{2}s [69]. Only the lithogenic fraction of sediments is simulated, which accounts about 75-80% of the total suspended load [27, 28]. Fluvial inputs of particles have been taken from the detailed study of Liquete et al. [29]. Seven rivers are included, being Guadiaro and Guadalhorce (Fig. 4) the most important ones.

Suspended particles (diameter < 63 μm) are characterized by a mean size that is considered to be representative of suspended matter in this environment. The particle size controls, through Stoke's law, the settling velocity of particles and, as a consequence, affects the sedimentation rate as well. A sensitivity analysis (described below) has been carried out, and the selected particle size is that for which the best agreement between measured and calculated sedimentation rates is obtained. The particle radius finally selected was 4.0 μm. This is a reasonable figure, given that sediments over all the Alborán Sea may be classified as silts [26, 27] and consequently particles have diameter in the range $4 - 63$ μm. The selected value is also consistent with Liu et al. [71], who used a radius of 2.5 μm to describe fine silts in their model. Of course, it would be convenient to have a detailed particle size spectra, but it is not available in current literature.

The computed steady distributions of suspended particles in the upper and lower layers may be seen in Figure 20. Particle concentrations in the upper layer are about 0.4 g/m^3 in the Strait of Gibraltar and in the Atlantic jet in the Alborán Sea. Concentration decreases in the WAG and in the south part of the sea to some 0.2 g/m^3. Slightly higher concentrations, about 0.4-0.5 g/m^3, are obtained in most of the bottom layer. These values are in agreement with measurements [11], which give suspended particle concentrations below 0.5 g/m^3 in several stations sampled in the Strait of Gibraltar and Alborán Sea (exact values of concentrations are not reported in such paper). The plumes of the two main rivers, Guadiaro and Guadalhorce (see Figure 4), may be appreciated in both layers. It seems that

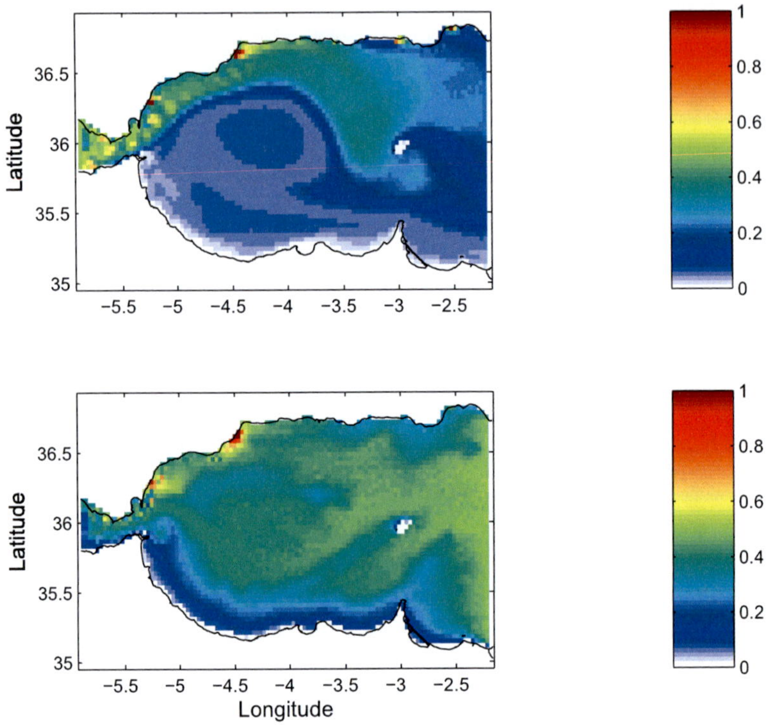

Figure 20. Computed suspended matter concentrations (g/m^3) in the upper and lower water layers (up and down respectively).

the distribution of particles in the upper layer is controlled by water circulation, with larger concentrations of particles along the Spanish coast, in the Atlantic jet, and a decrease in the region occupied by the WAG. Nevertheless, it is necessary to point out that these results correspond to a steady situation with constant water fluxes through the Strait and no winds. In practice, these particle distributions will be altered by changing water exchanges through the Strait, winds, changes in river particle supplies and eventual episodes of disappearance of the WAG.

Table 6. Measured and computed fluxes (g/m^{2}y) of the lithogenic particle fraction for the lower water layer. Measurements correspond to the average value from two sediment traps located in the lower water layer.

station	measured flux	computed flux
C	5.0×10^{-6}	7.1×10^{-6}
E	7.2×10^{-6}	7.4×10^{-6}

The vertical particle fluxes calculated by the model have been compared with those obtained from sediment traps [28] located at stations C and E (lower layer) for the lithogenic fraction of suspended sediments. Results are shown in Table 6. It may be seen that differences are, as maximum, a factor 1.5. Both measurements and the model give a higher flux in station E, which is more distant from the coast than C. This distribution is unusual for continental margins, where fluxes at a given depth generally decrease with distance from the coast. It has been attributed [28] to a funneling of particles from the periphery of the WAG towards its centre. However, it seems from Figure 20 that such process is not occurring in the upper water layer since suspended particle concentrations in the WAG are lower than in the Atlantic jet. It can be seen in Figure 20 that, in the lower layer, particle concentrations are slightly lower (and consequently also the fluxes) in the area of point C. Nevertheless, this effect is less apparent in the model results than in measurements and a clear explanation cannot be given.

A map of computed sedimentation rates (SR) is presented in Fig. 21. Except in the river plumes, they are of the order of 10^{-2} g/cm^{2}year and are obviously correlated to the suspended matter concentrations in the deep layer (see Fig. 20). The smallest SR is obtained in some areas of the Strait of Gibraltar and the Alborán Sea, where larger currents are present in the lower layer (Fig. 19) and/or stronger tidal currents (Fig. 18) exist. A comparison between measured [27] and computed SR may be seen in Table 7 (column corresponding to $R = 4$ μm). It seems, looking at Table 7, that essentially uniform SR are given by the model. However, it is evident from Fig. 21 that a clear structure is produced, with SR varying over one order of magnitude from the south to the north of the sea. The order of magnitude of the SR is correctly given by the model, although there is a general underestimation of SR over the Alborán Sea. This is not surprising given the approximations made in the model: episodes of heavy rains, atmospheric deposition and bedload are not included. Moreover, only the lithogenic particle fraction, that is about 80% of the total suspended load, is simulated. The high SR measured in point A has been attributed to the presence of a bottom nepheloid layer in this area [28], which may be generated in the Alborán Sea by resuspension induced by a internal wave focused along a canyon axis, as it is for instance the case with the Guadiaro submarine canyon [72]. Nepheloid layers cannot be reproduced by the model since it is depth-averaged in each layer. Consequently, the largest difference between measured and computed SR occurs in this point.

Masqué et al. [27] have concluded, from their set of measurements (Table 7), that there is a general trend by which SR decreases as the water column depth increases, specially in the area of Málaga. Model results show this effect; indeed it may be seen in Fig. 21 that higher SR in the Málaga area appears to be due to the influence of river discharges and the Atlantic jet. Nevertheless, the opposite effect is present along the south coast of the Alborán Sea: here SR increases with distance from the coast. Although there are not measurements in this area, model results are consistent with the fact that higher currents (and hence lower SR) exist close to the coast in the lower water layer. Also, it should be considered that this area is far from the particle sources included in the model (Spanish rivers and Atlantic inflow). Consequently, particles are deposited on the sea bed before reaching the south part of the Alborán Sea.

Generally speaking, the model gives a realistic representation of sedimentation process in the Alborán Sea. SR are mainly controlled by river inputs and water currents (both tidal and mean flow in the deep layer).

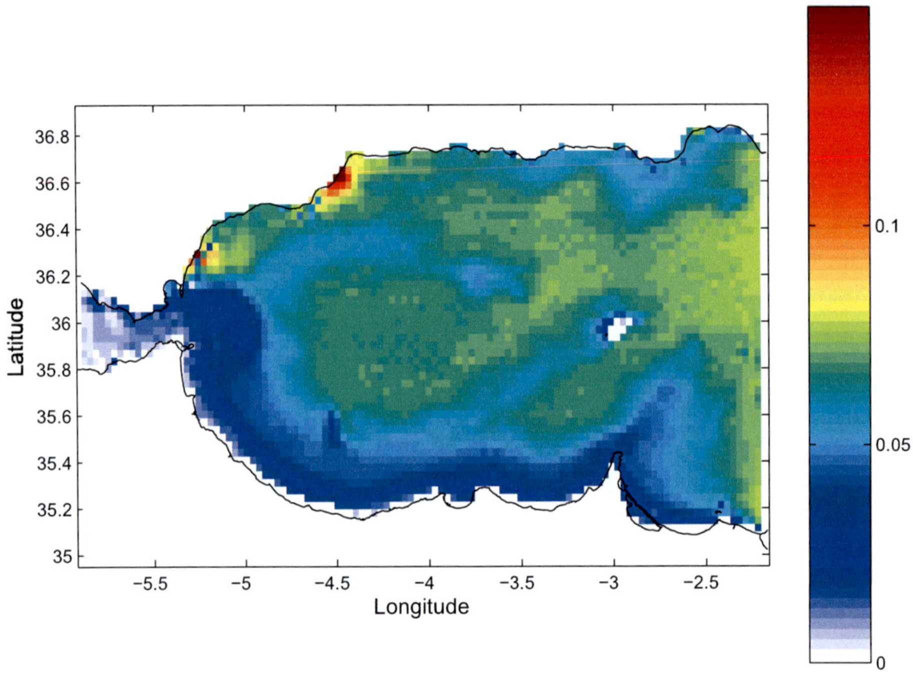

Figure 21. Computed sedimentation rates (g/cm^{2}y) over the model domain.

Table 7. Measured and computed sedimentation rates (SR, g/cm^{2}year) at 8 positions in the Alborán Sea and for three values of the mean particle radius, R, used in the model. The nominal value is $R = 4$ μm. The other results are discussed in Section 4.5.4..

station	measured SR	$R = 4$ μm	$R = 7.5$ μm	$R = 15$ μm
A	0.182 ± 0.011	0.055	0.076	0.071
B	0.101 ± 0.010	0.064	0.092	0.0092
C	0.079 ± 0.005	0.063	0.097	0.012
D	0.043 ± 0.002	0.065	0.11	0.018
E	0.058 ± 0.002	0.066	0.012	0.030
F	0.075 ± 0.007	0.067	0.14	0.052
G	0.084 ± 0.006	0.071	0.18	0.23
H	0.053 ± 0.002	0.061	0.014	0.089

4.5.3. Radionuclide Dispersion

The radionuclide dispersion model has been applied to two radioactive anthropogenic isotopes, ^{137}Cs and ^{239,240}Pu, which are introduced from atmospheric fallout as discussed in more detail below. Some simulations concerning the natural radionuclide ^{210}Pb have been carried out as well.

The parameters that describe the geochemical behaviour of the different radionuclides are the exchange velocity χ and the kinetic transfer coefficient for the desorption process k_2.

The correction factor ϕ has been fixed as 0.01 according to previous modelling works [58]. Sediment porosity is 0.6 according to measurements [26].

The so-called equilibrium distribution coefficient, k_d, describes the partition of a tracer between the dissolved and solid phases at equilibrium [60]:

$$k_d = \frac{C_s}{C_d} \tag{36}$$

where C_s and C_d are radionuclide concentrations in the solid and dissolved phases respectively. This parameter is of course depending on the tracer geochemical behaviour and on environmental conditions, thus it is a site-specific parameter.

In the case of ^{137}Cs, the kinetic coefficient k_2 is 1.16×10^{-5} s^{-1} according to the experiments in Nyffeler et al. [73], value that has been used in other modelling works concerning this radionuclide [49, 50, 56]. The exchange velocity has been obtained from the equation relating this parameter with k_2 and the equilibrium distribution coefficient, k_d [30, 58, 50]:

$$k_d = \frac{\chi}{k_2}\frac{3}{\rho R} \tag{37}$$

The k_d mean value recommended by the International Atomic Energy Agency is 4 m^3/kg [74] for Cs, although it may vary in more than one order of magnitude depending on environmental conditions. In the present application, the value that has been used is $k_d = 2$ m^3/kg since a better agreement with observations is obtained. Nevertheless, the model sensitivity to this parameter is studied in Section 4.5.4.. Since R, k_d and k_2 are already defined, the value of χ is determined from equation 37.

As described before [73], k_2 is very similar even for radionuclides with a rather different geochemical behaviour, being χ the essential parameter describing the tracer geochemical behaviour. Thus the value given to k_2 for ^{239,240}Pu and ^{210}Pb is the same as used for ^{137}Cs. Then the χ values for these radionuclides are obtained from equation 37 using their corresponding recommended k_ds, which are 100 and 200 m^3/kg for ^{239,240}Pu and ^{210}Pb respectively [74]. This procedure has been successfully used before [58, 30, 55, 56, 50].

Nevertheless, it has to be considered that the behaviour of Pu in aquatic systems is of considerable complexity due to the fact that can exist in different oxidation states simultaneously. The reduced Pu [Pu(III) and Pu(IV)] is highly particle reactive and has been shown to possess a k_d that is approximately two orders of magnitude higher than that of the more soluble oxidized Pu [Pu(V) and Pu(VI)]. Different exchange velocities should be used for oxidized and reduced Pu, as has already been done [75]. However, in this work different plutonium oxidation states have not been considered due to the lack of experimental data on Pu speciation in the area. Thus, measurements (of Pu concentrations and k_ds) represent the mixture of oxidation states that is present in the particular sample. A measured effective k_d, representative of such mixture, is used to obtain an effective exchange velocity. This approach for simulating Pu dispersion is used in other models [55, 76]. Moreover, the vertical transport of plutonium related to the settling of biogenic particles cannot be simulated by the model since only the lithogenic particle fraction is considered, as mentioned before. Thus, the results for plutonium have to be considered only as a first approach to a very complex problem.

Radionuclide distributions: ^{137}Cs and ^{239,240}Pu simulations The major source of the anthropogenic radionuclides ^{137}Cs and ^{239,240}Pu has been the atmospheric deposition due to nuclear weapon tests, as well as the Chernobyl accident fallout in the case of ^{137}Cs. Other sources as discharges from nuclear facilities, riverine contributions and run-off from land may be neglected in the area [10].

In the case of ^{137}Cs, the model has been run for the period 1985 (one year before Chernobyl accident) to 1997, when computed ^{137}Cs concentrations in the bed sediment are compared with measurements in [27]. Initial ^{137}Cs concentrations for both water layers are obtained from the long-term box model of Sánchez-Cabeza et al. [77] and are assumed to be homogeneous over the entire domain. They are 3.0 and 2.0 Bq/m^3 for the upper and lower layer respectively. Concentrations in suspended matter and bottom sediments are assumed to be initially in equilibrium with the dissolved phase, thus they are calculated from equation 36. Atmospheric and Chernobyl inputs are obtained from [77] as well.

Table 8. Measured and computed ^{137}Cs concentrations (Bq/kg) in surface sediments of the Alborán Sea at some positions and for several k_d values. The mean sediment inventory is given in Bq/m^2. The nominal simulation corresponds to $k_d = 2$ m^3/kg, the other results are discussed in Section 4.5.4..

station	measured	$k_d = 1$ m^3/kg	$k_d = 2$ m^3/kg	$k_d = 4$ m^3/kg
A	2.6 ± 1.2	1.92	3.85	7.75
C	2.5 ± 1.5	1.95	3.92	7.88
E	7.3 ± 2.2	1.93	3.91	7.83
G	3.2 ± 1.9	1.95	3.89	7.79
H	5 ± 3	2.00	3.99	7.99
J	2.8 ± 0.5	1.95	3.93	7.97
inventory	163 ± 10	79	158	318

The computed distribution of ^{137}Cs in the surface bed sediments at the end of the simulation may be seen in Fig. 22 and a comparison with measurements [27, 78] is presented in Table 8 (column corresponding to $k_d = 2$ m^3/kg). It seems from the table that the model gives uniform ^{137}Cs concentrations in the sediment. This is just an effect of the specific locations of sampling points and, indeed, a clear structure is produced (see Fig. 22). Computed concentrations are in the range $3 - 6$ Bq/kg (the color scale in Fig. 22 is limited to 5 Bq/kg for more clarity), in generally good agreement with measurements (Table 8). There are higher concentrations along both coasts of the Alborán Sea, particularly along the Spanish coast, where SR are higher due to the riverine input of particles. Thus, it could be deduced that ^{137}Cs incorporated to the surface water from the atmosphere is attached to suspended particles that are later deposited on the bed. Indeed, lower concentrations are in general obtained in the Strait of Gibraltar and along the main path followed by the deep

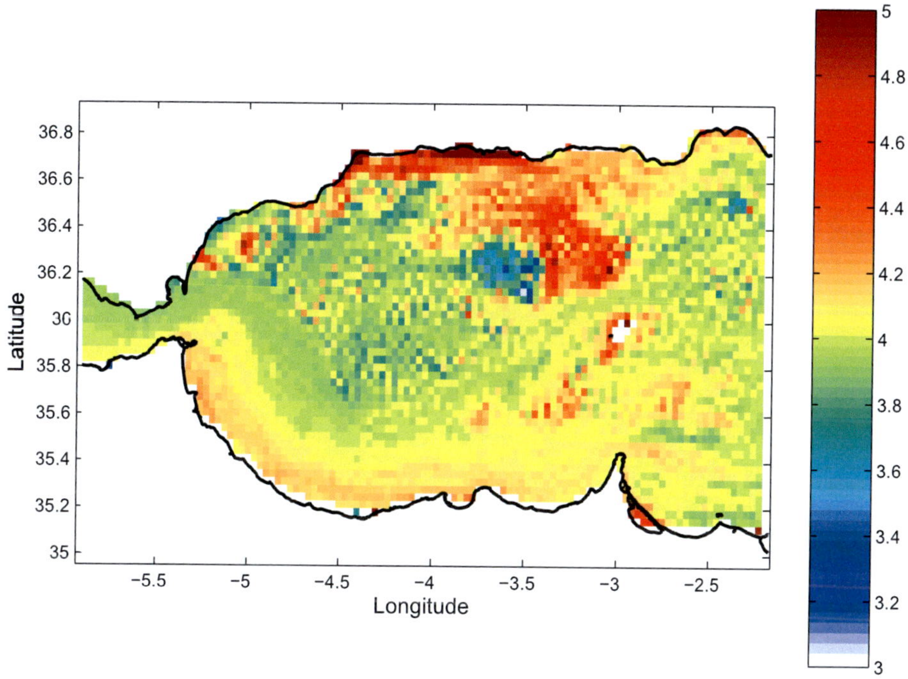

Figure 22. Computed distribution of ^{137}Cs (Bq/kg) in surface bed sediments of the Alborán Sea.

outflow current, where SR are smaller. Nevertheless, direct adsorption of ^{137}Cs from the dissolved phase to the bed sediment is occurring simultaneously and an absolutely clear correlation between SR and ^{137}Cs concentration in sediments cannot be established. Indeed, direct adsorption is higher in shallower areas, as can be seen from equation 30, and consequently the map in Fig. 22 is affected by topography with a general trend of increasing ^{137}Cs concentrations with decreasing water depth. Moreover, direct adsorption of radionuclides (and trace metals in general) is affected by the water-sediment contact time [79]. Thus, direct adsorption is smaller in regions of stronger currents in the bottom water layer, as is the case with SR. It may be concluded that a clear correlation between contamination of the bed sediment and SR cannot be established, although a correct estimation of SR is of course required for an appropriate simulation of particle-reactive radionuclide transport.

The mean inventory of ^{137}Cs in bed sediments over the Alborán Sea has been calculated as well. This inventory includes radionuclides in the surface sediments plus radionuclides that have migrated down the sediment core according to equations 27 and 34. The obtained value, 158 Bq/m^2 (Table 8) is in good agreement with the estimations carried out from measurements [78]: these authors have obtained an inventory of 163 ± 10 Bq/m^2 in the area of Málaga. The computed total inventory in bed sediments is 6.70 TBq. The total inventory in the water column (for both layers) has been calculated as well, resulting 75.2 TBq. However, there are not experimental estimations to compare with in this case.

Gascó et al. [10] have measured ^{137}Cs concentrations in unfiltered water samples from the Strait of Gibraltar (surface and deep water) in 1997. Model results have also been compared with these measurements. Results are presented in Table 9, where it may be seen that the model gives realistic results for the water column as well. Essentially the same ^{137}Cs concentrations are computed in surface and deep waters, in agreement with measurements [10].

Table 9. Computed and measured total concentrations in the water column (dissolved plus suspended matter) of ^{137}Cs (Bq/m^3) and ^{239,240}Pu (mBq/m^3) at several locations.

station		measured		computed	
		^{137}Cs	^{239,240}Pu	^{137}Cs	^{239,240}Pu
J	surface		10.0 ± 0.4		6.97
	deep		23 ± 5		12.9
L	surface	3.1 ± 0.3	14.1 ± 1.5	3.02	6.9
	deep	2.1 ± 0.3	18.0 ± 1.6	1.99	13.0
M	surface	2.62 ± 0.24	9.5 ± 1.2	3.02	6.9
	deep	2.2 ± 0.3	23.1 ± 1.7	1.98	13.7
N	surface		11.1 ± 0.4		7.7
	deep		21.5 ± 2.1		13.2
P	surface		12 ± 5		45.7
	deep		19.2 ± 2.1		14.9
Q	surface		4.8 ± 0.3		6.8
	deep		20.9 ± 0.7		12.6

In the case of ^{239,240}Pu, a simulation over 20 years has been carried out. The atmospheric deposition has been compiled from current literature [11, 80], and zero activity concentrations in all phases are considered as initial conditions for the simulation. The computed distribution of Pu in the sediment may be seen in Fig. 23. It can be clearly seen that this distribution is rather different to that of Cs, which is due to the different geochemical behaviour of both radionuclides. Only one measurement has been found to compare model results. Measured ^{239,240}Pu concentration in the surface sediment at point J is 0.550 Bq/kg [11], and the computed one is 0.379 Bq/kg.

The computed radionuclide concentrations corresponding to unfiltered water are presented for both water layers in Figs. 24 and 25, for ^{137}Cs and ^{239,240}Pu respectively. The distributions in the upper layer are similar for both radionuclides, showing higher concentrations in the area occupied by the WAG. This is due to the fact that some water is trapped in the gyre and thus its radionuclide concentration increases as a result of the atmospheric input. Nevertheless, simulations are made under steady flow conditions, thus this pattern will be destroyed by episodes of migration and disappearance of the WAG and winds. However, it is worth commenting that the pattern of higher radionuclide concentrations (for ^{210}Pb, introduced by atmospheric deposition as well) in the centre of the oceanic gyres (in both the Pacific and North Atlantic) has already been detected [81]. Consequently, the radionuclide enhancement in the WAG is a real process, although overestimated because of the simulation conditions (steady residual flow).

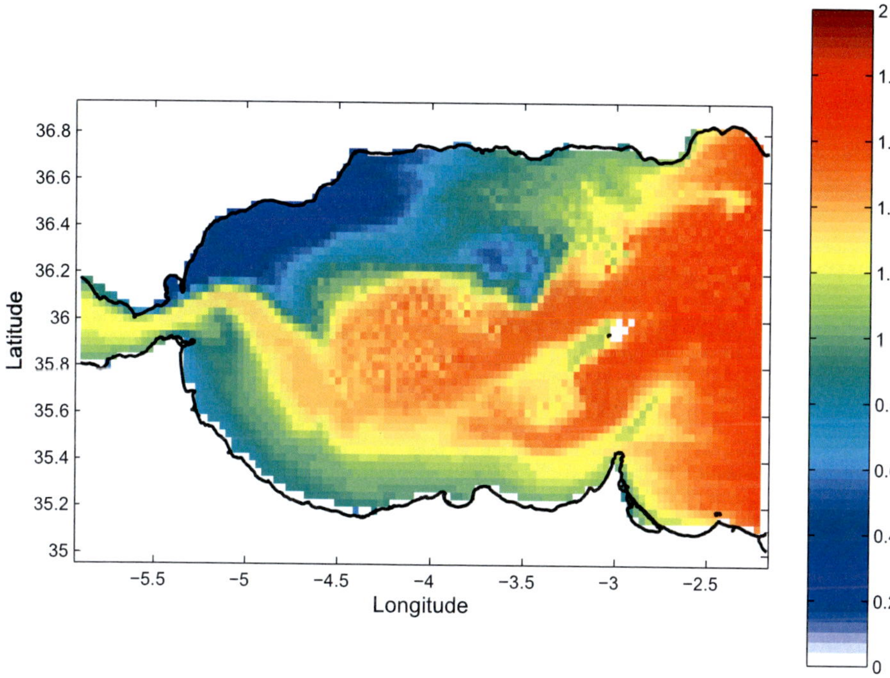

Figure 23. Computed distribution of $^{239,240}Pu$ (Bq/kg) in surface bed sediments of the Alborán Sea.

In the case of a highly reactive radionuclide as $^{239,240}Pu$, it is quickly fixed to suspended matter and then, by settling, introduced into the deep water layer. Finally radionuclides are deposited on the seabed. Although it cannot be clearly seen in Fig. 25 because of the scale in the colorbar, it is evident from Table 9 that computed concentrations in the deep layer are about a factor 2 higher than in the surface layer. This result is in agreement with the observations presented in Table 9, although the model is slightly underestimating concentrations.

Thus, in the lower layer, distributions are clearly different for both radionuclides due to the different efficiency of the removal process of dissolved radionuclides by settling suspended particles. In fact we can clearly see now that radionuclide distributions in sediments are correlated with those in the deep (unfiltered) water. This is not surprising since deposition of radionuclides on the sediment depend on the radionuclide concentration in suspended matter, and the direct adsorption of radionuclides by the sediment depends on the radionuclide concentration in dissolved form in the deep layer.

The computed average inventory of $^{239,240}Pu$ over the model domain in the sediment is 45.8 Bq/m^2. Gascó et al. [78] have obtained a value of 29 ± 2 Bq/m^2, although only one sediment core was analyzed. Consequently, it can just be said that the order of magnitude seems correct. The total inventory in the sediment results 1.9 TBq. The computed inventory of $^{239,240}Pu$ in the water column is 0.68 TBq, which agrees with the previous estimation of 0.64 TBq [11]. However, this agreement should be interpreted with care since we have seen

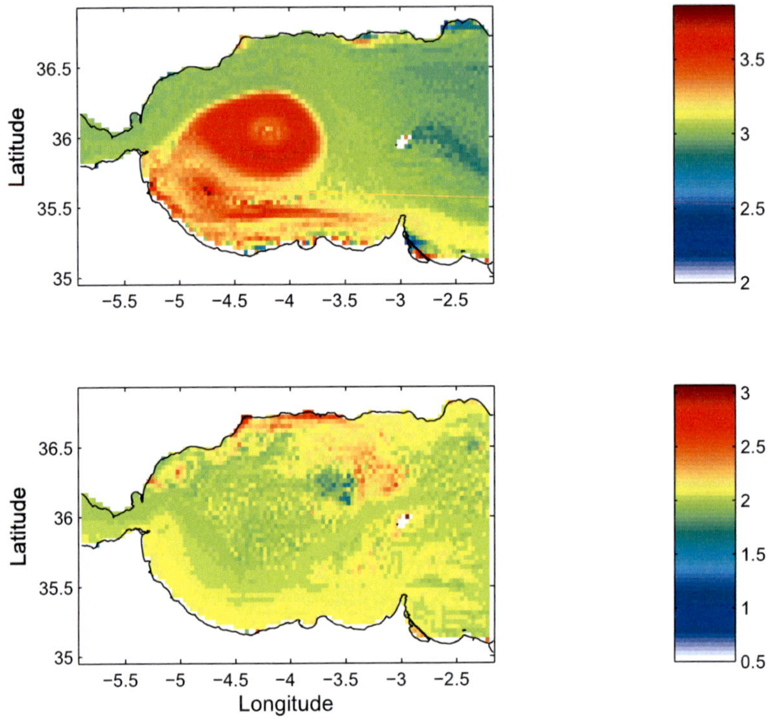

Figure 24. Computed ^{137}Cs concentrations in unfiltered water(Bq/m^3) in the upper and lower water layers (up and down respectively).

that the model tends to underestimate Pu concentrations in the water column. Thus, there must be a concentration overestimation somewhere in the sea. This area must be the WAG, because of the steady flow conditions used in the simulation as commented above. This seems confirmed by the surface sample at point P (see Table 9).

The computed fractions of ^{239,240}Pu in the dissolved phase may be seen in Fig. 26 for both water layers. Most of it remains dissolved in spite of the strong affinity of plutonium to be fixed to the solid phase, which is due to the low suspended particle concentrations in the Alborán Sea. Thus, in the surface layer the fraction of plutonium in particles ranges from values of the order of 1 % in the WAG to some 4-5 % in the Atlantic jet. In the deep layer this fraction is about 4 % for most of the sea, except in the plumes of the main rivers where suspended particle concentrations are higher. These results are in good agreement with León-Vintró et al. [11], who found that the percentage of ^{239,240}Pu in particulate form is in the range 2-4 %.

The model uses the distribution coefficient for calculating the exchange velocity for each radionuclide, as described above (equation 37). However, the partition of radionuclides between the liquid and solid phases has not to be in equilibrium necessarily. Indeed, the model can calculate the k_d that would be measured from a sample as the ratio between radionuclide concentrations in the solid and dissolved phases. As an example, the computed

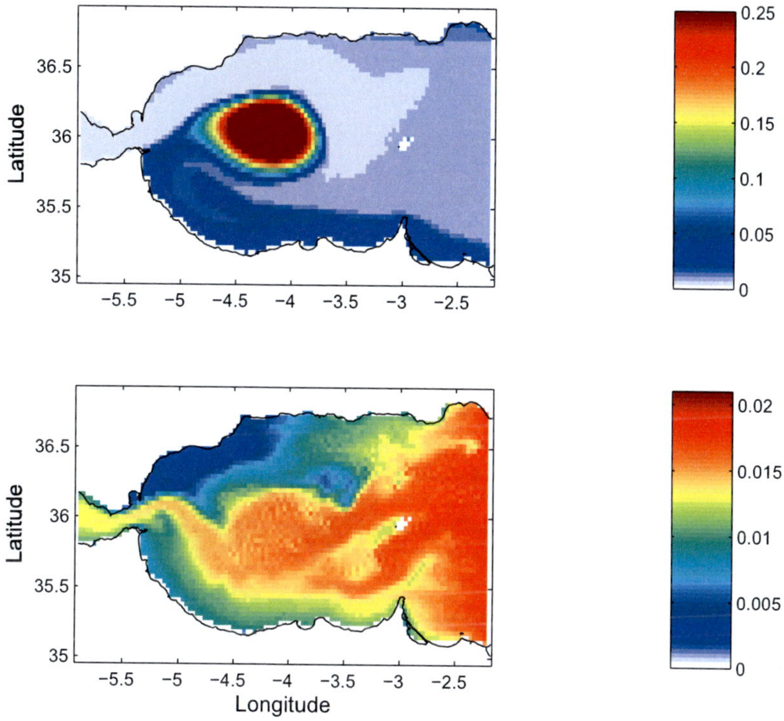

Figure 25. Computed ^{239,240}Pu concentrations in unfiltered water(Bq/m^3) in the upper and lower water layers (up and down respectively).

water-suspended matter k_ds for ^{137}Cs and ^{239,240}Pu in the upper water layer are presented in Fig. 27. In most of the sea radionuclide partition is at equilibrium since the computed k_d is essentially the same value used in equation 37 to obtain the exchange velocity (100 and 2 m^3/kg for Pu and Cs respectively, as discussed before). However, the system is slightly out from equilibrium in the area of the Strait of Gibraltar. This is probably due to the strong currents that are present here in the upper layer, which do not allow radionuclide partition to reach equilibrium. This effect is not observed in the lower water layer, where radionuclide partition is at equilibrium over all the Alborán Sea and the Strait. Currents are weaker in this layer.

^{210}Pb Simulations The case of ^{210}Pb is extremely complex. It is a member of the ^{238}U radioactive decay chain and is supplied to the sea water from the atmosphere, rivers and in-situ decay in the water column of its parent radionuclide ^{226}Ra. In the bed sediment, ^{210}Pb is also produced by decay of ^{226}Ra. Because of their desintegration periods and since both remain fixed to the sediment, these radionuclides are in secular equilibrium (their activities are the same) in the seabed. This amount of ^{210}Pb, which is in equilibrium with ^{226}Ra, is known as the supported fraction. Nevertheless, ^{210}Pb is also incorporated in the bed sediment due to deposition of ^{210}Pb containing particles from the water column. This is

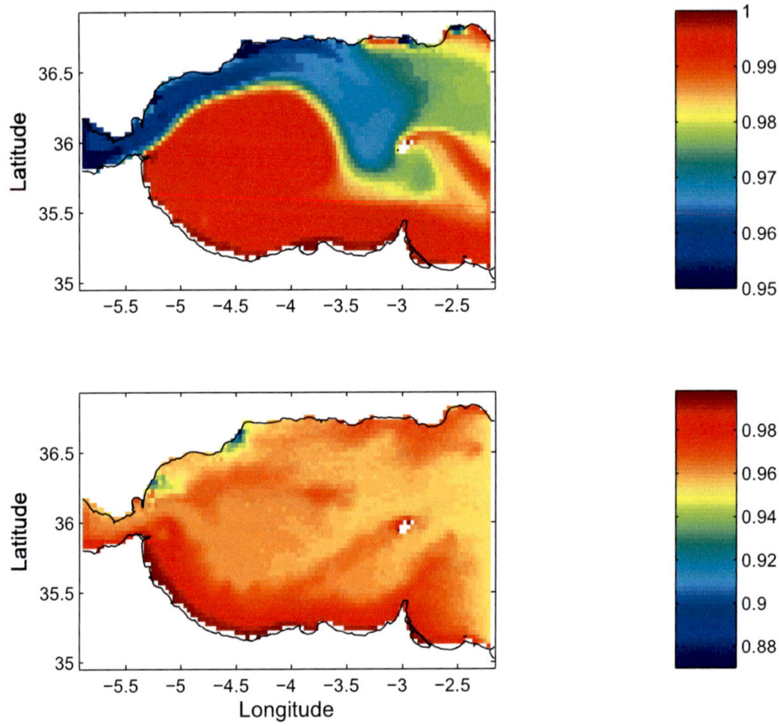

Figure 26. Computed fraction of dissolved $^{239,240}Pu$ in the upper and lower water layers (up and down respectively).

known as the excess ^{210}Pb fraction. In practice, the difference between total ^{210}Pb and ^{226}Ra activities in the sediment gives the ^{210}Pb in excess. Excess ^{210}Pb has been measured in sediments of the Alborán Sea [27] and will be used to test the model results, since all the ^{210}Pb is incorporated to the bed sediment from the water column in the model.

The atmospheric deposition of ^{210}Pb in the western Mediterranean is 81 Bq/m^{2}year [82]. In normal river water, ^{210}Pb concentrations are low, about 0.2 Bq/m^3 [83], and most of it is present in suspended particles. We have used this value for the ^{210}Pb supply from river waters, assuming that activity in suspended particles discharged by the rivers is in equilibrium with water (thus deduced from equation 36). It is not possible to calculate in-situ production of ^{210}Pb. The approach used has consisted of specifying ^{210}Pb concentrations at the open boundaries, where input of water occurs, from estimations of ^{210}Pb mean concentrations in the Atlantic inflow and Mediterranean outflow [10], and assuming that these concentrations contain the in-situ fraction of ^{210}Pb that is subsequently transported into the model domain. The limitations are obvious, since some of this ^{210}Pb will be lost by radioactive decay in the domain and its content in the water column will be underestimated. Moreover, in the case of the artificial ^{137}Cs and $^{239,240}Pu$ the model could be started from clean sediments, but this is obviously not the situation with the natural ^{210}Pb. One possibility consists of using an initial concentration of excess ^{210}Pb in both the mixed and deep

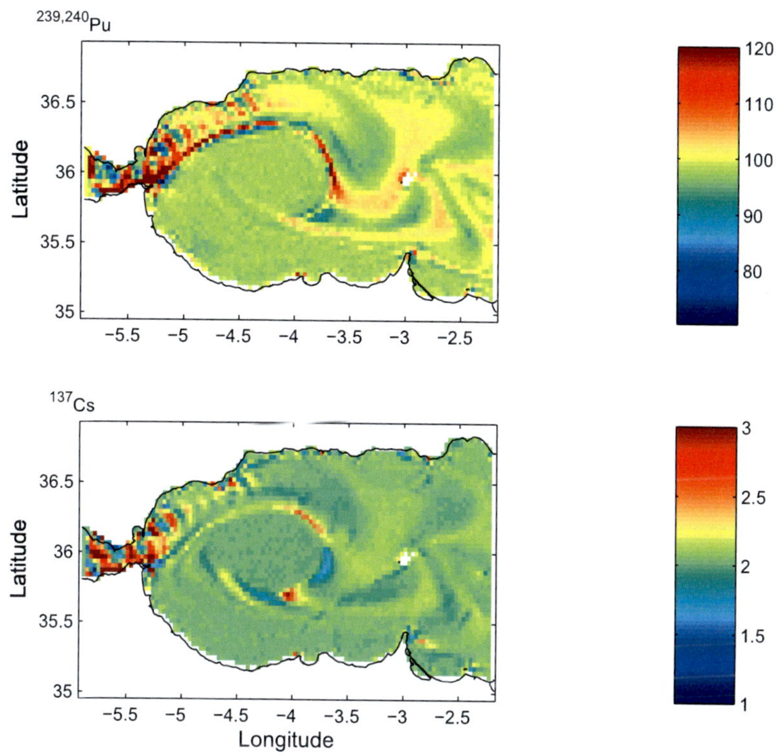

Figure 27. Computed ^{239,240}Pu and ^{137}Cs water-suspended matter distribution coefficient (m^3/kg) in the upper water layer.

sediment. However, this may look tricky since such initial concentrations could be fitted to obtain results that compare well with measurements. Thus, it was decided to start the model from a clean sediment, keeping in mind these limitations when analyzing the model results. Consequently results have to be interpreted with care and will be useful to have a general overview of the dynamics of ^{210}Pb in the system, but a good comparison with experimental data cannot be expected.

The dispersion model is run for 35 years, time in which a steady distribution of ^{210}Pb is achieved. Moreover, the effects of wrong initial conditions will have been reduced, given that the radioactive desintegration period of ^{210}Pb is 22 years. The computed inventories of excess ^{210}Pb may be seen in Fig. 28. Of course they include the mixed and the deep sediment. The distribution is much like that of ^{239,240}Pu (Fig. 23). This is not surprising since their geochemical behaviour is similar (k_ds of 1×10^2 and 2×10^2 m^3/kg for Pu and Pb respectively). It also seems that the contribution of rivers is negligible in comparison with atmospheric deposition. Finally, the ^{210}Pb distribution in the water column presents the same enhancement in concentrations in the area of the WAG as we have seen before and is not shown.

A comparison between measured and computed inventories in the sediments may be seen in Table 10. The inventories measured by Masqué et al. [27] are included in the table. However, these authors have found that they are larger than those expected from

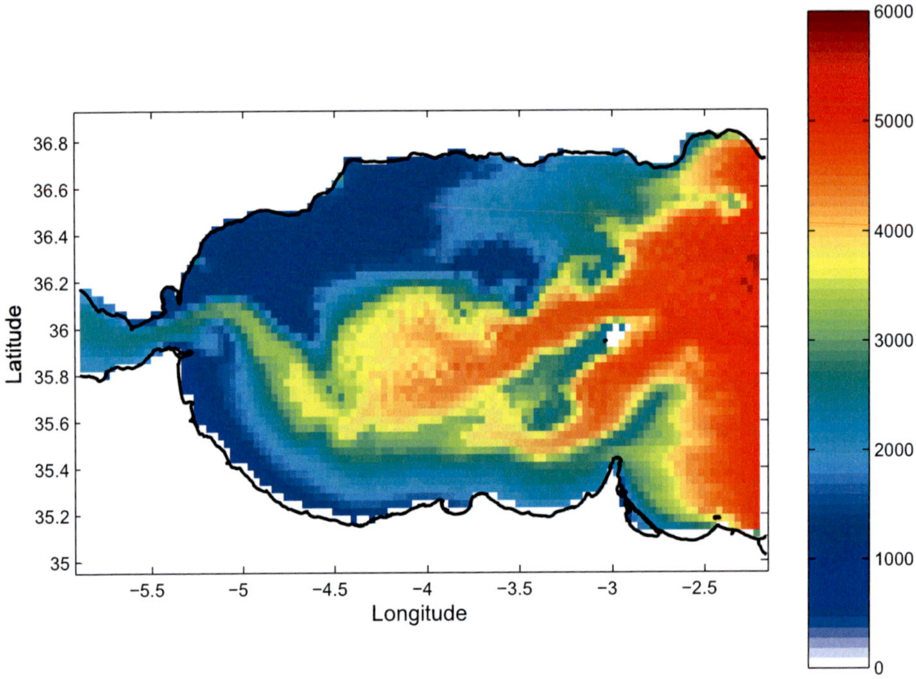

Figure 28. Computed distribution of excess ^{210}Pb inventories (Bq/m^2) in bed sediments of the Alborán Sea.

a balance model for the water column (details may be seen in the reference). Thus the expected inventories, deduced from the balance model by such authors, are also given in Table 10. Such authors have concluded that the discrepancy may be due, at least partially, to the presence of a bottom nepheloid layer and to the occurrence of turbiditic flows in some areas. This highlights the difficulties of the study of ^{210}Pb dispersion. Nevertheless, it may be seen in the Table that, at least, the order of magnitude of the calculated inventories seems reasonable. Moreover, Masqué et al. [27] state that there is a general trend of increasing excess ^{210}Pb inventories with water depth in the area of Málaga, and this trend is given by the model too, as may be seen in Fig. 28.

The computed mean inventory of excess ^{210}Pb in the sediments of the Alborán Sea is 2.98 kBq/m^2 and the total inventory is 126 TBq.

Radionuclide fluxes through the Strait of Gibraltar The fluxes of the two studied anthropogenic radionuclides through the Strait of Gibraltar (through a north-south section at the longitude of Tarifa) have been evaluated from the model results. Indeed, several estimations of fluxes for some radionuclides and metals have already been done from measurements [8, 12, 11, 10, 9].

The water transport used in the estimations ranged from 0.69/0.65 Sv (inflow/outflow) [84] to 1.77/1.73 Sv [85]. As a consequence of the variability of the water exchanges in

Table 10. Measured, computed and expected inventories (see text for explanation) of excess ^{210}Pb (in kBq/m^2) in bed sediments of the Alborán Sea.

station	measured	computed	expected
A	10.66 ± 0.17	1.15	3.05 ± 0.05
B	11.35 ± 0.14	1.43	3.24 ± 0.04
C	13.40 ± 0.14	1.81	3.27 ± 0.03
D	5.97 ± 0.10	2.65	3.32 ± 0.06
E	14.57 ± 0.15	3.73	3.55 ± 0.04
F	15.1 ± 0.2	4.66	3.36 ± 0.04
G	11.03 ± 0.17	4.75	3.94 ± 0.06
H	10.22 ± 0.18	4.72	3.79 ± 0.07

Table 11. Annual fluxes of radionuclides (TBq) in the Atlantic (F_{in}) and Mediterranean (F_{out}) waters through the Strait of Gibraltar. Water exchanges (inflow/outflow) are given in Sv.

^{137}Cs		^{239,240}Pu		inflow/outflow	source
F_{in}	F_{out}	F_{in}	F_{out}		
118	75	0.27	0.44	1.25/1.20	this model
72 ± 18	60 ± 13	0.28 ± 0.11	0.62 ± 0.16	0.90/0.90	[10]
		0.22	0.87	1.39/1.32	[11]
64	51	0.25	0.53	0.81/0.76	[9]

the literature, radionuclide and metal fluxes may present differences up to a factor 3 for the same concentration of the substance [9]. A comparison of the model calculations with previous estimations may be seen in Table 11, where the water exchanges used in each calculation are also given. In spite of the variability of the water exchanges all results are in reasonable agreement and differences reach a factor 2 as maximum. In general, it seems that the model calculation tends to overestimate the inflow of ^{137}Cs. It should be taken into account, however, that estimations are sometimes based on a single measurement of radionuclide concentration in the Strait.

Characteristic times of the system The turn-over-time is defined as the time in which the tracer inventory in the water column decreases in a factor *e* [86] and the sediment half-life is defined as the time in which the radionuclide inventory in the sediment decreases in a factor 2 [87]. These parameters are relevant for the water quality of a system, and it is important, from a management perspective, to know the time scale for a pollutant discharged into a water body to be transported out of the system [88]. Thus, flushing times (that are equivalent to turn-over-times in the case of constant flow) have been recently determined for a number of water bodies using numerical experiments [88, 89]. On the other hand, it is also known that a contaminated sediment may act as a long-term delayed source of previously released

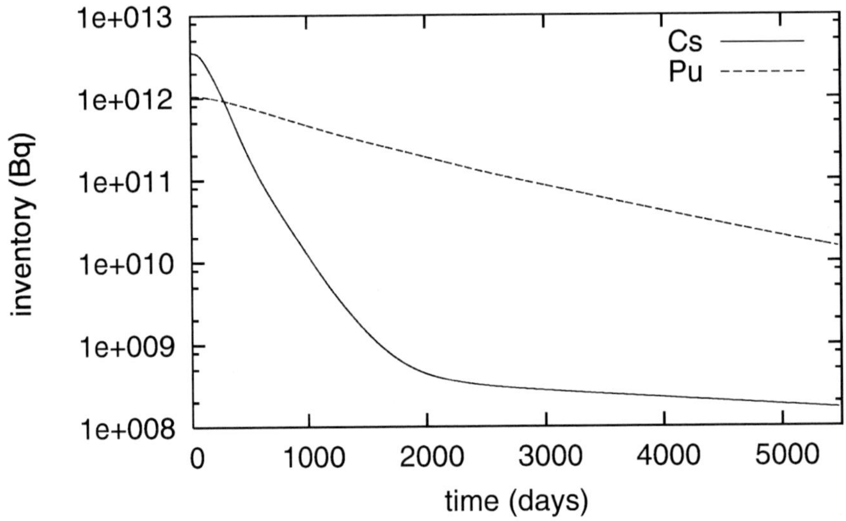

Figure 29. Computed temporal evolution of Cs and Pu inventories in the bed sediment (Bq) for the numerical experiments carried out to determine characteristic times of the system.

radionuclides [90, 91]. Consequently, it is also relevant to have estimations of the sediment half-life [87].

Two numerical experiments have been carried out to determine flushing times and sediment half-lives in the cases of ^{137}Cs and ^{239,240}Pu. The model is started, for each radionuclide, from the computed radionuclide distributions in water, suspended matter and bed sediments presented above and simulations over 15 years are carried out. Atmospheric radionuclide inputs are not considered and clean water, with zero radionuclide concentrations, is assumed to enter the domain through the open boundaries. The radionuclide inventories within the Alborán Sea in the region comprised between $-4.5°$ and $-2.7°$ longitude is evaluated each time step and results are written to a file. Thus, the system-wide (as defined by Choi and Lee [89]) flushing time has actually been determined. These authors found that this parameter is useful for determining the long-term water quality of a system.

As an example, the temporal evolution of the sediment inventories of ^{137}Cs and ^{239,240}Pu are shown in Fig. 29. From numerical fitting to exponential decay curves, sediment half-lives are 136 and 778 days for ^{137}Cs and ^{239,240}Pu respectively. Half-life for plutonium is about a factor 6 larger than that of caesium, revealing the higher affinity of plutonium to remain fixed to the solid phase. It is worth commenting that the model that has been used in this work to describe water-sediment interactions consists of a single reversible reaction (1-step model). Recent experiments [92, 93] have shown that a model involving the existence of two successive reversible reactions (2-step model) is more appropriate to simulate the release kinetics. Indeed, these two kinetic models were compared in the Irish Sea [75] for the case of ^{239,240}Pu and in English Channel for the case of ^{137}Cs [94]. In the case of plutonium, for instance, the 1-step model gave a sediment half-life that was essentially the same as in this study, 657 days. Nevertheless, it increased to 112 years when the 2-step model was used. Thus, the kinetic model clearly affects the sediment

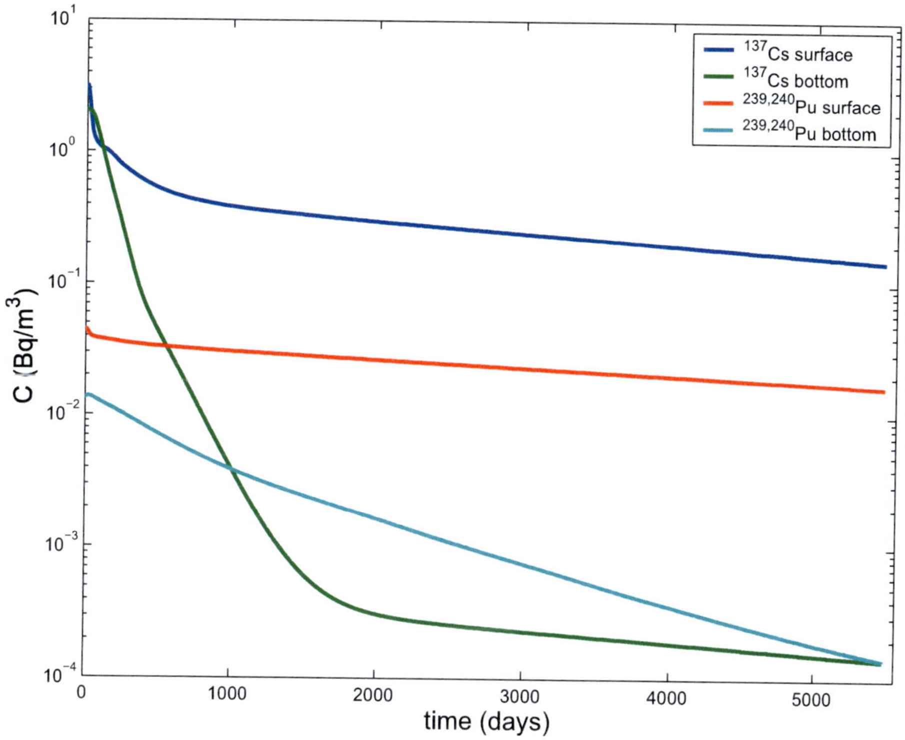

Figure 30. Computed temporal evolution of Cs and Pu mean dissolved concentrations (Bq/m^3) for the numerical experiments carried out to determine characteristic times of the system.

half-life and a larger value would be obtained if a 2-step model is applied. However, this is always difficult because of the lack of site-specific information regarding kinetic rates for both reactions. The application of more complex kinetic models is not the purpose of this work, although the model may of course be improved in the future. Consequently, we are giving conservative estimations from a management perspective.

Flushing times have been evaluated for each water layer separately and for the dissolved phase and suspended matter. The temporal evolution of the mean radionuclide concentration in each water layer is presented in Fig. 30 for both radionuclides (dissolved phase). Choi and Lee [89] have found that the system-wide flushing time is better obtained from a double exponential decay curve rather than a single exponential decay. The temporal evolution of the mass within the system, M, is thus given by the following equation:

$$\frac{M}{M_0} = (1+\beta)e^{-k_1 t} - \beta e^{-k_2 t} \tag{38}$$

where M_0 is the initial mass within the system. If the three parameters β, k_1 and k_2 are determined from numerical fitting to the decay curves in Fig. 30, the system-wide flushing time is given by [89]:

$$T_f = \frac{1+\beta}{k_1} - \frac{\beta}{k_2} \tag{39}$$

Table 12. Flushing times (days) for ^{137}Cs and ^{239,240}Pu in the surface and deep layers, for the dissolved phase and suspended matter. The regression coefficient of the numerical fitting to the double exponential decay function is also given.

	^{137}Cs		^{239,240}Pu	
	surface	bottom	surface	bottom
	water			
T_f (days)	292	121	3720	814
r^2	0.966	0.992	0.989	0.999
	SPM			
T_f (days)	248	123	4422	893
r^2	0.974	0.992	0.997	0.999

In the case of dissolved ^{137}Cs in the surface layer, for instance, a flushing time of 176 days is obtained if the corresponding curve in Fig. 30 is fitted to a single exponential decay, and the regression coefficient is $r^2 = 0.636$. However, if the double exponential is used (equation 38), the regression coefficient in this case increases to $r^2 = 0.966$ and the flushing time is 292 days.

Flushing times have been determined for ^{137}Cs and ^{239,240}Pu, for both water layers. Results are summarized in Table 12. It can be seen that Pu flushing time is about one order of magnitude larger than for Cs, specially for the surface layer. This is again due to the higher affinity of Pu to be fixed to the solid phase, which makes it less mobile than Cs. Essentially the same values are obtained for the dissolved phase and suspended matter for both radionuclides.

In the case of constant flow through a region, as is the case in our simulations, the flushing time can be obtained as V/q [86], where V is the region volume and q is the flow. A flushing time of some 80 days is obtained if this equation is applied to the surface water layer with $q = 1.25$ Sv. This is shorter than the flushing times presented in Table 12 because two reasons: firstly the fact that the WAG traps water and, secondly, radionuclide interactions with the solid phase make them less mobile, which is of course more evident for ^{239,240}Pu than for ^{137}Cs. The effect of water-sediment interactions on flushing times has not been investigated before.

4.5.4. Sensitivity Tests

Some sensitivity analysis have been carried out to study the model response to changes in parameters which are poorly defined. In particular, the cases of the mean particle size, R, and of the radionuclide distribution coefficient, k_d, are discussed here. The sensitivity analysis have been made for the case of ^{137}Cs since simulation times are shorter.

The sedimentation model has been tested for several values of the particle mean size. Results are shown in Table 7. The agreement between measured and computed SR is poorer as the mean particle size increases. An increase in the sediment size implies a higher settling velocity, thus, in principle, SR should increase as well. This is indeed the case for $R = 7.5$

μm in Table 7. Nevertheless, a higher settling velocity simultaneously implies that particles can remain suspended in the water column only for shorter times, thus they cannot travel too far from their source. As a consequence SR decreases well below the observed levels after a further increase of the particle size. It may be seen that, in general, the best results are obtained with a mean particle size of 4 μm.

The sensitivity of the ^{137}Cs transport model to the equilibrium distribution coefficient has been studied as well. Obviously, a higher k_d means that the substance has a higher affinity to be fixed to the solid phase and vice versa. Both the surface sediment concentrations and total inventories are linearly correlated with the k_d, as should be expected from the model formulation. It may be seen in Table 8 that computed ^{137}Cs concentrations in the surface sediment are well above the measured levels if the average k_d value given by the IAEA, 4 m^3/kg [74], is used. Computed concentrations diminish as the k_d is reduced and, in general, the best agreement with measurements is obtained with a k_d equal to 2 m^3/kg. Indeed, the sediment inventory is clearly underestimated if a k_d of 1 m^3/kg is used, although ^{137}Cs concentrations in the surface sediment are acceptable. A detailed study, by means of montecarlo methods, of the sensitivity of the description of the water-sediment interactions to the different parameters involved may be seen in an earlier work [95] and will not be repeated here.

5. Conclusion

Two examples of the kinds of models that may be developed for simulating radioactivity dispersion in the marine environment are presented. First, a rapid-response lagrangian, or particle-tracking, model for dissolved radionuclides in the Strait of Gibraltar. These models may be used to support the decision-making process after an emergency situation consisting of a radioactive spill in the sea. Then a example of a model developed for an environmental study is described (Alborán Sea model).

In a particle-tracking model the contamination release is simulated by a number of particles whose paths are computed. Diffusion and radioactive decay are simulated by a Monte Carlo method. The currents required to calculate the advective transport are obtained from a hydrodynamic model. Radionuclide concentrations are obtained from the density of particles per water volume unit. A 2D barotropic model is used to calculate tidal currents over the domain. This model provides the tidal constants that are used by the dispersion model to reconstruct the tidal current at any time and position in the model domain. This model has been validated through the comparison of computed tidal elevations, phases and currents with measurements in the area. Computed mean currents have also been compared with available data. Results are, in general, in good agreement with observations in the Strait.

Some dispersion calculations have been carried out as examples of model performance. In general, contaminants are flushed towards the east due to the mean current. Nevertheless, dominant east winds tend to retain contamination in the Strait and to enhance mixing. This is also the case if the release occurs close to the coast, where currents are weaker than in the central part of the Strait.

All the study of the Alborán Sea has been made in a sequential way. Thus, the hydrodynamic model is tested first. Once it is giving a correct representation of water circulation in the region, the sediment transport is included. Radionuclide dispersion is finally considered when the sediment model produces sedimentation rates and suspended matter concentrations in agreement with observations in the Alborán Sea.

Two models are used to describe water circulation: a 2D depth-averaged model is applied to tide computations (the same as in the Strait of Gibraltar) and a two-layer model is used for the residual circulation. Computed tide elevations and phases, as well as tidal ellipse parameters have been compared with observations in the Alborán Sea. The model gives a correct representation of tides, showing that tidal amplitudes and currents decrease quickly as going from the Strait of Gibraltar into the Alborán Sea. The 2-layer model also reproduces the dominant features of the residual circulation in the Alborán Sea, namely the Atlantic jet and Western Alborán Gyre in the upper layer and the fact that water in the bottom layer essentially flows to the west along the Morocco slope. The magnitudes of computed currents have been compared with observations and both set of data are, in general, in good agreement.

The sediment and radionuclide transport models are run off-line. Tidal analysis is used to calculate tidal constants for the two constituents included in the model, M_2 and S_2, which are stored in files that are later read by the dispersion models. This allows a fast determination of tidal currents.

The sediment model consists of an advection/diffusion equation to which terms describing sediment erosion, deposition and settling, and diffusion through the pycnocline are added. Sediment supplies from the main rivers discharging in the area are included. Computed suspended matter concentrations are low, below 1 g/m^3, which is in agreement with measurements. Also, computed sedimentation rates are in agreement, by order of magnitude, with those deduced from observations. They are of the order of 10^{-2} g/cm^{2}year, although the model generally underestimates them. This is due to model assumptions: only the lithogenic particle fraction is simulated, bedload is not included and episodes of high river discharge, when larger amounts of sediments are supplied, are not considered. The general trend of decreasing SR with water depth is given by the model.

The radionuclide dispersion model considers that radionuclides may be present in three phases: dissolved, suspended matter and bed sediments. Exchanges of radionuclides between the liquid and solid phases are described by means of kinetic transfer coefficients, assuming a single reversible reaction (1-step model). The migration of radionuclides from the sediment mixed layer to the deep sediment because of settling of suspended matter has also been described. The model has been applied to three radionuclides: ^{137}Cs, ^{239,240}Pu and ^{210}Pb. They are mainly introduced from atmospheric deposition. Pu and Pb interact strongly with sediments, thus a time-splitting method was required to solve the water-sediment interaction terms because of the highly restrictive stability conditions that these terms introduce.

Generally speaking, the model produces results in agreement with measurements. A correct estimation of ^{137}Cs levels in bed sediments and in the water column is made. Also, the model reproduces the observed fact that essentially the same Cs concentrations exist in both water layers. Although the model slightly underestimates Pu concentrations in the water column, it is obtained that concentrations in the deep water layer are essentially a factor 2 higher than in the surface layer because of the efficient removal of dissolved plutonium by settling suspended particles. Excess ^{210}Pb inventories in the sediment have been calculated. They are underestimated with respect to measurements, although this is due to the difficulties in modelling ^{210}Pb: this is a natural radionuclide which is also produced by in situ radioactive decay of its parent radionuclide ^{226}Ra. Also, initial conditions for the simulation cannot be appropriately defined.

An interesting effect has been observed. It consists of an enhancement of radionuclide concentrations in the WAG. This effect has also been observed in the centre of oceanic gyres in the Atlantic and Pacific. It is due to the fact that water remains trapped in the gyre region, remaining relatively isolated from the surroundings, and thus accumulating radionuclides introduced from the atmosphere. However, this effect is overestimated by the model due to the steady residual flow used in the simulations.

Radionuclide fluxes through the Strait of Gibraltar have been evaluated. They are in relative good agreement with previous estimations based upon radionuclide concentration measurements in the Strait. Finally, system time-scales that are useful from a management perspective have been calculated. They are the sediment half-life and the system-wide flushing time. Both time scales are significantly larger for ^{239,240}Pu than for ^{137}Cs, which is due to the fact that reactive elements are less mobile in the environment than those remaining dissolved. Thus, flushing time for ^{239,240}Pu in the surface water layer is one order of magnitude larger than that of ^{137}Cs, for instance.

References

[1] Scott, E.M. *Modelling Radioactivity in the Environment*. Elsevier, Amsterdam, The Netherlands, 2003, pp. 1-427.

[2] Cabello, J. it Parque Natural del Estrecho, un nuevo futuro. *Revista Medio Ambiente* 2003, 43. Available on-line at http://www.juntadeandalucia.es/medioambiente (in Spanish).

[3] Autoridad Portuaria de la Bahía de Algeciras, 2006. Web site http://www.apba.es (in English and Spanish).

[4] Nav42. Report to the Maritime Safety Committee. International Maritime Organization, 1998. Available on line at http://www.navcen.uscg.gov/marcomms/imo/document.htm

[5] Tsimplis, M.N. Bryden, H.L. Estimations of the transports through the Strait of Gibraltar. *Deep Sea Res.* 2000, 47, pp. 2219-2242.

[6] Bryden, H.L., Candela, J. and Kinder, T.H. Exchange through the Strait of Gibraltar. *Prog. Oceanog.* 1994, 33, pp. 201-248.

[7] Lacombe, H., Richez, C. The regime of the Strait of Gibraltar. In: *Hydrodynamics of Semi-Enclosed Seas*, Editor J.C.J. Nihoul; Elsevier, Amsterdam, 1982, pp. 13-73.

[8] Van Geen, A., Boyle, E. Variability of trace-metal fluxes through the Strait of Gibraltar. *Palaeogeog. Palaeoclim. Palaeoecol.* 1990, 89, pp. 65-79.

[9] Gómez, F. The role of exchanges through the Strait of Gibraltar on the budget of elements in the western Mediterranean Sea: consequences of human-induced modifications. *Mar. Pol. Bull.* 2003, 46, pp. 685-694.

[10] Gascó, C., Antón, M.P., Delfanti, R., González, A.M., Meral, J., Pappuci, C. Variation of the activity concentrations and fluxes of natural (^{210}Po, ^{210}Pb) and anthropogenic (^{239,240}Pu, ^{137}Cs) radionuclides in the Strait of Gibraltar (Spain). *J. Environ. Radioact.* 2002, 62, pp. 241-262.

[11] León-Vintró, L., Mitchell, P.I., Condren, O.M., Downes, A.B., Papucci, C., Delfanti, R. Vertical and horizontal fluxes of plutonium and americium in the western Mediterranean and the Strait of Gibraltar. *Sci.Tot. Environ.* 1999, 237, pp. 77-91.

[12] Elbaz-Poulichet, F., Morley, N.H., Beckers, J.M., Nomerange, P. Metal fluxes through the Strait of Gibraltar: the influence of the Odiel and Tinto Rivers (SW Spain). *Mar. Chem.* 2001, 73, pp. 193-213.

[13] Criado-Aldeanueva, F., García-Lafuente, J., Vargas, J.M., del Río, J., Vázquez, A., Reul, A. Sánchez, A., Distribution and circulation of water masses in the Gulf of Cádiz from in situ observations. it Deep Sea Res. II 2006, 53, pp. 1144-1160.

[14] Gascard, J.C., Richez, C. Water masses and circulation in the western Alborán Sea and in the Straits of Gibraltar. *Prog. Oceanog.* 1985, 15, pp. 157-216.

[15] Parrilla, G., Kinder, T.H., Preller, R.H. Deep and intermediate Mediterranean water in the western Alborán Sea. *Deep Sea Res.* 1986, 33, pp. 55-88.

[16] Izquierdo, A., Tejedor, L., Sein, D.V., Backhaus, J.O., Brandt, P., Rubino, A., Kagan, B.A. Control variability and internal bore evolution in the Strait of Gibraltar: a 2D two-layer model study. *Estuar. Coastal Shelf Sci.* 2001, 53, pp. 637-651.

[17] Preller, R.H. A numerical model study of the Alborán Sea gyre. *Progr. Oceanog.* 1986, 16, pp. 113-146.

[18] Heburn, G.W., La Violette, P. Variations in the structure of the anticyclonic gyres found in the Alborán Sea. *J. Geophys. Res.* 1990, 95C2, pp. 1599-1613.

[19] Perkins, H., Kinder, T., La Violette, P. The Atlantic inflow in the western Alborán Sea. *J. Phys. Oceanog.* 1990, 20, pp. 242-263.

[20] Vargas-Yáñez, M., Plaza, F., García-Lafuente, J., Sarhan, T., Vargas, J.M., Velez-Belchi, P. About the seasonal variability of the Alborán Sea circulation. *J. Mar. Syst.* 2002, 35, pp. 229-248.

[21] Vélez-Belchí, P., Vargas-Yáñez, M., Tintoré, J. Observation of a western Alborán gyre migration event. *Progr. Oceanog.* 2005, 66, pp. 190-210.

[22] Bryden, H.L., Stommel, H.M. Origin of the Mediterranean outflow. *J. Mar. Res.* 1982, 40, pp. 55-71.

[23] Mañanes, R., Bruno, M., Alonso, J., Fraguela, B., Tejedor, L. Non-linear interaction between tidal and subinertial barotropic flows in the Strait of Gibraltar. *Oceanol. Acta* 1998, 21, pp. 33-46.

[24] Tejedor, L., Izquierdo, A., Kagan, B.A.,Sein, D.V. Simulation of the semidiurnal tides in the Strait of Gibraltar. *J. Geophys. Res.* 1999, 104, pp. 13541-13557.

[25] Tsimplis, M.N., Proctor, R., Flather, R.A. A two dimensional tidal model for the Mediterranean Sea. *J. Geosphys. Res.* 1995, 100, pp. 16223-16239.

[26] Auffret, G.A., Pastouret, L., Chamley, H., Lanoix, F. Influence of the prevailing current regime on sedimentation in the Alborán Sea. *Deep Sea Res.* 1974, 21, pp. 839-849.

[27] Masqué, P., Fabres, J., Canals, M., Sanchez-Cabeza, J.A., Sanchez-Vidal, A., Cacho, I., Calafat, A.M., Bruach, J.M. Accumulation rates of major consituents of hemipelagic sediments in the deep Alborán Sea: a centennial perspective of sedimentary dynamics. *Mar. Geol.* 2003, 193, pp. 207-233.

[28] Fabres, J., Calafat, A., Sanchez-Vidal, A., Canals, M., Heussner, S. Composition and spatio-temporal variability of particle fluxes in the western Alborán gyre, Mediterranean Sea. *J. Mar. Syst.* 2002, 33-34, pp. 431-456.

[29] Liquete, C., Arnau, P., Canals, M., Colas, S. Mediterranean river systems of Andalusia, southern Spain, and associated deltas: a source to sink approach. *Mar. Geol.* 2005, 222-223, pp. 471-495.

[30] Periáñez, R. *Modelling the Dispersion of Radionuclides in the Marine Environment*, Springer-Verlag, Heidelberg, Germany, 2005, pp. 1-201.

[31] Pugh, D.T. *Tides, Surges and Mean Sea Level*, Wiley, Chichester, UK, 1987, pp. 1-472.

[32] Candela, J., Winant, C., Ruiz, A. Tides in the Strait of Gibraltar. *J. Geophys. Res.* 1990, 95, pp. 7313-7335.

[33] García-Lafuente, J. Variabilidad del nivel del mar en el estrecho de Gibraltar: mareas y oscilaciones residuales (Sea level variability in the Strait of Gibraltar: tides and residual oscillations). Ph.D. Dissertation, Instituto Español de Oceanografía, Málaga, Spain (in Spanish), 1986, pp. 1-154.

[34] Jensen, T.G. Open boundary conditions in stratified ocean models. *J. Mar. Syst.* 1998, 16, pp. 297-322.

[35] Delhez, E.J.M. On the residual advection of passive constituents. *J. Mar. Syst.* 1996, 8, pp. 147-169.

[36] Sannino, G., Bargagli, A., Artale, V. Numerical modelling of semidiurnal tidal exchange through the Strait of Gibraltar. *J. Geophys. Res.* 2004, 109, pp. C05011.

[37] Riddle, A.M. The specification of mixing in random walk models for dispersion in the sea. *Cont. Shelf Res.* 1998, 18, pp. 441-456.

[38] Proctor, R., Flather, R.A., Elliott, A.J. Modelling tides and surface drift in the Arabian Gulf: application to the Gulf oil spill. *Cont. Shelf Res.* 1994, 14, pp. 531-545

[39] Elliott, A.J. Shear diffusion and the spread of oil in the surface layers of the North Sea. *German J. Hydrogr.* 1986, 39, pp. 113-137.

[40] Periáñez, R., Elliott, A.J. A particle tracking method for simulating the dispersion of non-conservative radionuclides in coastal waters. *J. Environ. Radioact.* 2002, 58, pp. 13-33.

[41] Periáñez, R. Modelling surface radioactive spill dispersion in the Alborán Sea. *J. Environ. Radioact.* 2006, 90, pp. 48-67.

[42] Dick, S., Schonfeld, W. Water transport and mixing in the North Frisian Wadden Sea. Results of numerical investigations. *German J. Hydrogr.* 1996, 48, pp. 27-48.

[43] Schonfeld, W. Numerical simulation of the dispersion of artificial radionuclides in the English Channel and the North Sea. *J. Mar. Syst.* 1995, 6, pp. 529-544.

[44] Elliott, A.J. Simulations of atmospheric dispersion following a spillage of petroleum at sea. *Spill Sci. Tech. Bull.* 1999, 5, pp. 39-50.

[45] Elliott, A.J., Wilkins, B.T., Mansfield, P. On the disposal of contaminated milk in coastal waters. *Mar. Poll. Bull.* 2001, 42, pp. 927-934.

[46] Echevarría, F., García-Lafuente, J., Bruno, M., Gorsky, G., Goutx, M., González, N., García, C.M., Gómez, F., Vargas, J.M., Picheral, M., Striby, L., Varela, M., Alonso, J.J., Reul, A., Cozar, A., Prieto, L., Sarhan, T., Plaza, F., Jiménez-Gómez, F. Physical-biological coupling in the Strait of Gibraltar. *Deep Sea Res. II* 2002, 49, pp. 4115-4130.

[47] Periáñez, R., Pascual-Granged, A. Modelling surface radioactive, chemical and oil spills in the Strait of Gibraltar. *Comp. Geosci.* in press.

[48] Werner, F.E., Cantos-Figueroa, A., Parrilla, G., A sensitivity study of reduced-gravity flows with applications to the Alborán Sea. *J. Phys. Oceanog.* 1988, 18, pp. 373-383.

[49] Periáñez, R. Modelling the transport of suspended particulate matter by the Rhone River plume (France). Implications for pollutant dispersion. *Environ. Poll.* 2005, 133, pp. 351-364.

[50] Periáñez, R. Modelling the dispersion of radionuclides by a river plume. Application to the Rhone River. *Cont. Shelf Res.* 2005, 25, pp. 1583-1603.

[51] Liu, W.C., Hsu, M.H., Kuo, A.Y. Modelling of hydrodynamics and cohesive sediment transport in Tanshui River estuarine system, Taiwan. *Mar. Poll. Bull.* 2002, 44, pp. 1076-1088.

[52] Lumborg, U., Windelin, A. Hydrography and cohesive sediment modelling: application to the Romo Dyb tidal area. *J. Mar. Syst.* 2003, 38, pp. 287-303.

[53] Cancino, L., Neves, R. Hydrodynamic and sediment suspension modelling in estuarine systems. Part I: description of the numerical models. *J. Mar. Syst.* 1999, 22, pp. 105-116.

[54] Monte, L., Periáñez, R., Kivva, S., Laptev, G., Angeli, G., Barros, H., Zheleznyak, M. Assessment of state-of-the-art models for predicting the remobilisation of radionuclides following the flooding of heavily contaminated areas: the case of the Pripyat River floodplain. *J. Environ. Radioact.* 2006, 88, pp. 267-288.

[55] Periáñez, R. Three dimensional modelling of the tidal dispersion of non conservative radionuclides in the marine environment. Application to $^{239,240}Pu$ dispersion in the eastern Irish Sea. *J. Mar. Syst.* 1999, 22, pp. 37-51.

[56] Periáñez, R. Modelling the tidal dispersion of ^{137}Cs and $^{239,240}Pu$ in the English Channel. *J. Environ. Radioact.* 2000, 49, pp. 259-277.

[57] Periáñez, R. The enhancement of ^{226}Ra in a tidal estuary due to the operation of fertilizer factories and redissolution from sediments: experimental results and a modelling study. *Estuar. Coastal Shelf Sci.* 2002, 54, pp. 809-819.

[58] Periáñez, R., Martínez-Aguirre, A. U and Th concentrations in an estuary affected by phosphate fertilizer processing: experimental results and a modelling study. *J. Environ. Radioact.* 1997, 35, pp. 281-304.

[59] Periáñez, R., Abril, J.M., García-León, M. Modelling the dispersion of non conservative radionuclides in tidal waters. Part 1: conceptual and mathematical model. *J. Environ. Radioact.* 1996, 31, pp. 127-141.

[60] Duursma, E.K., Carroll, J. *Environmental Compartments*, Springer, Berlin, Germany, 1996, pp. 1-277.

[61] Vested H.J., Baretta J.W., Ekebjærg, L.C. Labrosse, A. Coupling of hydrodynamical transport and ecological models for 2D horizontal flow. *J. Mar. Syst.* 1996, 8, pp. 255-267.

[62] Kowalick, Z., Murty, T.S. *Numerical Modelling of Ocean Dynamics*, World Scientific, Singapore, pp. 1-481.

[63] Tappin, A.D., Burton, J.D., Millward, G.E., Statham, P.J. A numerical transport model for predicting the distributions of Cd, Cu, Ni, Pb and Zn in the southern North Sea: the sensitivity of model results to the uncertainties in the magnitudes of metal inputs. *J. Mar. Syst.* 1997, 13, pp. 173-204.

[64] Manzella, G.M.R. and Elliott, A.J. EUROSPILL: Mediterranean tidal and residual databases. *Mar. Poll. Bull.* 1991, 22, pp. 553-558.

[65] García-Lafuente, J.M., Cayo-Lucana, N. Tidal dynamics and associated features of the northwestern shelf of the Alborán Sea. *Cont. Shelf Res.* 1994, 14, pp. 1-21.

[66] Béranguer, K., Mortier, L., Crépon, M., Seasonal variability of water transport through the Straits of Gibraltar, Sicily and Corsica, derived from a high-resolution model of the Mediterranean circulation. *Progr. Oceanog.* 2005, 66, pp. 341-364.

[67] Sannino, G., Bargagli, A., Artale, V. Numerical modelling of the mean exchange through the Strait of Gibraltar. *J. Geophys. Res.* 2002, 107, pp. 3094-3118.

[68] Nikiema, O., Devenon, J.L., Baklouti, M. Numerical modelling of the Amazon River plume. *Cont. Shelf Res.* 2007, 27, pp. 873-899.

[69] Tattersall, G.R., Elliott, A.J., Lynn, N.M., 2003. Suspended sediment concentrations in the Tamar estuary. *Estuar. Coastal Shelf Sci.* 2003, 57, pp. 679-688.

[70] Uncles, R.J., Stephens, J.A. Distribution of suspended sediments at high water in a macrotidal estuary. *J. Geophys. Res.* 1989, 94, pp. 14395-14405.

[71] Liu, J.T., Chao, S., Hsu, R.T. Numerical modeling study of sediment dispersal by a river plume. *Cont. Shelf Res.* 2002, 22, pp. 1745-1773.

[72] Palanques, A., El Khatab, M., Puig, P., Masqué, P., Sánchez-Cabeza, J.A., Isla, E. Downward particle fluxes in the Guadiaro submarine canyon depositional system (north-western Alborán Sea), a river flood dominated system. *Mar. Geol.* 2005, 220, pp. 23-40.

[73] Nyffeler, U.P., Li, Y.H., Santschi, P.H. A kinetic approach to describe trace element distribution between particles and solution in natural aquatic systems. *Geochim. Cosmochim. Acta* 1984, 48, pp. 1513-1522.

[74] IAEA *Sediment distribution coefficients and concentration factors for biota in the marine environment* Technical Reports Series 2004, 422, Vienna, Austria, pp. 1-93.

[75] Periáñez, R. Kinetic modelling of the dispersion of plutonium in the eastern Irish Sea: two approaches. *J. Mar. Syst.* 2003, 38, pp. 259-275.

[76] Aldridge, J.N., Kershaw, P., Brown, J., McCubbin, D., Leonard, K.S., Young, E.F. Transport of plutonium ($^{239/240}$Pu) and caesium (^{137}Cs) in the Irish Sea: comparison between observations and results from sediment and contaminant transport modelling. *Cont. Shelf Res.* 2003, 23, pp. 869-899.

[77] Sánchez-Cabeza, J.A., Ortega, M., Merino, J., Masqué, P. Long-term box modelling of ^{137}Cs in the Mediterranean Sea. *J. Mar. Syst.* 2002, 33/34, pp. 457-472.

[78] Gascó, C., Antón, M.P., Pozuelo, M., Meral, J., González, J.M., Papucci, C., Delfanti, R. Distributions of Pu, Am and Cs in margin sediments from the western Mediterranean (Spanish coast). *J. Environ. Radioact.* 2002, 59, pp. 75-89.

[79] Benes, P., Cernik, M., Lam-Ramos, P. Factors affecting the interaction of radiocesium with freshwater solids. Contact time, concentrations of solids and temperature. *J. Radioanal. Nucl. Chem.* 1992, 159, pp. 201-218.

[80] Hirose, K., Igarashi, Y., Aoyama, M., Kim, C.K., Kim, C.S., Chang, B.W. Recent trends of plutonium fallout observed in Japan: plutonium as a proxy for desertification. *J. Environ. Monitor.* 2003, 5, pp. 302-307.

[81] Ivanovich, M., Harmon, R.S. *Uranium-series Disequilibrium: Applications to Earth, Marine and Environmental Sciences*, Clarendon Press, Oxford, UK, 1992.

[82] Sánchez-Cabeza, J.A., Masqué, P., Martínez-Alonso, M., Mir, J., Esteve, I. ^{210}Pb atmospheric flux and growth rates of a microbial mat from the northwestern Mediterranean Sea (Ebro River delta). *Environ. Sci. Tech.* 1999, 33, pp. 3711-3715.

[83] Ivanovich, M., Harmon, R.S. *Uranium-series Disequilibrium: Applications to Environmental Problems*, Clarendon Press, Oxford, UK, 1982.

[84] Sarmiento, J.L., Herbert, T., Toggweiler, J.R. Mediterranean nutrient balance and episodes of anoxia. *Glob. Biogeochem. Cycles* 1988, 14, pp. 669-681.

[85] Greze, V.N., Kovalev, A.V., Baldina, E.P., Bileva, O.K. Shmeleva, A.A. Zooplankton transfer through the Strait of Gibraltar and peculiarities of its taxonomic composition and distribution in adjacent areas. *Investigaciones Pesqueras* 1985, 49, pp. 3-13.

[86] Prandle, D. A modelling study of the mixing of ^{137}Cs in the seas of the European Continental Shelf. *Phil. Transac. R. Soc. London* 1984, A310, pp. 407-436.

[87] Periáñez, R. Redissolution and long-term transport of radionuclides released from a contaminated sediment: a numerical model study. *Estuar. Coastal Shelf Sci.* 2003, 56, pp. 5-14.

[88] Shen, J., Haas, L. Calculating age and residence time in the tidal York River using three dimensional model experiments. *Estuar. Coastal Shelf Sci.* 2004, 61, pp. 449-461.

[89] Choi, K.W., Lee, J.H.W. Numerical determination of flushing times for stratified water bodies. *J. Mar. Syst.* 2004, 50, 263-281.

[90] Cook, G.T., MacKenzie, A.B., McDonald, P., Jones, S.R. Remobilization of Sellafield derived radionuclides and transport from the north east Irish Sea. *J. Environ. Radioact.* 1997, 35, pp. 227-241.

[91] Mitchell, P.I., Condren, O.M., León-Vintró, L. and McMahon, C.A. Trends in plutonium, americium and radiocaesium accumulation and long-term bioavailability in the western Irish Sea mud basin. *J. Environ. Radioact.* 1999, 44, pp. 223-251.

[92] Ciffroy, P., Garnier, J.M., Pham, M.K. Kinetics of the adsorption and desorption of radionuclides of Co, Mn, Cs, Fe, Ag and Cd in freshwater systems: experimental and modelling approaches. *J. Environ. Radioact.* 2001, 55, pp. 71-91.

[93] El-Mrabet, R., Abril, J.M., Manjón, G., García-Tenorio, R. Experimental and modelling study of plutonium uptake by suspended matter in aquatic environments from southern Spain. *Wat. Res.* 2001, 35, pp. 4184-4190.

[94] Periáñez, R. Testing the behaviour of different kinetic models for uptake/release of radionuclides between water and sediments when implemented in a marine dispersion model. *J. Environ. Radioact.* 2004, 71, pp. 243-259.

[95] Periáñez, R. On the sensitivity of a marine dispersion model to parameters describing the transfers of radionuclides between the liquid and solid phases. *J. Environ. Radioact.* 2004, 73, pp. 101-115.

In: Atlantic and Indian Oceans
Editors: E. S. Askew and J. P. Bromley, pp. 77-114
ISBN 978-1-60692-475-4

Chapter 3

MARINE NATURAL PRODUCTS RESEARCH IN THE WESTERN INDIAN OCEAN

Anne Gauvin-Bialecki
Laboratoire de Chimie des Produits Naturels et des Sciences des Aliments,
Faculté des Sciences et Technologies, Université de la Réunion,
15 Avenue René Cassin, BP 7151, 97 715 St Denis,
Messag Cedex 9, La Réunion, France

ABSTRACT

Marine invertebrates, algae and microorganisms produce a considerable number of structurally unique and biologically active compounds, also called secondary metabolites. The ecological roles of these natural products, although not completely understood, range from chemical defense against predation to intra-specific cues for larval settlement. Since the late 1960's marine natural product bioprospecting has demonstrated that most of these molecules may be drug candidates.

In addition to their pharmaceutical applications, secondary metabolites may also be employed as chemotaxonomic markers useful for elucidating classification problems and phylogenetic relationships. Drug discovery and chemotaxonomy are two main tracks in marine natural products chemistry which have attracted the attention of numerous biologists and chemists throughout the world. However, despite this intense global interest, the natural products chemistry of marine flora and fauna of the western Indian Ocean islands is still largely ignored.

This paper will provide therefore an outline of the contribution made by the Chemistry Laboratory of Natural Substances and Food Sciences (LCSNSA) at Reunion Island University to the isolation and identification of natural bioactive molecules as well as chemotaxonomic markers from marine organisms of Mayotte, a French Overseas Territory of the Comoros archipelago located in the western Indian Ocean.

INTRODUCTION

Marine invertebrates, algae and microorganisms produce a considerable number of structurally unique and biologically active compounds, also called secondary metabolites. The ecological roles of these natural products, although not completely understood, range from chemical defense against predation to intra-specific cues for larval settlement. Since the late 1960's marine natural product bioprospecting has demonstrated that most of these molecules may be drug candidates. In addition to their pharmaceutical applications, secondary metabolites may also be employed as chemotaxonomic markers useful for elucidating classification problems and phylogenetic relationships. Drug discovery and chemotaxonomy are two main tracks in marine natural products chemistry which have attracted the attention of numerous biologists and chemists throughout the world. However, despite this intense global interest, the natural products chemistry of marine flora and fauna of the western Indian Ocean is still largely ignored. This chapter will provide therefore an outline of the contribution made by the Chemistry Laboratory of Natural Substances and Food Sciences (LCSNSA) at Reunion Island University to marine natural products research in the west part of Indian Ocean.

Our marine natural products research and development program in the western Indian Ocean, was first concentrated on the marine invertebrates from Mayotte, a French Overseas Territory of the Comoros archipelago situated in the Mozambique Channel between Madagascar and the African mainland (Latitude: -12.8°, Longitude: 45.1°) (figure 1). Mayotte encompasses one of the largest coral lagoons in the world (1286 Km^2), cut by several channels (figure 1) and showing an exceptional biodiversity. Among the numerous marine invertebrates encountered in this area, sponges have predominated in all our large-scale collections and have therefore received special attention from our research group. Thus, in this chapter, we wish to use our investigation mainly focused on sponge extracts from Mayotte to illustrate the two approaches of marine natural products chemistry:

(1) *New drug research* with our investigation on secondary metabolites from the sponge *Luffariella variabilis*.
(2) *Chemotaxonomic study* with our investigation on sterols from two sponges of the genus *Haliclona*.

I. Drug Research in Oceans

I.1. Significance of the Study

Even today, after more than 100 years of research in pharmaceutical industries, there is still a great need for innovative drugs. Traditionally plant and terrestrial microorganisms have been in the focus for the search of new drug candidates from nature. However, since the sixties, the increasing needs for drugs able to control new illnesses or resistant strains of microorganisms stimulated to look for unconventional new sources of bioactive natural products. The oceans turned out to be an attractive field.

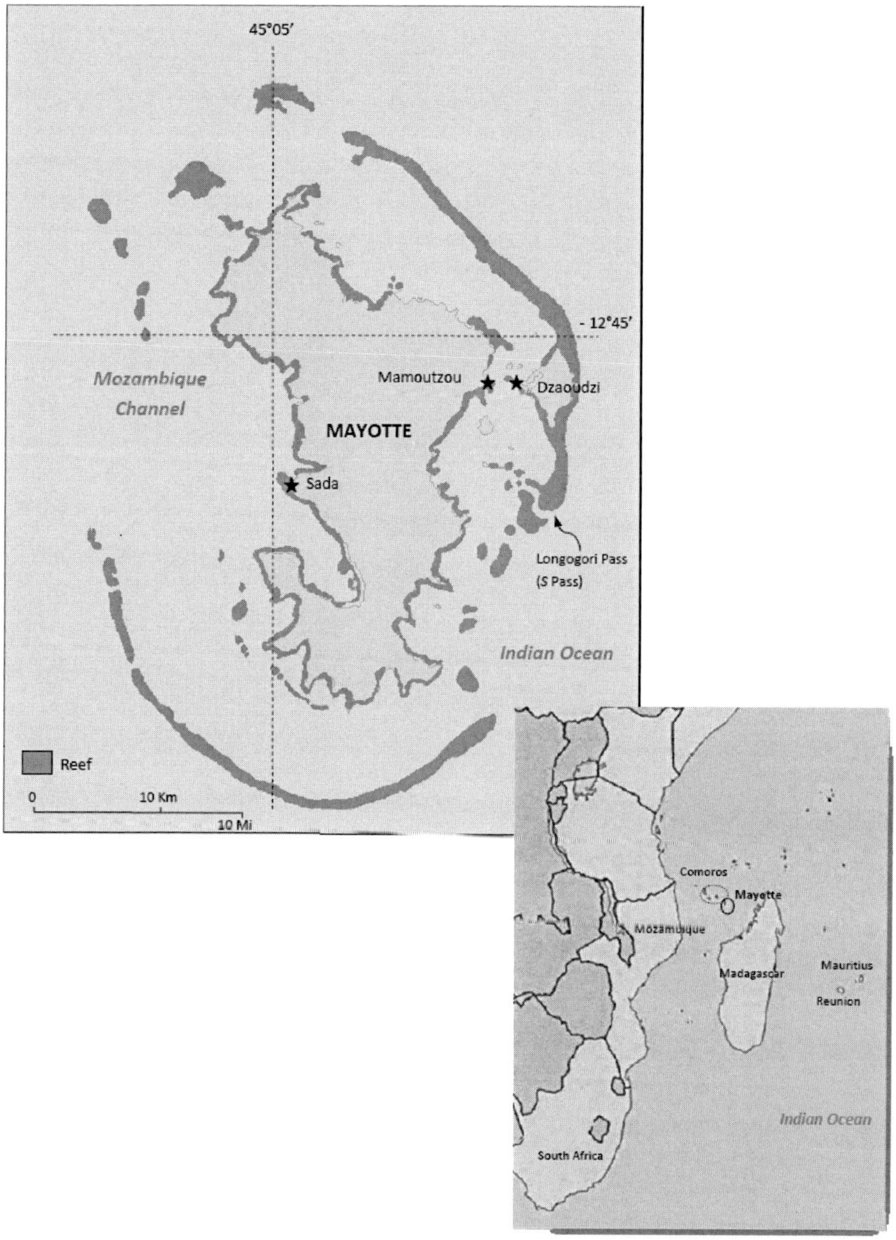

Figure 1. Mayotte - Coral reef map.

The marine environment, particularly in tropical areas, offers indeed a rich diversity of species, which is in many ways comparable to that of tropical rain forests. This environment also contains a wide number of organisms for which there are no terrestrial counterparts. Besides, ecological pressures on marine organisms, which include competition for space, deterrence of predation and pathogens, ability to successfully reproduce may have led to the evolution of unique chemical components responsible for these actions and interactions. Such compounds, having no role in the basic life process i.e. being not essential for the growth and the survival of the organism, are called secondary metabolites and their great diversity of structures in marine organisms is one of the most fascinating aspects of marine natural products chemistry. To date, after 40 years of intensive research, chemistry of marine natural products has become a mature field. The search for new biomedicals from marine micro- and macroorganisms resulted in the isolation of more than 17 000 new secondary metabolites, many of which endowed of pharmacodynamic properties. A broad spectrum of biological activities has been detected, such as antibiotic, antifungal, toxic, cytotoxic, neurotoxic, antimitotic, antiviral, antineoplastic. In more recent years, new targets have been added to the general screening, for example AIDS, immunosupression, anti-inflammation, Alzheimer disease, ageing process and some tropical disease. Among the various marine organisms, sponges have provided a seemingly endless parade of novel structures with a great occurrence of potential pharmaceuticals [Faulkner, 1998 and subsequent annual plublications of these series].

Sponges (phylum Porifera) are the oldest group of multicellular animals (appeared on earth more than 500 million years ago).

Their basic mode of organization and their biochemical properties place them at the very base of the kingdom Metazoa. Sponges come in a variety of shapes (massive, globular, tubular, incrusting, branching, vase- or barrel-shaped …), colors and sizes and their consistency range from extremely soft and delicate to hard and rugged. Sponges are able to filter large amounts of water with a very good efficiency both in water transport and in the retention of a large variety of particles from small colloids to unicellular algae of 50 μm. This ability to feed on very small particulate organic matter and, perhaps, on dissolved organic matter may be one of the reasons for the dominance of sponges over other filter feeders. Sponges also contribute significantly to the biomass of the marine ecosystems. They have a world-wide distribution and are found in a great diversity of habitats: not only in the Oceans from intertidal zones to water thousands of meters deep but also in the fresh water (lakes and rivers). It is generally assumed that this persistent ecological success is due not only to well adapted and adaptable reproductive and physiological strategies, but also to effective defense mechanisms. Being sessile, sponges have to face a large array of dangers: attacks from mobile predators, overgrowing by space competitors, fouling by epizoic organisms and infection by external pathogenic microorganisms [Bakus et al. 1986; Wulff 2006]. A few sponges seem to derive protection from the presence of sharp or special cortical spicules[1] providing a superficial armour. Others lives in crevices or holes, or are covered by epibiotic organisms which can ensure them an effective camouflage. However, physical or behavioral defenses are rare among sponges. For these marine invertebrates, chemical defense (secretion of toxic molecules) seems then to be an efficient alternative to discourage attack of mobile predators like fishes, to kill potential fouling or contaminating organisms, and also to prevent competitors from intruding into their living space. This explains why sponges are the source of the greatest diversity of marine natural products. About one-third of all marine natural products have been isolated from sponges, which make them currently the most popular source of novel compounds. To date, some of these molecules are still in preclinical or early clinical development. Typical examples are discodermolide isolated from *Discodermia dissoluta* (Phase I-anticancer), manoalide isolated from *Luffariella variabilis* (Phase II-antipsoriatic), KRN 7000 a synthetic analog of agelasphins isolated from *Agelas mauritianus*

[1] Spicules are needle-like rods. The rods are individual elements, but they may interweave to form a strong support system. Calcite and Silica both are known as spicule components.

(Phase I-anticancer), IPL 576,092 isolated from *Petrosia contignata* (Phase II – antiantiasmathic and anti-inflammation) and isohomohalichondrin B isolated from *Halichondria okadai* (Phase I - anticancer) [Kornprobst, 2005a]. Today, the pioneering countries in sponge natural products discovery such as USA, Spain, Japan and Australia are benefiting from the great potential and social value of such research. This has again led to significant recognition of sponge bio-prospecting word.

I.2. Strategy and Approach of the Research

Generally, strategies for drug discovery can be separated into three categories: chemically driven, biologically driven and a combination of both [McConnell et al. 1994].

In the chemically driven approach, which has been pursued mainly by academic research groups, the object of the search was to find novel compounds from marine sources. Hence, crude extracts obtained by maceration using organic solvents are screened by Thin Layer Chromatography, ^{1}H and ^{13}C Nuclear Magnetic Resonance for unusual and interesting patterns. The next step for this approach is then finding biological properties for purified compounds.

The biologically driven strategy is the bioassay-guided approach. This method which involves screening crude extracts for biological activity, followed by the crucial work of backtracking the active compounds from the "hit"-extracts dominate natural products research up to the present. However, a lot of experience is required to exclude both false positive and false negative results. False positives may occur when the activity shown by an extract is attributed to a synergistic effect of more than one constituent in the extract. False negatives may be obtained for crude extracts when the screening assays are not sensitive enough. It seems then more advantageous to perform a screening with pure compounds rather than with crude extracts. Moreover, this method required considerable effort to get access to sufficient quantities of raw material for reproduction, isolation, structure elucidation and subsequent verification of biological activity. The complete process proved to be highly time and capacity consuming.

The combination of the chemically and biologically driven approaches means that selecting extracts for chemical fractionation is based on the biological activity profile of the crude extract. However, instead of using a bioassay-guided approach to purify the compounds responsible for the activity of the extract, Nuclear Magnetic Resonance and some chromatographic techniques are used to isolate the chemically most interesting substances. Ideally the structurally unusual or novel compounds are also responsible for the activity of the extract. This approach works well when the active compounds are present in high concentration.

The strategy adopted by our research team mainly focused on the discovery of antitumor and antiviral molecules was closed to the combination of both chemically and biologically approaches. However this technique was slightly modified by considering that the methods for the detection of biological activity of natural product mixtures can be divided into two groups for screening purposes: general screening bioassays and specialized screening bioassays. The search for specific pharmacologic activities often overlooks other useful activities which are not detected or are ignored, in the screening process. Furthermore specific test methods are often cumbersome and expensive. Hence, for bioassay-guided isolation of actives from natural product mixtures single, inexpensive and general bioassays for the rapid screening of extracts and fractions have to be employed. Brine-shrimp assay for example which is rapid, reliable, inexpensive and simple (e.g. no aseptic techniques are required) is clearly suitable. This technique is an *in vivo* lethality test in a tiny crustacean, the brine shrimp (*Artemia salina* Leach). It has been previously utilized in various bioassay systems including in the analysis of pesticide residues, mycotoxins, stream pollutants, anesthetics, dinoflagellate toxins, morphine-like compounds, toxicity of oil dispersants, cocarcinogenicity of phorbol esters and toxicants in marine environments [Meyer et al. 1982]. This test takes into account the basic premise that pharmacology is simply toxicology at a lower dose, and that toxic substances might indeed elicit, at a lower non-toxic dose, interesting pharmacologic effects [McLaughlin 1991]. The procedure enabled us to screen and select the more active lipophilic sponge extracts. Once the crude extracts selected, some chromatographic techniques were used to isolate the potent active components. A battery of specific and sophisticated bioassays were then performed on the pure isolated compounds to test their antiviral (HTLV1: *Human T cell leukemia/lymphoma virus type 1*, HIV: Human immunodeficiency virus) and antitumor properties (human breast carcinoma cells McF_7WT).

I.3. Application to the Sponge *Luffariella Variabilis* from Mayotte: Research of Potent Antitumor and Antiviral Secondary Metabolites

1. Introduction

Luffariella variabilis (Poléjaeff 1884) is a cryptic, coral reef sponge distributed widely through the Indo-Pacific [Bergquist 1980, 1995], in particular in Palau (originally known as the Western Caroline), where it was estimated to be one of the 10 most common sponges on the reefs around the islands [Faulkner et al. 2004]. *Luffariella variabilis* produces a range of related bioactive sesterterpenes (C_{25} terpenes) including manoalide (1) which was the first of this series of compounds, isolated from a Palauan specimen, by de Silva and Scheuer [1980]. This compound was first reported as an antimicrobial agent but its most important activity was as an analgesic and anti-inflammatory agent. The groups of Jacobs [de Freitas et al. 1984; Glaser and Jacobs 1986, 1987; Jacobs et al. 1985] and Dennis [Deems et al. 1987; Lombardo and Dennis 1985] independently established that manoalide inhibits the enzyme phospholipase A_2, which is involved in the initial step of the anti-inflammatory response. Since manoalide was present at a level of about 1% dry weight from a common sponge, it was regarded as a good candidate for drug development and its use as an anti-inflammatory agent was patented by the University of California. It was tested by Allergan Corp. as a topical treatment for psoriasis but the formulation used did not allow sufficient drug to pass through the skin. The apparent failure of manoalide as a potential drug did not discourage additional research on details of its mechanism as a phospholipase A_2 (PLA_2) inhibitor [Bennett et al. 1987; Glaser and Jacobs 1987; Jacobson et al. 1990; Ortiz et al. 1993; Reynolds et al. 1991]. PLA_2 is the enzyme response for the hydrolysis of membrane-bound phospholipids to release arachidonic acid that is subsequently converted into the prostaglandins and leukotrienes that ultimately cause the pain experienced in bee stings, psoriasis, arthritis and other inflammatory conditions. It was shown that manoalide undergoes an irreversible chemical reaction with lysine units on the interfacial binding site on PLA_2 and thus prevents the correct binding between PLA_2 and membrane-bound phospholipids [Potts et al. 1992a]. Structure-activity studies of manoalide and its derivatives have been reported [Glaser et al. 1989; Reynolds et al. 1988, 1991] together with two reviews of the work [Potts et al. 1992b; Soriente 1999]. Since the discovery of manoalide in 1980, marine sponges of the genus *Luffariella* and more generally of the family Thorectidae have proved to be a rich source of bioactive manoalide-related compounds (Table 1 and Figure 2).

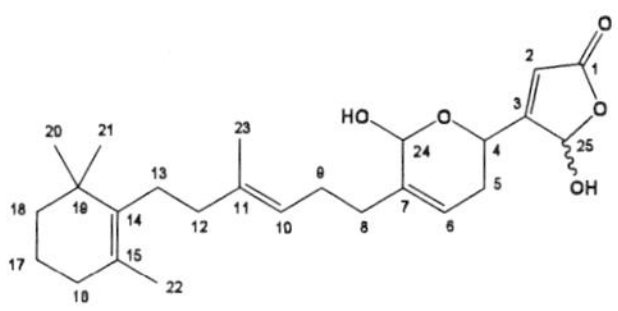

(**1**) Manoalide

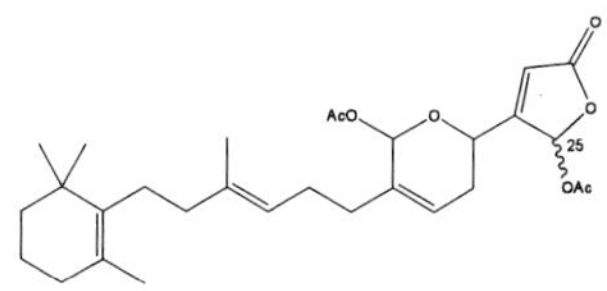

(**1a, 1b**) Manoalide diacetate (C-25 epimers)

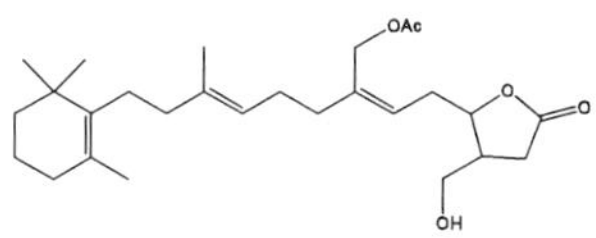

(**2**) (6Z)-24-Acetoxy-2,3-dihydroneomanoalide

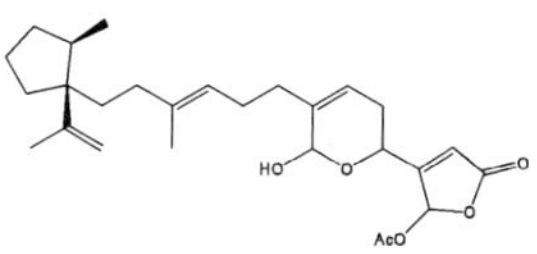

(**3**) 25-Acetoxyluffariellin A

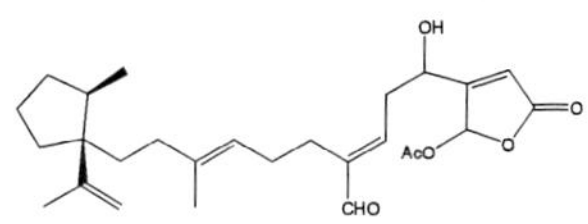

(**4**) 25-Acetoxyluffariellin B

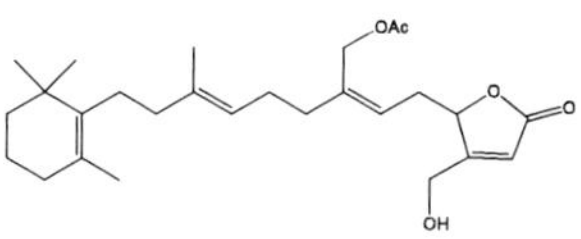

(**5**) (6Z)-24-Acetoxyneomanoalide

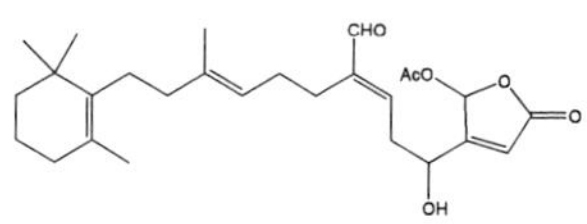

(**6**) 25-Acetoxyseco-manoalide

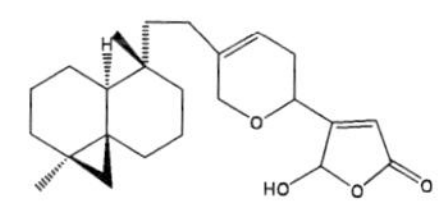

(**7**) Cacospongionolide

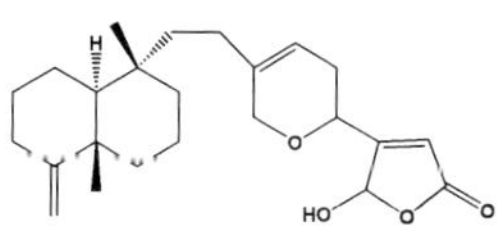

(**8**) Cacospongionolide B

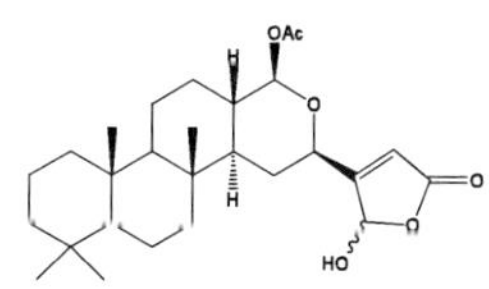

(**9**) Cavernosolide

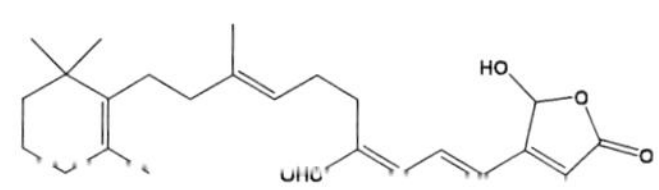

(**10**) (4E,6E)-Dehydromanoalide

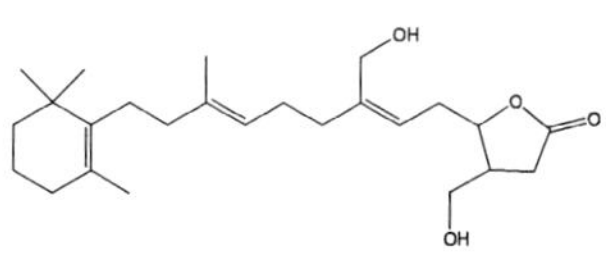

(**11**) (6Z)-2,3-Dihydroneomanoalide

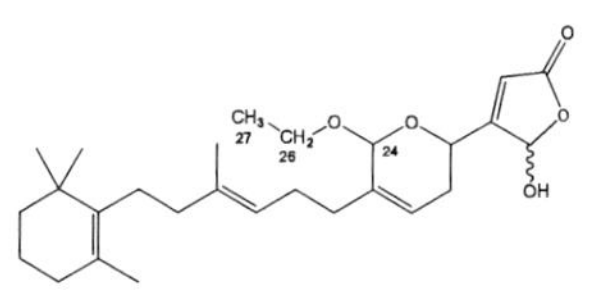

(**12**) 24-Ethyl-*O*-manoalide

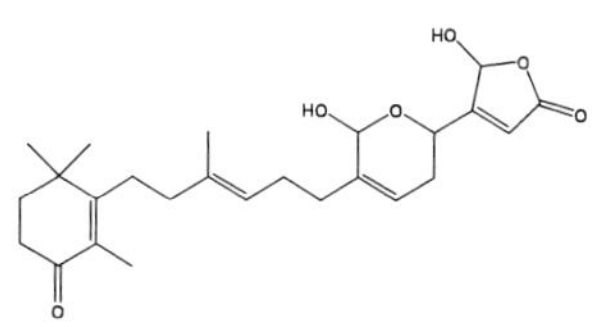

(**13**) Fasciospongide A

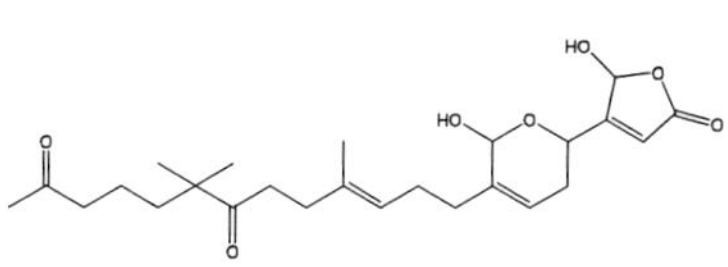

(**14**) Fasciospongide B

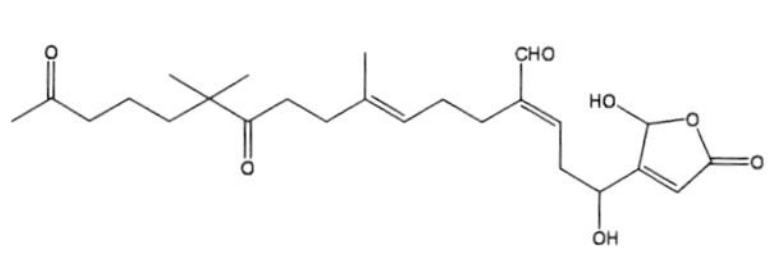

(**15**) Fasciospongide C

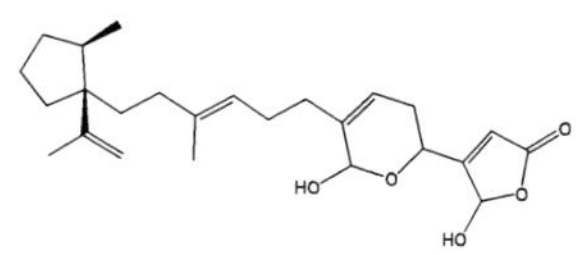

(**16**) Luffariellin A

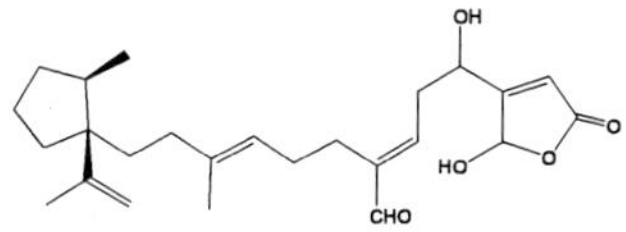

(**17**) Luffariellin B

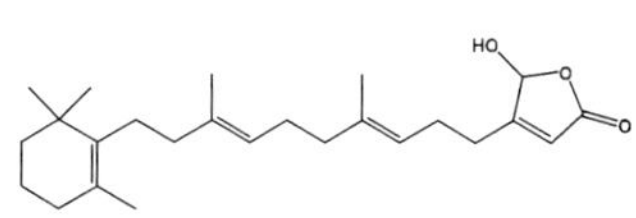

(**18**) Luffariellolide

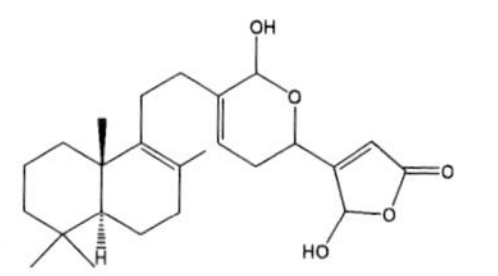

(**19**) Luffarin A

Figure 2. Structures of manoalide-related sesterterpenes.

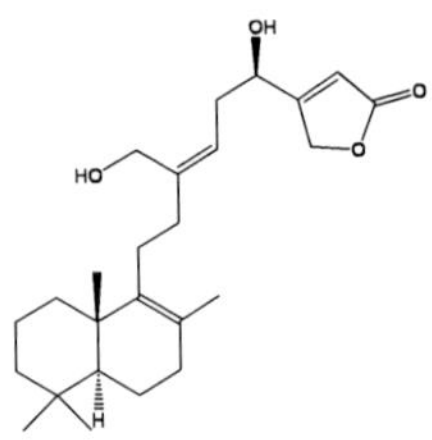

(**27**) Luffarin I

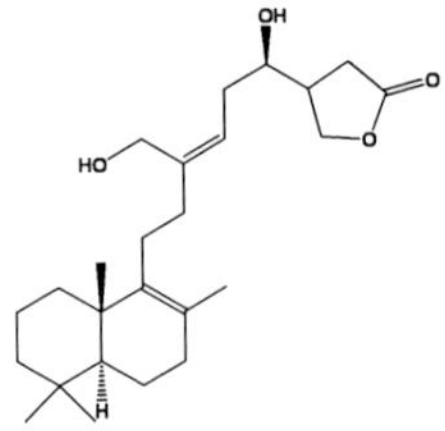

(**28**) Luffarin J

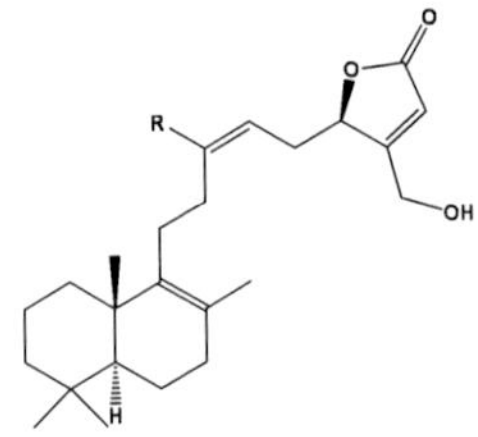

(**29**) R = CH_2OH Luffarin K
(**31**) R = CHO Luffarin M

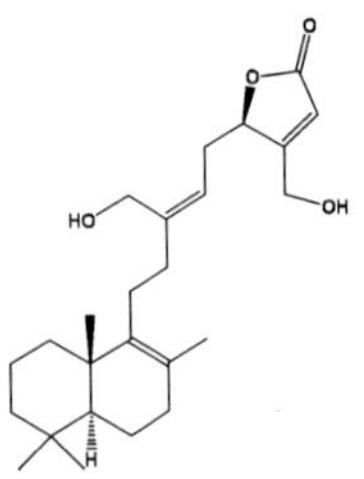

(**30**) Luffarin L

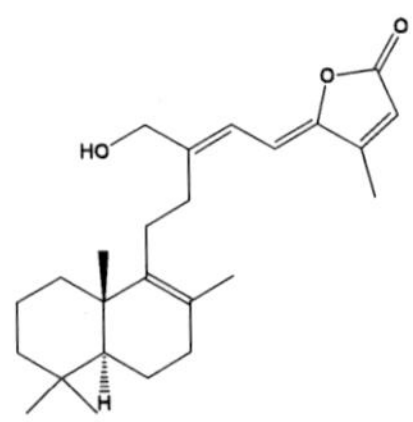

(**32**) Luffarin N

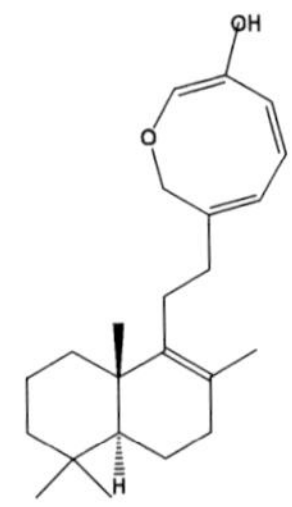

(**33**) Luffarin O

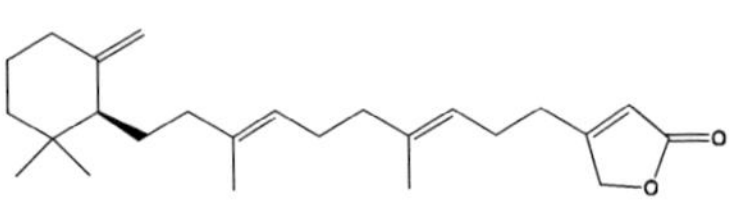

(**34**) Luffarin P

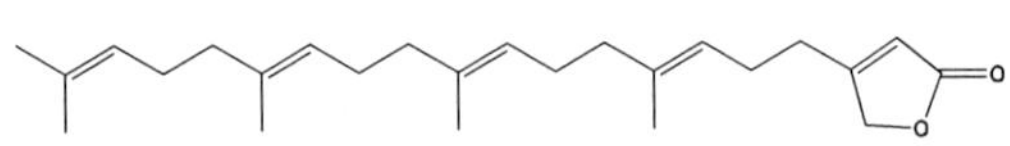

(**35**) Luffarin Q

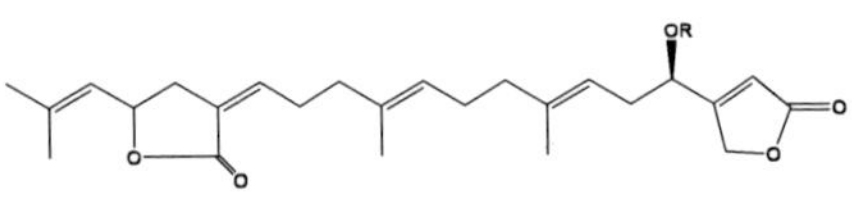

(**36**) R = H Luffarin R
(**37**) R = Ac Luffarin S

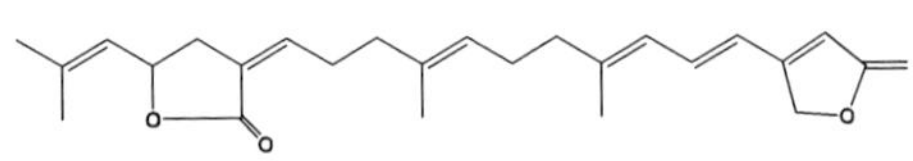

(**38**) Luffarin T

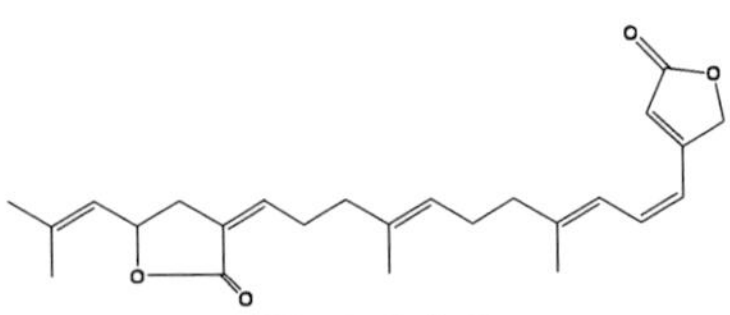

(**39**) Luffarin U

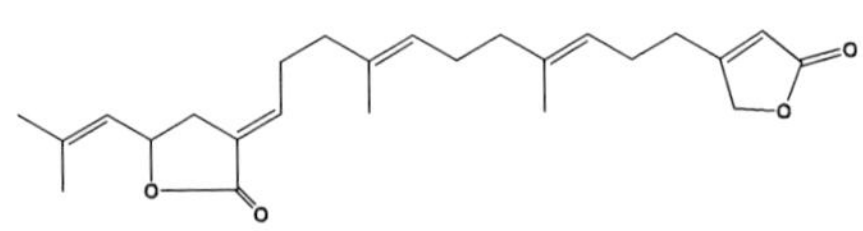

(**40**) Luffarin V

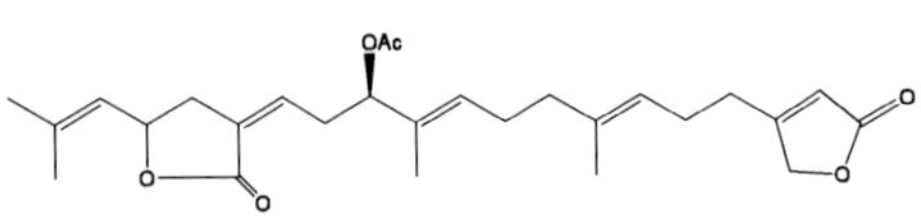

(**41**) Luffarin derivative 1

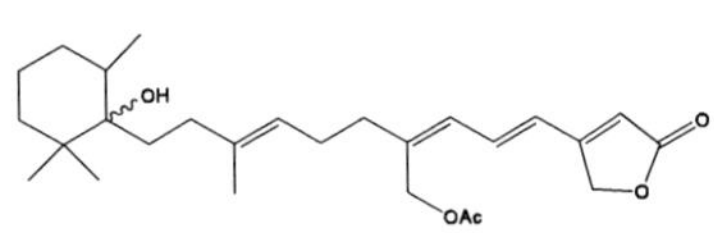

(**42**) Luffarin derivative 2

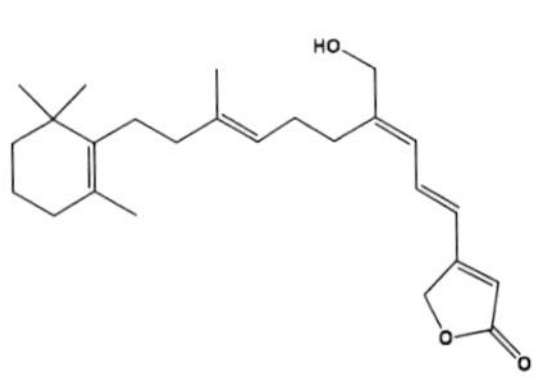

(**43**) Luffariolide A

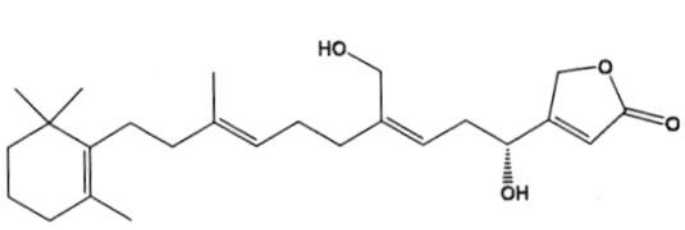

(**44**) Luffariolide B

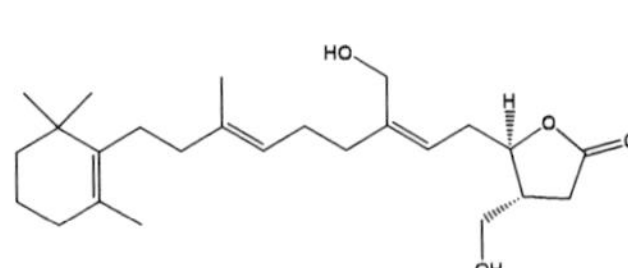

(**45**) Luffariolide C

Figure 2. (continued).

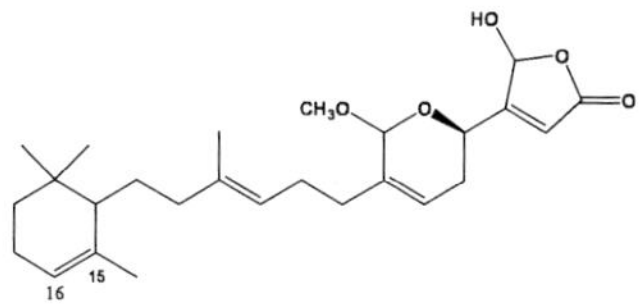

(**50**) Luffariolide H

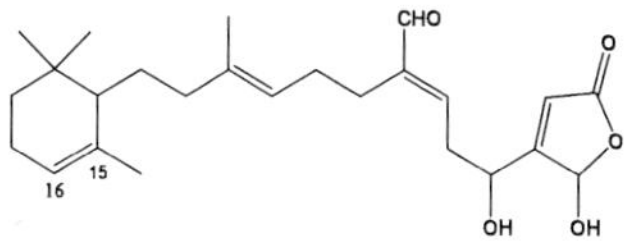

(**51**) Luffariolide J

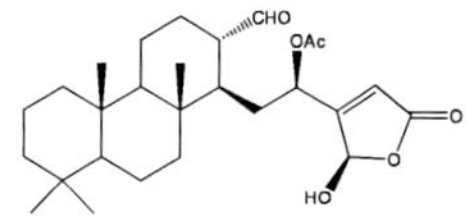

(**52**) Luffolide

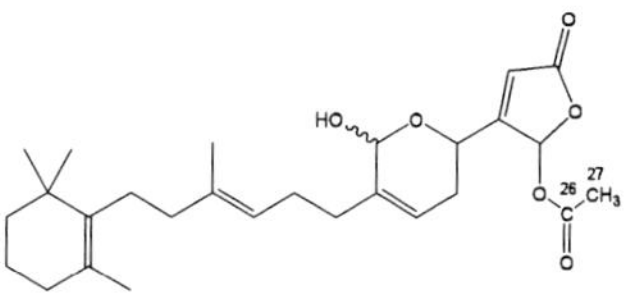

(**53**) Manoalide monoacetate

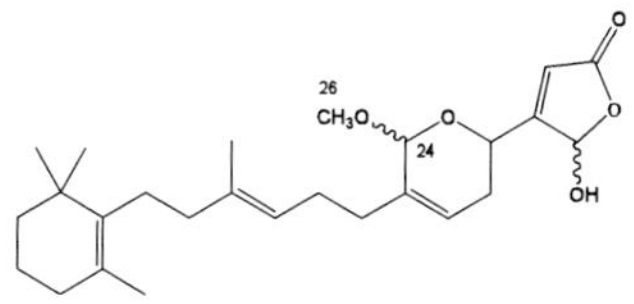

(**54**) 24-Methyl-*O*-manoalide

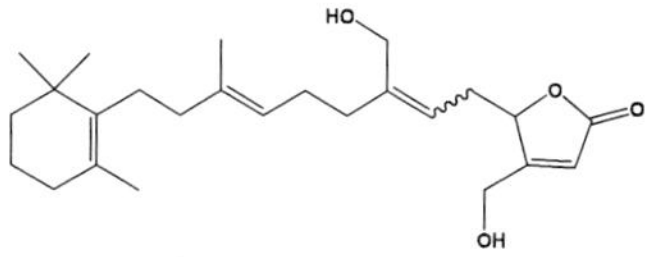

(**55**) (6E)-Neomanoalide
(**56**) (6Z)-Neomanoalide

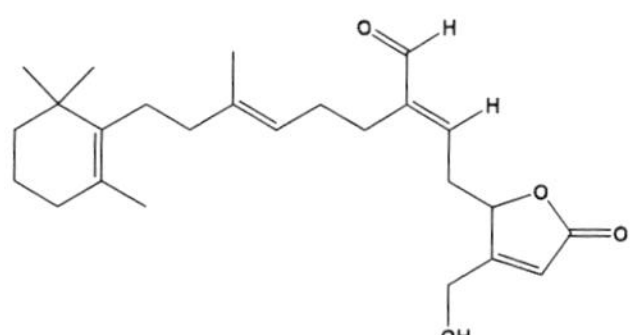

(**57**) (6E)-Neomanoalide-24-al

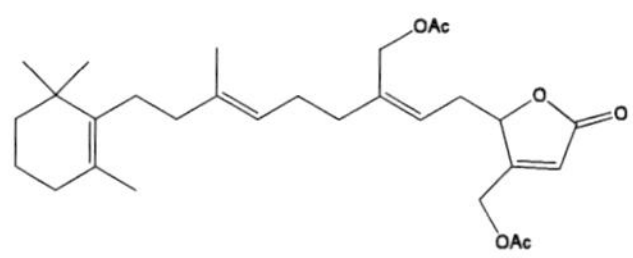

(**58**) (6Z)-Neomanoalide-24,25-diacetate

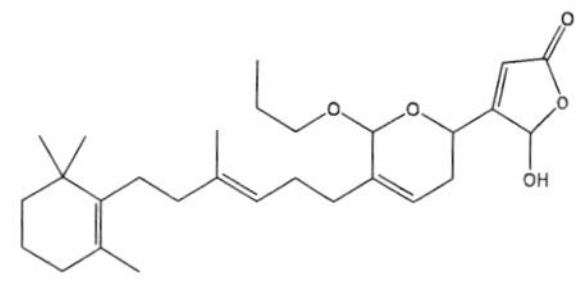

(**59**) 24-Propyl-*O*-manoalide

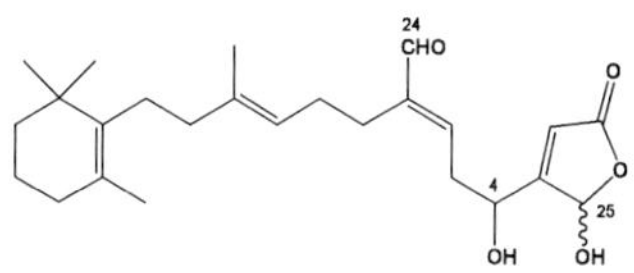

(**60**) Seco-manoalide

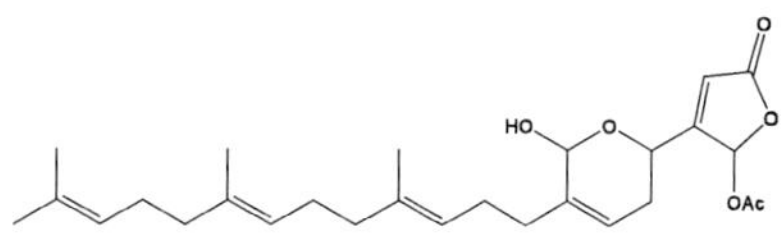

(**61**) R = H Thorectolide
(**62**) R = Ac Thorectolide monoacetate

Figure 2. (continued).

More than 60 natural manoalide derivatives have been described during the last three decades and most of them were reported to be potent anti-inflammatory agents. In our ongoing effort to discover and develop new marine natural product biomedicinals, we investigated between 1995 and 1998, *Luffariella variabilis* collected off the coast of Mayotte [Gauvin 1998, 2000].

Table 1. Some manoalide-related sesterterpenes from the marine sponges of the family Thorectidae

Manoalide derivatives		**Origin**		**Bioactivity**	**References**
n°	Usual name *(in alphabetical order)*	Sponges	Locations		
2	(6*Z*)-24-Acetoxy-2,3-dihydroneomanoalide	*Luffariella* sp. *Luffariella* sp.	Australia (Great Barrier Reef, Pelorus Island) *Pacific Ocean* Okinawa (Kerama Islands) *Pacific Ocean*	*Antibacterial activity* towards *Bacillus subtilis, Micrococcus luteus* *Cytotoxic activity* towards murine lymphoma L1210 cells and human epidermoid KB cells	[Kobayashi et al. 1993a, König et al. 1992]
3, 4	25-Acetoxyluffariellins A, B	*Luffariella variabilis*	Australia (Great Barrier Reef, Orpheus Island) *Pacific Ocean*		[Ettinger-Epstein et al. 2007]
5	(6*Z*)-24-Acetoxyneomanoalide	*Luffariella* sp. *Luffariella* sp.	Australia (Great Barrier Reef, Pelorus Island) *Pacific Ocean* Okinawa (Kerama Islands) *Pacific Ocean*	*Antibacterial activity* towards *Bacillus subtilis, Micrococcus luteus* *Cytotoxic activity* towards murine lymphoma L1210 cells and human epidermoid KB cells	[Kobayashi et al. 1993a, König et al. 1992]
6	25-Acetoxyseco-manoalide	*Luffariella variabilis*	Australia (Great Barrier Reef, Orpheus Island) *Pacific Ocean*		[Ettinger-Epstein, et al. 2007]
7	Cacospongionolide	*Fasciospongia cavernosa* (erroneously classified as *Cacospongia mollior*)	Croatia (Rovinj) *Adriatic sea*	*Cytotoxic activity* on *Artemia salina* (brine shrimp)	[de Rosa et al 1988, 1995]
8	Cacospongionolide B	*Fasciospongia cavernosa*	Croatia (Rovinj) *Adriatic sea*	*Antibacterial activity* towards *Bacillus subtilis, Micrococcus luteus* *Anti-inflammatory activity* *Cytotoxic activity* on *Artemia salina* (brine shrimp)	[de Rosa M et al. 1998, de Rosa S et al. 1995]

Manoalide derivatives		Origin		Bioactivity	References
n°	Usual name *(in alphabetical order)*	Sponges	Locations		
9	Cavernosolide	*Fasciospongia cavernosa*	Italy (Bay of Naples) *Thyrrenian sea*		
10	(4E,6E)-Dehydromanoalide	*Luffariella variabilis*	Palau *Pacific Ocean*	*Cytotoxic activity* towards murine lymphoma L1210 cells, human epidermoid KB cells and human colon tumor HCT-116 cells.	[Kobayashi et al. 1993a, Namikoshi et al. 2004, Potts et al. 1992c, Zhou & Molinski 2006]
		Luffariella sp.	Okinawa (Kerama Islands) *Pacific Ocean*		
		Luffariella sp.	Federated States of Micronesia (Pohnpei) *Pacific Ocean*		
		Luffariella sp.	Palau *Pacific Ocean*		
11	(6*Z*)-2,3-Dihydroneomanoalide	*Luffariella* sp.	Australia (Great Barrier Reef, Pelorus Island) *Pacific Ocean*	*Antibacterial activity* towards *Escherichia coli, Micrococcus luteus*	[König et al. 1992]
12	24-Ethyl-*O*-manoalide	*Luffariella variabilis*	Mayotte (Longogori Pass) *Indian Ocean*		[Gauvin 1998]
13 - 15	Fasciospongides A-C	*Fasciospongia sp.*	New Caledonia (Ouvéa) *Pacific Ocean*		[Montagnac et al. 1994]
16	Luffariellin A	*Luffariella variabilis*	Palau *Pacific Ocean*	*Anti-inflammatory activity*	[Ettinger-Epstein et al. 2007, Kernan et al. 1987]
		Luffariella variabilis	Australia (Great Barrier Reef, Orpheus Island) *Pacific Ocean*		

Table 1. (Continued)

Manoalide derivatives		Origin		Bioactivity	References
n°	Usual name *(in alphabetical order)*	Sponges	Locations		
17	Luffariellin B	*Luffariella variabilis*	Palau *Pacific Ocean*	*Anti-inflammatory activity*	[Kernan et al. 1987]
18	Luffariellolide	*Luffariella* sp.	Palau *Pacific Ocean*	*Anti-inflammatory activity*	[Albizati et al. 1987, Kernan et al. 1989]
		Luffariella sp.	Palau *Pacific Ocean*		
19-40	Luffarins A-V	*Luffariella geometrica*	Australia (Great Australian Bight) *Indian Ocean*		[Butler & Capon 1992]
41, 42	Luffarin derivatives 1, 2	*Fasciospongia cavernosa*	Croatia(Rovinj) *Adriatic sea*		[de Rosa & Carbonelli 2006]
43-47	Luffariolides A-E	*Luffariella* sp.	Okinawa *Pacific Ocean*	*Cytotoxic activity* towards murine lymphoma L1210 cells	[Tsuda et al. 1992]
48, 49	Luffariolides F, G	*Luffariella* sp.	Okinawa (Kerama Islands) *Pacific Ocean*	*Cytotoxic activity* towards murine lymphoma L1210 cells and human epidermoid KB cells	[Kobayashi et al. 1993]
50, 51	Luffariolides H, J	*Luffariella* sp.	Okinawa (Nakijin) *Pacific Ocean*	*Antibacterial activity* towards *Bacillus subtilis, Micrococcus luteus, Staphylococcus aureus* *Cytotoxic activity* towards murine lymphoma L1210 cells and human epidermoid KB cells	[Tsuda et al. 2002]
52	Luffolide	*Luffariella* sp.	Palau *Pacific Ocean*	*Anti-inflammatory activity*	[Kernan et al. 1989]

Manoalide derivatives		Origin		Bioactivity	References
n°	Usual name *(in alphabetical order)*	Sponges	Locations		
53	Manolide monoacetate	*Fasciospongia sp.*	New Caledonia (Ouvéa) *Pacific Ocean*		
		Hyrtios erecta	Japan (Amami Island) *Pacific Ocean*		
		Luffariella sp.	Palau *Pacific Ocean*		
		Luffariella variabilis	Australia (Great Barrier Reef, Orpheus Island) *Pacific Ocean*		
		Luffariella variabilis	Mayotte (Longogori Pass) *Indian Ocean*		
		Thorectandra excavatus	Australia (Darwin) *Indian Ocean*		
54	24-Methyl-*O*-manoalide	*Luffariella variabilis*	Mayotte (Longogori Pass) *Indian Ocean*	*Antibacterial activity* towards *Staphylococcus aureus*	[Gauvin 1998, Namikoshi et al. 2004, Zhou & Molinski 2006]
		Luffariella sp.	Palau *Pacific Ocean*	*Cytotoxic activity* towards human colon tumor HCT-116 cells.	
		Luffariella sp.	Federated States of Micronesia (Pohnpei) *Pacific Ocean*		

Table 1. (Continued)

Manoalide derivatives		Origin		Bioactivity	References
n°	Usual name *(in alphabetical order)*	Sponges	Locations		
55	(6*E*)-Neomanoalide	*Hyrtios erecta*	Japan (Amami island) *Pacific ocean*	*Cytotoxic activity* towards murine lymphoma L1210 cells and human epidermoid KB cells	[Kobayashi et al. 1993a, Kobayashi et al. 1994, de Silva & Scheuer 1981, Tsuda et al. 1992]
		Luffariella variabilis	Palau *Pacific Ocean*		
		Luffariella sp.	Okinawa (Kerama Islands) *Pacific Ocean*		
		Luffariella sp.	Okinawa *Pacific Ocean*		
56	(6*Z*)-Neomanoalide	*Hyrtios erecta*	Japan (Amami island) *Pacific Ocean*	Cytotoxic activity towards murine lymphoma L1210 cells and human epidermoid KB cells	[Kobayashi et al. 1993a, Kobayashi et al. 1994, de Silva & Scheuer 1981, Tsuda et al. 1992]
		Luffariella variabilis	Palau *Pacific Ocean*		
		Luffariella sp.	Okinawa (Kerama Islands) *Pacific Ocean*		
		Luffariella sp.	Okinawa *Pacific Ocean*		
57	(6*E*)-Neomanoalide-24-al	*Luffariella* sp.	Australia (Great Barrier Reef, Pelorus Island) *Pacific Ocean*	*Antibacterial activity* towards *Bacillus subtilis*, *Micrococcus luteus*	[Kobayashi et al. 1993a, König et al. 1992]
		Luffariella sp.	Okinawa (Kerama Islands) *Pacific Ocean*	*Cytotoxic activity* towards murine lymphoma L1210 cells and human epidermoid KB cells	
58	(6*Z*)-Neomanoalide-24,25-diacetate	*Luffariella* sp.	Palau *Pacific Ocean*		[Namikoshi et al. 2004, Zhou & Molinski 2006]

Manoalide derivatives		Origin		Bioactivity	References
n°	Usual name *(in alphabetical order)*	Sponges	Locations		
59	24-Propyl-*O*-manoalide	*Luffariella* sp.	Federated States of Micronesia (Pohnpei) *Pacific Ocean*		
60	Seco-manoalide	*Hyrtios erecta*	Japan (Amami island) *Pacific Ocean*	*Antibacterial activity* towards *Staphylococcus aureus.*	[Ettinger-Epstein et al. 2007, Gauvin 1998, Glaser et al. 1989, Kobayashi et al. 1994, Namikoshi et al. 2004, Reddy et al. 1997, de Silva & Scheuer 1981, Zhou & Molinski 2006]
		Luffariella spp.	Palau *Pacific Ocean*	Antifungal activity towards *Candida albicans, Candida glabrata, Candida krusei.*	
		Luffariella sp.	Federated States of Micronesia (Pohnpei) *Pacific Ocean*	*Anti-inflammatory activity*	
		Luffariella variabilis	Palau *Pacific Ocean*	*Cytotoxic activity* towards murine lymphoma L1210 cells, human epidermoid KB cells and human colon tumor HCT-116 cells.	
		Luffariella variabilis	Australia (Great Barrier Reef, Orpheus Island) *Pacific Ocean*		
		Luffariella variabilis	Mayotte (Longogori Pass) *Indian Ocean*		
61	Thorectolide	*Hyrtios* sp.	New Caledonia (Walpole) *Pacific Ocean*	*Antiviral activity* : inhibitor of both HIV-1 nucleocapside and integrase *Cytotoxic activity* towards human epidermoid KB cells	[Bourguet-Kondracki et al. 1996]
62	Thorectolide monoacetate	*Hyrtios* sp.	New Caledonia (Walpole) *Pacific Ocean*	*Anti-inflammatory activity*	[Bourguet-Kondracki et al. 1996, Cambie et al. 1998]
		Thorectandra excavatus	Australia (Darwin) *Indian Ocean*	*Cytotoxic activity* towards human epidermoid KB cells	

Analysis of the secondary metabolites from this sponge resulted in the isolation of two new manoalide-type sesterterpenes named 24-methyl-*O*-manoalide (54) and 24-ethyl-*O*-manoalide (12) together with the previously reported manoalide (1), seco-manoalide (60) and manoalide monoacetate (53). This report details the isolation, structure elucidation and cytotoxicity data of these five marine natural products. The possible artifact formation of 24-methyl-*O*-manoalide (54) and 24-ethyl-*O*-manoalide (12) during solvent extraction is also discussed.

2. Isolation of the Secondary Metabolites

The sponge *Luffariella variabilis* was collected by hand using scuba at depths between 10 and 30 m in *Longogori* Pass (or *S* pass), a strict fishing reserve located at the East of Mayotte (Figure 1). The animal was kept frozen at - 20 °C for transport to the laboratory and subsequently extracted with a mixture of methanol-chloroform (2:1 v/v). The crude lipid extract obtained showing significant level of toxicity (DL_{50} = 50 µg/mL) in our brine shrimp lethality test was then subjected to solvent partitioning, as outlined in the "Materials and Methods" section. The *n*-hexane fraction was repeatedly fractionated by silica gel column chromatography, followed by HPLC using normal phase silica gel column to afford 24-ethyl-*O*-manoalide (**12**), manoalide monoacetate (**53**) and 24-methyl-*O*-manoalide (**54**). The carbon tetrachloride and chloroform fractions were combined and chromatographed on a silica gel column to furnish manoalide (**1**) and seco-manoalide (**60**). Compound 60 was further purified by HPLC over normal phase silica gel.

Excluding manoalide monoacetate (**53**), all the metabolites (**1**, **12**, **54**, **60**) which behaved as single compounds on TLC and HPLC (ordinary phase and reverse phase) with several solvents systems, had ^{1}H and ^{13}C NMR spectra showing two sets of signals (Table 2). This indicated the compounds to be a mixture of epimers in approximately 1:1 ratio for manoalide (**1**), 24-ethyl-*O*-manoalide (**12**) and 24-methyl-*O*-manoalide (**54**) and in 7:3 ratio for seco-manoalide (**60**).

Table 2. ^{13}C NMR[a] and ^{1}H NMR[b] data of compounds manoalide, seco-manoalide, manoalide monoacetate, 24-methyl-*O*-manoalide and 24-ethyl-*O*-manoalide in $CDCl_3$

	manoalide		seco-manoalide		manoalide monoacetate		24-methyl-*O*-manoalide		24-ethyl-*O*-manoalide	
position	^{13}C	^{1}H(*J*, Hz)	^{13}C	^{1}H(*J*, Hz)	^{13}C	^{1}H(*J*, Hz)	^{13}C	^{1}H(*J*, Hz)	^{13}C	^{1}H(*J*, Hz)
1	171.94, 171.72		170.74, 170.68		169.11[c]		170.33, 170.29		170.38, 170.31	
2	117.98, 116.87	6.10, 5.99 s	118.91, 118.77	6.11, 6.03 s	118.95	6.08 s	118.46, 117.54	6.07 s	118.43, 117.49	6.19, 6.02 s
3	169.08, 168.32		169.85, 168.79		165.85		167.58, 167.35		167.73, 167.43	
4	63.24, 62.37	4.86, 4.79 m	67.13, 66.79	4.80 m	61.86	4.74 dd (9.8, 4.6)	62.29, 63.20	4.77 m	63.17, 62.31	4.78, 4.86 dd (3.4, 10.5)
5	29.21, 28.65	2.20 m	26.98	2.80 m	28.50	2.30 m	29.12, 28.68	2.20 m	29.14, 28.78	2.20 m
6	121.70, 121.01	5.66 m	148.32, 147.85	6.56 m	120.70	5.68	120.85, 120.71	5.67 m	120.77, 120.57	5.66 m
7	137.46, 137.15		146.16, 146.05		137.56		137.12, 136.92		137.12, 136.81	
8	32.57	2.10 m	34.97, 34.65	2.30 m	32.66	2.12 m	32.62	2.10 m	32.66	2.10 m
9	25.96	2.11 m	24.75	2.30 m	26.04	2.12 m	26.06	2.10 m	26.13	2.10 m
10	122.83	5.10 m	122.52	5.10 t (6.7)	122.84	5.10 m	122.90	5.12 t (6.5)	122.94	5.12 t (6.1)
11	137.09		137.77		137.14		137.12		137.12	
12	40.30	1.99 m	40.43	2.10 m	40.36	1.99 m	40.32	2.00 m	40.31	2.00 m
13	27.91	2.00 m	28.01	2.10 m	27.99	1.99 m	27.95	2.00 m	27.94	2.00 m
14	136.90		137.21		137.08		136.92		136.90	
15	126.27		127.27		127.10		127.06		127.06	
16	32.76	1.84 t (6.3)	32.97	1.86 t (6.4)	32.83	1.87 t (6.0)	32.81	1.88 t (6.2)	32.81	1.88 t (6.2)
17	19.56	1.52 m	19.75	1.54 m	19.62	1.52 m	16.90	1.54 m	19.60	1.53 m
18	39.85	1.38 m	40.05	1.38 m	39.91	1.38 m	39.89	1.39 m	39.89	1.39 m
19	34.98		35.20		35.06		35.04		35.04	
20	28.65	0.95 s	28.85	0.96 s	28.71	0.96 s	28.69	0.97 s	28.69	0.97 s
21	28.65	0.95 s	28.85	0.96 s	28.71	0.96 s	28.69	0.97 s	28.69	0.97 s
22	19.85	1.56 s	20.04	1.56 s	19.82	1.57 s	19.88	1.58 s	19.88	1.58 s
23	16.11	1.61 s	16.28	1.58 s	16.18	1.62 s	16.15	1.62 s	16.14	1.62 s
24	91.47, 91.24	5.30, 5.26 s	195.19, 195.05	9.38 s	91.68	5.33 s	98.58, 98.43	4.79 s	97.17, 97.13	4.92, 4.89 s
25	98.34, 97.48	6.22, 6.03 s	97.84, 97.07	6.09, 6.25 s	92.92	6.17 s	97.74, 97.15	6.17 s	97.71, 97.13	6.23, 6.09 s
26					169.60[c]		55.94, 55.90	3.44 s	64.26, 64.01	3.83, 3.55 m
27					20.77	2.14 s			15.40, 15.29	1.24, 1.23 t (7.0)

[a] δ(ppm) 100.11 MHz. [b] δ(ppm) 400.13 MHz; s, singlet; d, doublet; t, triplet; m, multiplet. [c] assignments may be interchanged.

3. Structure Elucidation of the Secondary Metabolites

Manoalide (**1**) was obtained as an optically active pale yellow glass $[\alpha]^{25}_{D}$ +79 (*c* 0.5, MeOH). Its structure was unambiguously established by comparison of the IR, ^{1}H and ^{13}C NMR and eiMS data with those earlier reported [de Silva and Scheuer 1980; Kobayashi et al. 1994]. As regards its stereochemistry, **1** contains three asymmetric carbons at C-4, C-24 and C-25. While the absolute configurations at C-24 and C-25 could never been examined in detail since the hemiacetal rings seem to readily open and recyclize, the C-4 absolute configuration was shown to be (R) either by synthesis [Ekow Amoo et al. 1988] or by applying the Horeau's method after chemical conversion [Kobayashi et al. 1994]. Besides the (R) configuration at C-4, Kobayashi et al. [1994], who were the first to isolate **1** as a mixture of epimers from the marine sponge *Hyrtios erecta*, showed that the mixture consisted of C-25 epimers. This was elucidated after acetylation of the mixture and separation by HPLC to provide two diacetates **1a** and **1b**. Following the same method, our manoalide (**1**) was acetylated and the diacetates were separated. By comparison of the spectral data including the optical activity ($[\alpha]^{25}_{D}$ +80 (*c* 0.5, $CHCl_3$) and $[\alpha]^{25}_{D}$ +32 (*c* 0.5, $CHCl_3$)) of each epimer with those given by the literature ($[\alpha]^{25}_{D}$ +81 (*c* 1.0, $CHCl_3$) and $[\alpha]^{25}_{D}$ +32 (*c* 0.7, $CHCl_3$)), we may presume that we have isolated C-25 epimers with the (R) configuration at C-4.

Manoalide monoacetate (**53**) crystallized from *n*-hexane as white flakes, mp 117-118 °C, $[\alpha]^{25}_{D}$ +33 (*c* 0.5, $CHCl_3$). The ^{1}H and ^{13}C NMR as well as the eiMS data were in good agreement with the published values [Cambie et al. 1988; de Silva and Scheuer 1980; Kobayashi et al. 1994]

Seco-manoalide (**60**) was obtained as an unstable colorless glass $[\alpha]^{25}_{D}$ +15 (*c* 0.5, $CHCl_3$). Its structure was established by comparison of the IR, ^{1}H and ^{13}C NMR and eiMS data with those earlier reported [de Silva and Scheuer 1981; Kobayashi et al. 1994]. Besides, taken into account that the C-4 absolute configuration of manoalide-related sesterterpenes among which seco-manoalide, was always assigned as (R) [Bourguet-Kondracki et al. 1996; Kobayashi et al., 1994; Tsuda et al. 1992], we may suppose that the mixture of epimers obtained consisted of 4(R),25(R) and 4(R),25(S) epimers.

24-Ethyl-O-manoalide (**12**) was obtained as a colorless glass, $[\alpha]^{25}_{D}$ +63 (*c* 0.5, $CHCl_3$). The IR spectrum contained three bands at 3410, 1790 and 1762 cm^{-1} which were assignable to a γ-hydroxybutenolide group and a band at 1098 cm^{-1} that indicated the presence of an ether function. The eiMS showed a molecular peak at m/z 444. This datum together with ^{1}H and ^{13}C NMR spectra suggested the formula of $C_{27}H_{40}O_5$. The mass spectrum showed an intense peak at m/z 137 and fragments ions at m/z 121, 107 and 95 derived from the m/z 137, implying the presence of the alkylated cyclohexenyl end group $C_{10}H_{17}$ commonly generated by manoalide-related sesterterpenes. The ^{1}H and ^{13}C NMR showed the characteristic signals of an additional ethoxy group at δ 1.23 – 1.24 (H-27, *t*, J = 7Hz, 3H), δ 64.01 -34.26 (C-26) and δ 15.29 – 15.40 (C-27). The ether linkage between C-24 and C-26 was suggested by the ^{13}C NMR chemical shift of C-24 which resonated at lower field (δ 97.13 – 97.17) than the C-24 (δ 91.24 – 91.47) of **1** bearing an hydroxyl group. These data suggested structure **12** for 24-ethyl-*O*-manoalide.

24-Methyl-O-manoalide (**54**) $[\alpha]^{25}_{D}$ +48 (*c* 0.5, $CHCl_3$), was isolated as a colorless glassy solid. The IR spectrum contained three bands at 3380, 1790 and 1765 cm^{-1} typical of a γ-hydroxybutenolide moiety and an absorption at 1097 cm^{-1} supporting the presence of an ether group. The molecular formula of **54** was suggested as $C_{26}H_{38}O_5$ by ^{1}H and ^{13}C NMR together with eiMS. Like compound **12**, the fragment ions m/z 137, 121, 107 and 95 typical of the alkylated cyclohexenyl end group were observed. The ^{1}H and ^{13}C NMR of **54** were almost identical with those of manoalide (1) (Table 2). However, its ^{1}H NMR spectrum contained an additional methoxy signal at δ 3.44 (H-26, *s*, 3H). This was confirmed by the ^{13}C NMR spectrum which exhibited a peak at δ 55.90 – 55.94 for C-26. The methoxy function was clearly positioned at C-24, as the C-24 carbon resonance was now at δ 98.43 – 98.58 compared with δ 91.24 – 91.47 in **1**. Thus the structure of 24-methyl-*O*-manoalide was concluded to be **54**. Once the structure of this sesterterpene had been established, it was found that 54 was already produced from natural manoalide (**1**) or seco-manoalide (**60**) by de Silva and Scheuer [1981]. 54 is also an intermediate of the synthesis of manoalide [Katsumura et al., 1985].

The stereochemistry of the two new sesterterpenes **12** and **54** remains undetermined.

However it should be point out that 24-*O*-ethylmanoalide (**12**) and 24-*O*-methylmanoalide (**54**) may be suspected to be artifacts derived from manoalide (**1**) during solvent extraction. Manoalide is indeed a hemiacetal and its extraction under some particular conditions would be expected to produce compounds **12** and **54**. The conversion of **1** into 24-*O*-methylmanoalide (**54**) may be explained by acid-catalyzed conversion with the MeOH used in the process of extraction [Namikoshi et al., 2004]. The conversion of **1** into 24-*O*-ethylmanoalide (**12**) may be explained by acid-catalyzed conversion with EtOH present as a preservative in $CHCl_3$ also used in extraction.

4. Bioactivity of the Secondary Metabolites

Antiviral activities. The five pure sesterterpenes were evaluated for their anti-HIV and anti-HTLV-1 activities. All the compounds were found to be ineffective.

Cytotoxicity activity. Manoalide (**1**) and monoacetate manoalide (**53**) were tested for growth inhibition using the McF_7WT human breast cancer cell line. Both were found to show appreciable cytotoxicity toward McF_7WT cells: 98% and 96% inhibitions respectively for **1** and 53 at 100 μM. The limited quantities of seco-manoalide (**60**), 24-methyl-*O*-manoalide (**54**) and 24-ethyl-*O*-manoalide (**12**) after the two antiviral tests prohibited further investigation of their potential as molecular-targeted antitumor agents.

5. Materials and Methods

General experimental procedures. Nuclear Magnetic Resonance (NMR) spectra were recorded on a Brucker AMX-400, in $CDCl_3$, at 400.13 MHz for ^{1}H spectra and at 100.11 MHz for ^{13}C spectra with TMS as internal standard. Chemical shifts are reported in ppm and coupling constants (J) are reported in Hz. Electron Impact Mass spectra (eiMS) were obtained on a Jeol AX-500 mass spectrometer. Infra-Red (IR) spectra were determined on a Perkin-Elmer 1600 FT-IR spectrometer and optical rotations $[\alpha]_D$ on a Perkin-Elmer 341 polarimeter. The melting points (mp) were determined with a Kofler hot bench type WME. High Performance Liquid Chromatography (HPLC) was performed on a spectraseries P100 equipped with a differential refractometer (Thermoseparation products – Refractomonitor). A Merck Lichrospher Si-60 column (25 cm x 10 mm i.d.) was used. Thin Layer Chromatography (TLC) was carried out on silica Kieselgel 60 F254 plates, and TLC plates were visualized by spraying with 50% H_2SO_4 aqueous solution followed by heating.

Sponge material. Luffariella variabilis (class Demospongiae, order Dictyoceratida, family Thorectidae) was collected by hand using scuba at depths between 10 and 30 m in *Longogori* Pass (or *S* pass), a strict fishing reserve located at the East of Mayotte (Figure 1). The sponge has a dark grey to black exterior with a cream to orange-brown interior. It is massive, moderately firm and less compressible with many different shapes. It was identified by Dr Nicole Boury-Esnault (Station Marine D'Endoume, Marseille, France) and Pr Patricia Bergquist (School of Biological Sciences, Auckland, New Zealand). A voucher specimen (reference AGL-2-97M) is on deposit at the Chemistry Laboratory of Natural Substances and Food Chemistry, University of Reunion island, France.

Brine Shrimp assay [Meyer et al. 1982]. *Sample preparation.* Samples were prepared by dissolving 50 mg of extract in 5 mL of methanol (solution A). Solution B were prepared by diluting 0.5 mL of A to 10 mL with methanol. Appropriate amounts of solution (100 µL B, 50 µL A and 500 µL A for 10, 100 and 1 000 µg/mL, respectively) were transferred to 1.25 cm disc of filter paper (Schleicher and Schuell, n° 740-E). The discs were dried under nitrogen. Control discs were prepared using only methanol. Five replicates were prepared for each dose level.

Hatching the Shrimp. Brine shrimp eggs (*Artemia salina*) were hatched in shallow rectangular dish (22 x 32 cm) filled with sea water. A plastic divider with several 2 mm holes was clamped in the dish to make two unequal compartments. The eggs (ca. 50 mg were sprinkled into the larger compartment which was darkened, while the smaller compartment was illuminated. After 48 hours the phototropic nauplii were collected by pipette from the lighted side, having been separated by divider from their shells.

Bioassay. Ten shrimp were transferred to each sample vial using a disposable pipette (Dispo Short Pipet) and sea water was added to make 5 mL. The nauplii can be counted macroscopically in the stem of the pipette against a lighted background. The vials were maintained under illumination. Survivors were counted, with the aid of a 3 x magnifying glass after 6 and 24 hours, and the percents deaths at each dose and control were determined. The 24 hours counts were more useful. In cases where control deaths occurred, the data were corrected using Abbot's formula: % deaths = [(test – control)/control] x 100.

LD_{50} determination. LD50's were determined from the 24 hour counts according to the method of Hafner et al. [1977].

Extraction and isolation. Frozen sponge tissue (1343 g dry weight after extraction) was cut up and homogenized in a waring-blender in methanol-chloroform 1:2 (v/v). After filtration, the solvent was removed under reduced pressure to give the crude material (33.4 g), which was successively partitioned between equal volumes of aqueous methanol, percent adjusted to produce a biphasic solution, and a solvent series of *n*-hexane (yield 5.71 g), carbon tetrachloride (yield 11.95 g) and chloroform (yield 7.44 g). A portion of the *n*-hexane fraction (2.98 g) was then repeatedly subjected to silica gel columns using eluants of increasing polarity from 5% ethyl acetate in *n*-hexane to 10% ethyl acetate in *n*-hexane, to afford a mixture of 12, 53 and 54. The resulting material was further purified by semi-preparative High Performance Liquid Chromatography over normal phase silica with *n*-hexane-ethyl acetate 75:20 (v:v) to yield pure 12 (19 mg, 0.0027%), 53 (18 mg, 0.0026% wet weight) and 54 (13 mg, 0.0019%). Carbon tetrachloride and Chloroform solubles were combined, and a 4.38 g portion was fractionated by silica gel column chromatography eluted with *n*-hexane-ethyl acetate using a step gradient of increasing ethyl acetate (90:10 to 70:30 v/v) to afford pure 1 (99 mg, 0.033%) and impure 60. Final purification via semipreparative

High Performance Liquid Chromatography using Si gel column with *n*-hexane-ethyl acetate 20:30 (v/v) gave 60 (76 mg, 0.025%).

24-Methyl-O-manoalide (**54**). Colourless glass; $[\alpha]^{25}_{D}$ + 48 (*c* 0.5, $CHCl_3$); $C_{26}H_{38}O_5$; EIMS m/z (%) 430 (M^+,13), 412(4), 398(8), 380(2), 260(5), 203(3), 177(3), 137(100), 123(10, 121(10), 107(8), 95(22), 81(22); IR ν max/cm^{-1} ($CHCl_3$) 3380, 2925, 1790, 1765, 1097, 1040; ^1H and ^{13}C NMR see Table 2.

24-Ethyl-O-manoalide (**12**). Colourless glass; $[\alpha]^{25}_{D}$ + 63 (*c* 0.5, $CHCl_3$); $C_{27}H_{40}O_5$; EIMS m/z (%) 444 (M+,22), 426(3), 398(9), 380(2), 261(5), 203(4), 177(6), 137(100), 123(12), 121(12), 107(9), 95(26), 81(22); IR ν max/cm^{-1} ($CHCl_3$) 3410, 2925, 1790, 1762, 1098, 1040; ^1H and ^{13}C NMR see Table 2.

*Acetylation of **1***. Acetylation of **1** giving the diacetates **1a** and **1b** was performed as described by Kobayashi et al. [1994].

Diacetates (**1a**, **1b**). Colourless oils; $[\alpha]^{25}_{D}$ + 80 (*c* 0.5, $CHCl_3$); $[\alpha]^{25}_{D}$ + 32 (*c* 0.5, $CHCl_3$); $C_{29}H_{40}O_7$; EIMS m/z (%) 500 (M^+, 17), 440 (M-AcOH, 16), 380 (M-2 AcOH, 11).

Acknowledgment. We are grateful to N. Boury-Esnault (Station Marine d'Endoume, Marseille, France) and P. Bergquist (University of Auckland) for taxonomic identification of sponge materials. We thank H. Fleury and M. Ventura (Laboratoire de virology de l'hôpital Pellegrin, Bordeaux, France) for anti-HIV tests; B. Guillemain, T. Astier-Gin and J.P. Portail (Fondation Bergognié, INSERM, Bordeaux, France) for anti-HTLV-1 tests; J. Robert (Laboratoire de Pharmacologie des medicaments anticancéreux, Fondation Bergognié, Bordeaux, France) for bioassays on McF_7WT cells. This work was supported by the Regional council of Reunion Island.

II. Sponge Chemosystematics

II.1. Significance of the Study

Sponge taxonomy and classification are inherently difficult as sponges possess rather few morphological characters. Expertise and long experience are therefore usually needed for a good approximation to the problem. Hence, finding new sets of characters, e.g. those of chemical nature, to aid in classification and phylogeny of sponges has quickly become obvious. Secondary metabolites with their great diversity in sponges have been suggested as a possible alternative to morphological characters, first because they increase in number from year to year [van Soest and Braekman, 1999], after discovery of sponge-produced bioactive and pharmaceutically valuable molecules, and then because the structural complexity of the molecules promises a large source of new characters. This complementary approach called "chemotaxonomy" or "chemosystematics" is defined as a classification method in relation to the chemical composition of the organisms or more precisely as a classification according to shared compounds or compound groups (synapomorphies). Such compounds also called "chemotaxonomic markers" are ideally not only present in all members of the considered taxon (classes, orders, families or genera) but also absent in others (the importance of the latter is frequently underestimated!).

II.2. Strategy and Approach of the Research

1. Use of Sterols as Biomarkers: Advantages

Among the numerous sponge natural products considered as suitable chemotaxonomic markers, sterols were the first group to be used in biochemical taxonomic studies by Bergmann [Bergmann 1949, 1962]. In his extensive work, it was shown that all sponges contain sterols and that the sterol composition is different for each sponge. Some of Bergmann's finding were inaccurate due to the lack of modern analytical instruments at that time, however, his idea led to numerous investigations on sponge sterols for chemotaxonomic purposes [see e.g. Bergquist et al. 1980, 1986, 1991; Carballeira and Shalabi, 1989; Castellanos et al. 2003; Erpenbeck and van Soest 2005; Fromont et al. 1994; Gauvin et al. 1998, 2004, de Rosa et al. 1973]. These innumerable researches on sponge sterols have shown in particular that along with conventional sterols (i.e. a C-19 tetracyclic nucleus with the normal C-8 side-chain which may be modified by the addition of one or two carbon atoms at C-24, (figures 3, 4)), a diverse and biogenetically unprecedented array of unconventional steroids with no terrestrial counterpart can occur in these primitive invertebrates [Aiello et al. 1999; Kerr and Baker 1991; Kornbropst 2005b; Sheikh and Djerassi 1974]. Commonly encountered features in unconventional steroids include additional oxygenation of both the nucleus and the side chain, side chains extensively modified by alkylation and degradation, and structural modification of the nucleus skeleton.

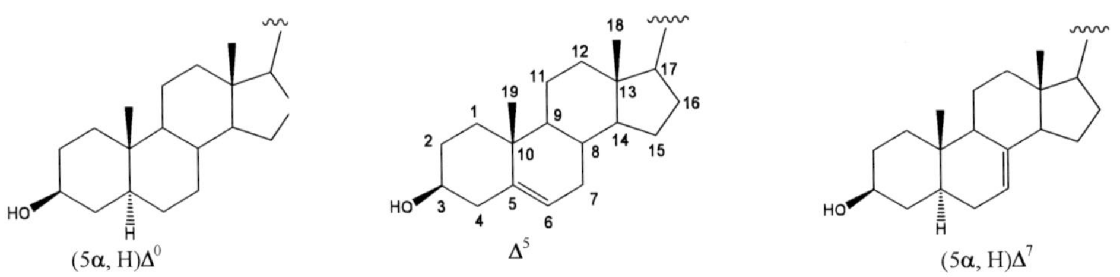

Figure 3. Conventional nuclei of 3β-hydroxysterols encountered in sponges.

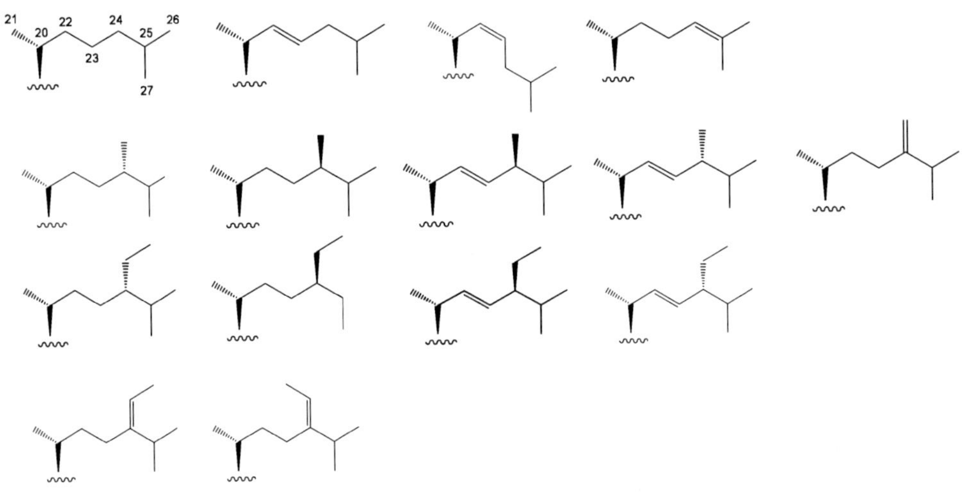

Figure 4. Conventional side chains of sterols encountered in sponges.

The occurrence of sulfate esters of polyoxygenated sterols in sponges has also been reported. A number of representative examples of unconventional sterols which with few exceptions, are unique to the phylum Porifera, are shown in figures 5-9.

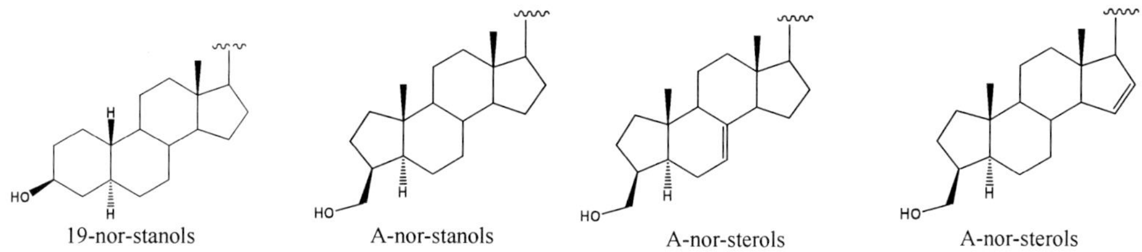

Figure 5. Selected unconventional nuclei of stanols/sterols encountered in sponges (from Kornprobst [2005b]).

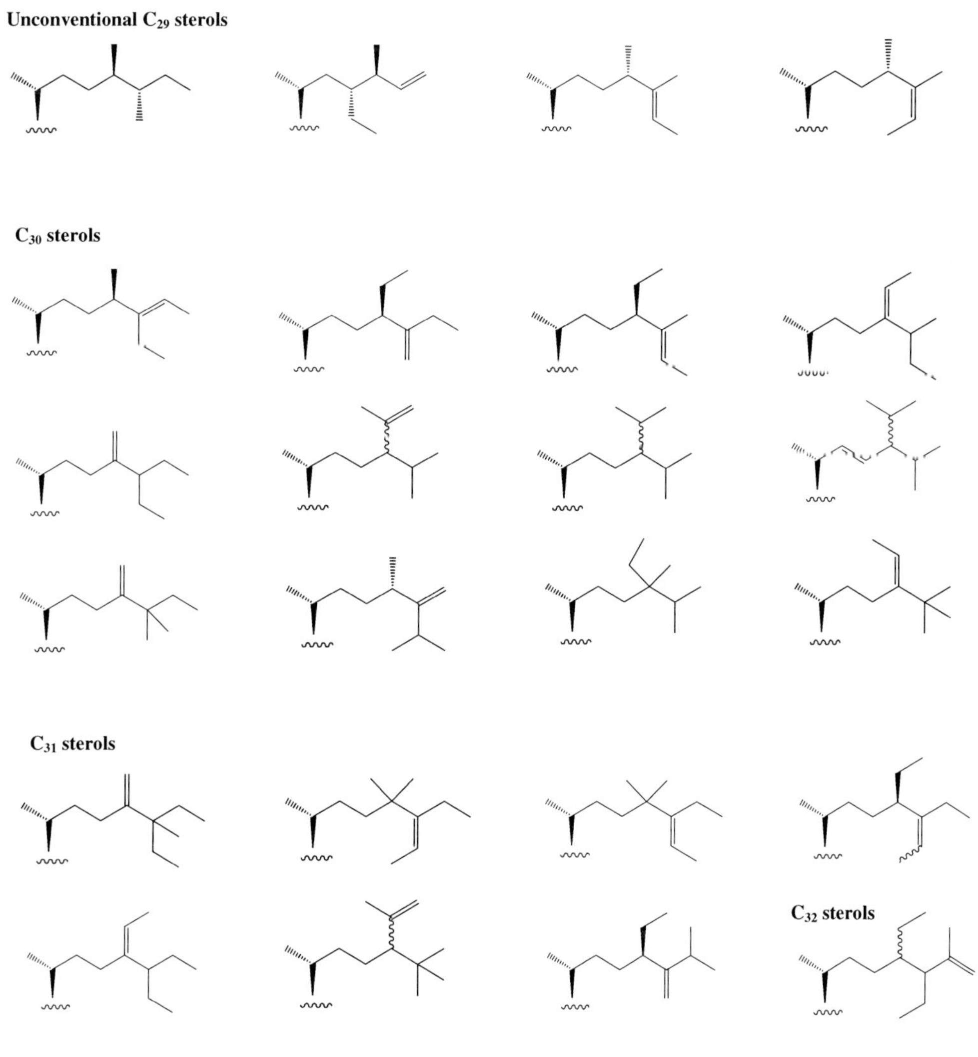

Figure 6. Selected highly alkylated side chains of sterols encountered in sponges (from Kornprobst [2005b])

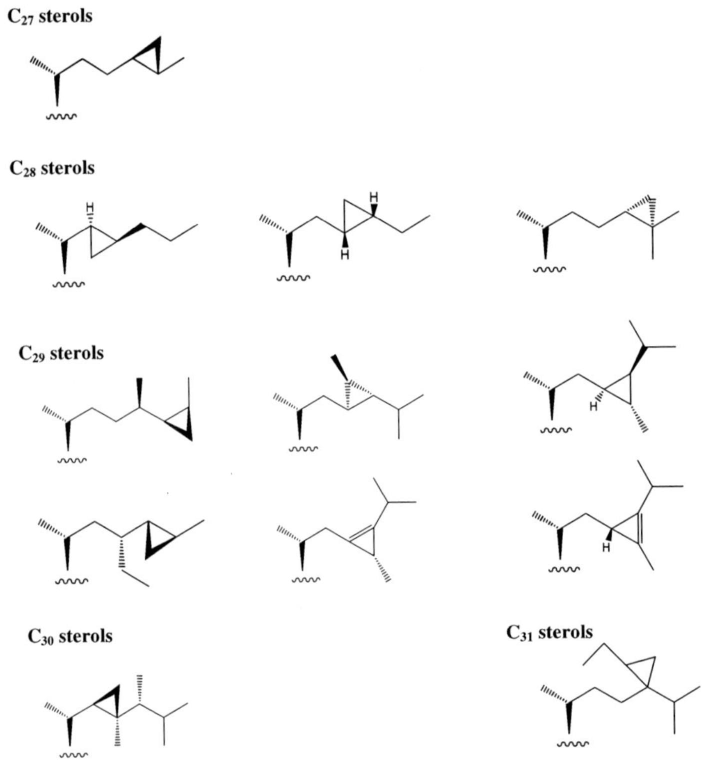

Figure 7. Selected cyclopropane/cyclopropene-containing side chains of sterols encountered in sponges (from Kornprobst 2005b).

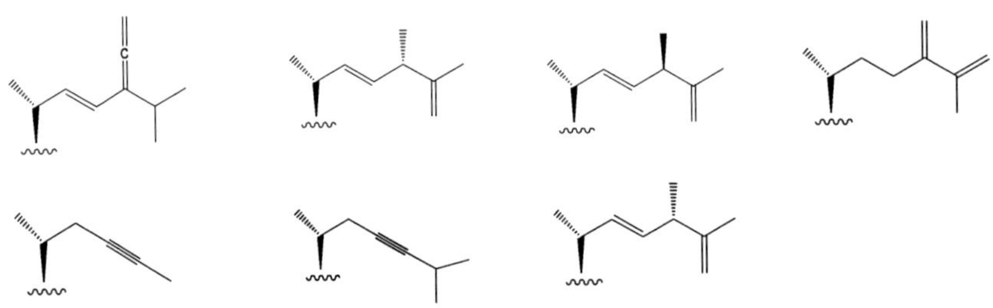

Figure 8. Selected diunsaturated side chains of sterols encountered in sponges (from Kornprobst 2005b).

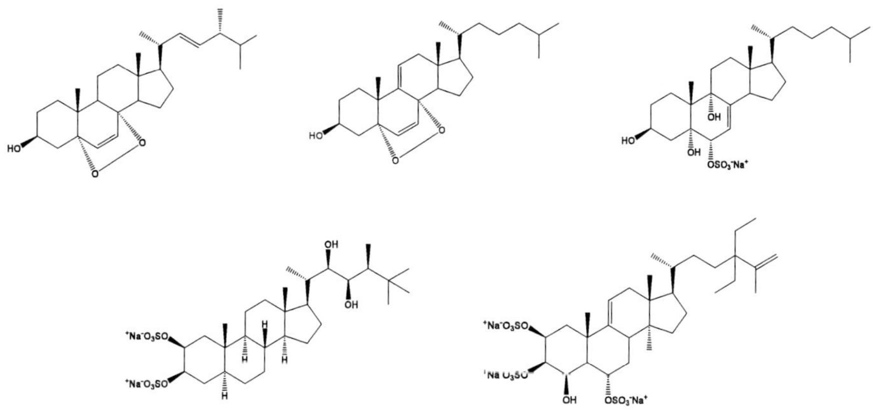

Figure 9. Selected oxygenated and/or sulfated sterols encountered in sponges (from Kornprobst 2005b).

Steroids with high degree of skeletal specificities often co-occur with conventional ones and are sometimes present in small amounts. However, many exceptions are reported for some sponges that are found with unusual structures as the predominant steroids rather than cholesterol or the conventional 3β-hydroxy sterols. In such a case, it has been suggested that these intriguing sterols very likely play a functional (rather than metabolic) role in maintaining the integrity of membranous structures [Bloch 1983]. It has also been hypothesized that the uniqueness of sterols in cell membranes of sponges is accurately related to the other membrane components, particularly the phospholipids.

As these latter have been found to be very different from those of higher animals, the structural modifications exhibited by the sponge sterols may be a sort of structural adjustments for a better fit with other membrane components [e.g. Carballeira et al. 1986; Dembitsky et al. 2003; Djerassi and Lam 1991; Hahn et al. 1988; Lam et al. 1989].

In addition to the great diversity and atypical framework encountered among sponge sterols, it was further shown that the overall sterol content of sponges remains relatively constant over time despite season and location [Kerr and Baker 1991]; one additional criterion for the choice of sterols as target for chemotaxonomic analyses. Suitable chemotaxonomic markers require indeed stability in their occurrence [Erpenbeck and van Soest 2007].

2. Use of Sterols as Biomarkers: Weakness

As very other phylogenetic reconstruction methods, sponge chemotaxonomy reveals peculiar pitfalls [Erpenbeck and van Soest 2007] among which "*the true origin of the compounds*". Sponges, efficient filter feeders, are indeed very fertile host animals for diverse symbiotic microorganisms including a large range of archea, heterotrophic bacteria, cyanobacteria, green algae, red algae, cryptophytes, dinoflagellates and diatoms [Lee et al. 2001; Wilkinson 1978a,b,c, 1987]. Able to resist the sponge digestive process and immune response, the symbionts can successfully inhabit sponges both intra- and extra cellularly. They can be very numerous, constituting for some sponges a substantial part of the total cell material, up to 60% of the sponge volume in cases. To explain this peculiarity, some putative benefits to symbiotic microorganisms or their hosts have been evoked in the literature. The surfaces or internal spaces of marine sponges are more nutrient-rich than sea water and sediments; therefore sponges offer nourishment and a safe habitat to their symbionts [Bultel-Poncé et al. 1999]. On the other hand, symbiotic microorganisms help in the nutritional process, either by intracellular digestion or by translocation of metabolites including nitrogen fixation, nitrification and photosynthesis [Wilkinson and Fay 1979; Wilkinson and Garrone 1980]. Microorganisms also stabilize the sponge skeleton [Wilkinson et al. 1981] and participate in the host's chemical defense system against predators and biofouling [Bakus et al. 1986; Proksch 1994; Unson et al. 1994]. It is thus important to consider the role of sponge symbionts as potential sources of bioactive compounds isolated from sponges since these symbionts can comprise a substantial part of the total cell material that is extracted. In some cases, there is good circumstantial or even direct evidence that some metabolites are actually derived from symbiotic microorganisms than being produced by the sponge from which the particular compound was isolated [Kobayashi et al. 1993b; Shigemori et al. 1992; Unson et al. 1994]. As a consequence, this compound production by sponge associates generates severe difficulties for chemotaxonomy. First because the assignment of a compound as sponge or symbiont compound is difficult. Then because symbiont compounds cannot be used as taxon markers as long as the symbionts varies by sponge, by season, by reproductive state of the host sponge, or simply at random. Successful applications of symbiont compounds in chemotaxonomy are then dependent on the stability of the host-symbiont relationship. Some symbionts might display such high host specificity that their compounds could be considered as suitable markers.

Concerning more specifically sterols biosynthesis in sponges, it is usually admit that there are four possible pathways: (1) de novo, (2) dietary uptake, (3) dietary uptake with modification or (4) synthesis from sponge associates [Erpenbeck and van Soest 2005; Kerr and Baker 1991]. So, chemotaxonomic statements using sterols as chemosystematic tool have to be made with caution.

Since 1990, to investigate taxonomy of some sponges from Indian Ocean, our research team conducts studies on sterol biomarkers complementary to the classical morphological approach [Gauvin 1998; Gauvin et al. 1998, 2000, 2004]. Thus, in this chapter, by investigating the sterols of two *Haliclona* species (Phylum Porifera, order *Haplosclerida*, family *Haliclonidae*) from Mayotte and using the literature available about both the genera *Haliclona* and *Reniera*, we attempted to illustrate the possibility to use sterol patterns to characterize sponge species of the two closely related genera *Haliclona* and *Reniera*.

II.3. Application to Sponges of the Genus *Haliclona* from Mayotte: Sterol Composition and Chemotaxonomic Considerations

1. Introduction

One disagreement between taxonomists concerns sponges belonging to genera *Haliclona* and *Reniera*. In the literature, opinions vary as to whether *Haliclona* (order Haplosclerida, family Haliclonidae) and *Reniera* (order Haplosclerida, family Renieridae) should be maintained as separate genera or whether they should be merged [Bergquist et al. 1980]. The confusion has its origin in the differing historical usages of the generic names, and persists today because of the inadequate criteria for separating the skeletally simple representatives of each genus. As a matter of fact, taxonomists are faced with great morphological variability within species and a lack of macroscopic characters to work with. In order to support the existence of two different genera instead of only one, Griessinger [1971] attempted to distinguish between species of *Reniera* and *Haliclona*, which possess a simple skeletal network, by utilizing criteria such as spicule size and the arrangement of the aquiferous system. A few years later, Rovirosa et al., a chilean group of workers [1990], confirmed this point of view by a chemical approach. Using the literature available about both genera *Haliclona* [Ballantine and Williams 1977; Bergquist et al. 1980; Findlay et al. 1985; de Rosa et al. 1973; Seldes et al. 1985; Sheikh and Djerassi 1974; Zielinski et al. 1982] and *Reniera* [Dini et al. 1985; Lawson et al. 1988; de Rosa et al. 1973] as well as their own results, they tried to determine the general sterol pattern for each genus. As a result of this work, they found that the *Reniera* species have mainly C_{27}, C_{28} and C_{29} usual sterols with Δ^0 and Δ^5 skeletons. In the *Haliclona* species, C_{27}, C_{28} and C_{29} sterols abound but they possess more quantities of the Δ^5 skeleton.

In order to confirm and to generalize their findings, we have decided to analyze the sterol composition of two specimens of *Haliclona*, collected on the coasts of Mayotte Island.

2. Results and Discussion

The twelve sterols isolated from the two *Haliclona* species (*Haliclona* sp. 1 and *Haliclona* sp. 2) from Mayotte are listed in Table 3, together with molecular weight, percentage of sterol components in the mixture and main fragments of the mass spectra. Table 3 contains also the Gas Chromatographic - Relative Retention Time (GC-RRT) data observed for the acetylated sterols. The relative configuration at the asymmetric centers of the side-chain has not been determined because side-chain epimeric sterols show virtually identical Gas Chromatographic (GC) and Mass Spectral (MS) data.

The gas chromatogram of *Haliclona* sp. 1 indicates the presence of eight C_{27}, C_{28} and C_{29} sterols (**S1**; **S2**; **S4**; **S6-S8**; **S10**; **S12**), all with a Δ^5 nuclear unsaturation. The most abundant components of the mixture are cholest-5-en-3β-ol (**S4**) which represent 40.2% and 22-*trans*-24-methylcholesta-5,22-dien-3β-ol (**S6**) accounting for 21.0% of the total mixture.

The results obtained from *Haliclona sp*. 2 show that this sponge contains seven Δ^5 sterols (**S3**; **S4**; **S6**; **S8**; **S10-S12**), which represents 76.6% of the total mixture. The remaining 20.8% corresponds to sterols of the 5α-stanols series (**S5**; **S9**). The major sterol (53.1% of the total sterol composition) is 24-ethylcholest-5-en-3β-ol (**S12**), and among the stanols, 5α-cholestan-3β-ol (**S5**) is the most abundant (20.8% of the total). The presence of two minor unknown sterols (**S3**; **S11**) and of 24-methyl-5α-cholestan-3β-ol (**S9**) has been observed solely on the GC-chromatogram of the steryl acetates. It was therefore only possible to achieve the GC-mass spectra of their acetylated derivatives. The identification of **S3** as $C_{27}\Delta^5$ and of **S11** as $C_{29}\Delta^5$ must be considered as tentative only.

In order to complete the earlier work of Rovirosa et al. [1990], we compared the sterol pattern of *Haliclona* sp. 1 and *Haliclona* sp. 2 with those given by the literature published before and after Rovirosa et al. study, for different species of *Haliclona* and *Reniera*. The percentage composition of the sterol mixture in relation to sterol size and nucleus unsaturation for each species is presented in Table 4.

Table 3. Sterols isolated from *Haliclona* sp. 1 and *Haliclona* sp. 2 : molecular weight (Mol. wt), composition (%), relative retention times (RRT), and main fragments of mass spectra (MS)

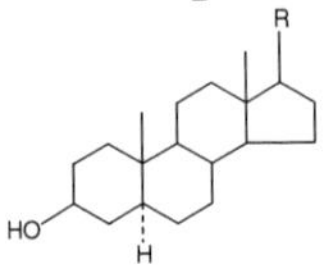

(5α,H) Δ^{0}

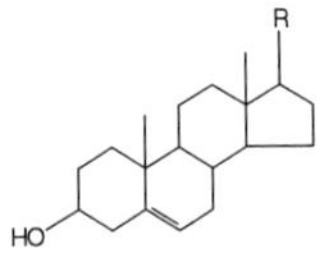

Δ^{5}

Sterol n°	Structure	Mol. wt	*Haliclona* sp. 1 %[a]	Haliclona sp. 2 %[a]	RRT[b]	MS[c]
S1	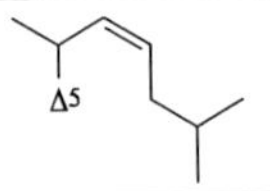 Δ^{5}	384	1.7	-	0.89	384, 369, 366, 351, 324, 300, 285, 282, 273, 271, 255, 231, 213
S2	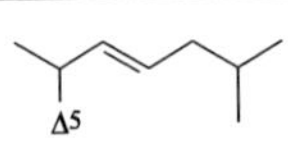 Δ^{5}	384	11.1	-	0.91	384, 369, 366, 351, 324, 300, 285, 282, 273, 271, 255, 213
3S	Unknown $C_{27}\Delta^{5}$ [d]	384	-	2.6	0.93	428, 416, 353, 344, 329, 315, 302, 281, 257, 231, 213
S4	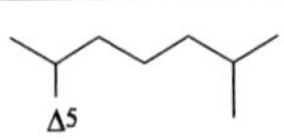 Δ^{5}	386	40.2	5.8	1.00	386, 371, 368, 353, 326, 301, 275, 255, 247, 231, 213
S5	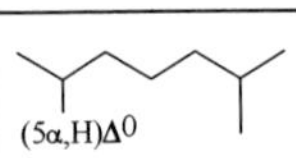 (5α,H)Δ^{0}	388	-	20.8	1.03	388, 373, 355, 331, 262, 234, 233, 215
S6	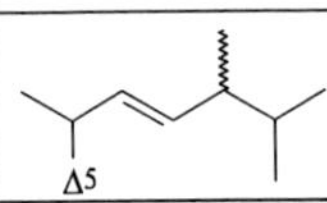 Δ^{5}	398	21.0	8.7	1.11	398, 383, 380, 365, 337, 314, 300, 271, 255, 232, 229, 215, 213
S7	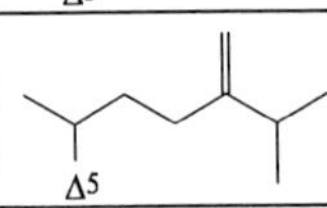 Δ^{5}	398	trace	-	1.26	398, 380, 314, 299, 281, 271, 255, 253, 229, 213
S8	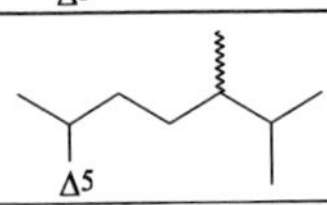 Δ^{5}	400	11.8	5.2	1.28	400, 385, 382, 367, 340, 315, 289, 273, 261, 255, 231, 213
S9	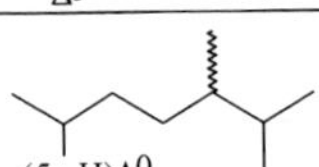 (5α,H)Δ^{0}	402	-	Trace	1.34	444, 429, 384, 369, 330, 315, 276, 275, 230, 215
S10	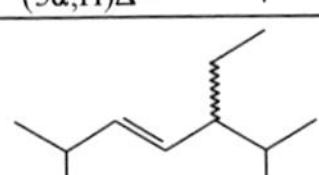 Δ^{5}	412	3.1	2.1	1.39	412, 397, 394, 379, 368, 351, 314, 301, 300, 299, 271, 255, 229, 213
S11	Unknown $C_{29}\Delta^{5}$ [d]	412	-	1.7	1.52	394, 379, 341, 271, 255, 253, 213
S12	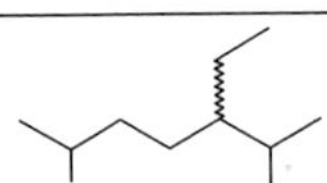 Δ^{5}	414	11.1	53.1	1.57	414, 399, 396, 381, 329, 314, 303, 273, 255, 231, 213

[a] Quantification of the sterols was based on the area of the GC peaks. [b] Gas chromatographic retention times of acetylated derivatives are relative to cholesteryl acetate (1.00); column OV-1; oven temperature 260°C. [c] Mass spectra of the free sterols except for compounds *S3, S9, S11* identified by the mass spectrum of their acetylated derivative. [d] Tentative assignment.

Table 4. Sterol composition (%) in relation to sterol size and nucleus unsaturation of *Haliclona* and *Reniera* species

Sponges		Sterol size					Nucleus unsaturation				References
Species	Location	C_{26}	C_{27}	C_{28}	C_{29}	C_{30}	Δ^0	Δ^5	Δ^7	$\Delta^{7,9(11)}$	
H. aglutinatta	Chile (Easter island) *Pacific Ocean*	4.2	65.1	15.9	14.7	-	12.0	85.8	2.2	-	[Rovirosa et al. 1990]
H. chilensis	Chile (Chiloé) *Pacific Ocean*	2.4	30.6	59.9	4.7	0.2	6.8	89.2	1.5	-	[Seldes et al. 1985a]
H. cinerea	Bulgaria (Pomorie) *Black Sea*	3.0	47.0	38.0	11.0	-	60.0	-	47.0	-	[Elenkov et al. 1999]
H. cinerea	Bulgaria (Nesebar) *Black Sea*	2.0	21.0	68.0	6.0	-	75.0	-	22.0	-	[Elenkov et al. 1999]
H. flavescens	Bulgaria *Black Sea*	2.7	50.7	39.8	6.7	-	60.3	1.8	37.6	0.2	[Zielinski et al. 1982]
H. flavescens	Bulgaria (Nesebar) *Black Sea*	2.0	37.0	50.0	10.0	-	60.0	-	39.0	-	[Elenkov et al. 1999]
H. oculata	Canada (New, Brunswick, Passamaquoddy Bay) *Atlantic Ocean*	2.3	-	87.4	2.9	-	-	92.6	-	-	[Findlay & Patil 1985]
H. permollis	Central California (Monterey) *Pacific Ocean*	0.5	30.0	47.0	18.0	-	-	95.5	-	-	[Sheikh & Djerassi 1974]]
H. rubens	Jamaica *Caribbean Sea*	+[a]	+[a]	+[a]	+[a]	-	29.1	65.6	-	-	[Ballantine et al. 1977]
H. simulans	France (Roscoff) *Atlantic Ocean*	trace	63.0	19.0	15.0	-	-	97.0	-	-	[de Rosa et al. 1973]

Table 4. (Continued)

Sponges		Sterol size					Nucleus unsaturation				References
Species	Location	C_{26}	C_{27}	C_{28}	C_{29}	C_{30}	Δ^0	Δ^5	Δ^7	$\Delta^{7,9(11)}$	
H. sp.	New Zealand *Pacific Ocean*	trace	41.0	26.0	25.0	-	-	92.0	-	-	[Berquist et al. 1980]
H. sp.	Central California (Monterey) Pacific Ocean	2.0	51.0	34.0	3.0	-	-	90.0	-	-	[Sheikh & Djerassi 1974]]
H. sp.	Australia (North-Western coat) *Pacific Ocean*	tr	35.4	43.7	17.3	0.4	0.9	95.9	-	-	[Santalova et al. 2004]
H. sp. 1	Mayotte *Indian Ocean*	53.0	32.8	14.2	-	-	-	100	-	-	[Gauvin, 1998]
H. sp. 2	Mayotte *Indian Ocean*	29.2	13.9	56.9	-	-	20.8	76.6	-	-	[Gauvin, 1998]
R. cratera	Italy (Gulf of Naples) *Thyrrenian Sea*	2.0	46.0	37.0	14.0	-	+[a]	+[a]	-	-	[de Rosa et al. 1973]
R. macropora	Chile (Chiloé) *Pacific Ocean*	2.7	29.3	54.9	7.8	-	22.9	71.6	0.2	-	[Rovirosa et al. 1990]
R. sarai	Italy (Gulf of Naples) *Thyrrenian Sea*	0.6	46.9	9.8	41.2	0.9	48.8	50.6	-	-	[Dini et al. 1985]
R. sp.	-	+[a]	+[a]	+[a]	+[a]	+[a]	-	100	-	-	[Lawson et al. 1988]

[a] % unspecified.

Sterols with additional oxygen functionalities in the ring or in the side chain which have been isolated from *Haliclona* sponges [Ballantine and Williams 1977; Findlay et al. 1985; Seldes et al. 1985a,b] are not discussed in this paper.

According to Rovirosa *et al.*, both genera present the sterol profile of most the Ceractinomorpha [Bergquist et al. 1980] where C_{27}, C_{28} and C_{29} typical sterols (alkylated at C-24) predominate. Another interesting finding is that the difference between the two genera may come from the nucleus unsaturation. A very high content of sterols with Δ^5 skeleton seems to be specific to *Haliclona* species. In contrast, species belonging to *Reniera* genus, may contain not only Δ^5 sterols but a high percentage of Δ^0 derivatives as well. Based on these hypotheses, the group of Chilean workers suggested that *Haliclona rubens* [Ballantine and Williams 1977], which presents a chemical composition similar to *Reniera* species, should be considered as *Reniera* instead of *Haliclona*. Similarly, *Reniera sp.* [Lawson et al. 1988], which only has Δ^5 sterols, should belong to *Haliclona* genus. However four sponges representatives of the genus *Haliclona* investigated by Zielinski et al. in 1982 (before Rovirosa et al. study) and Elenkov et al. in 1999 (after Rovirosa et al. study) show a sterol composition quite distinct from the *Haliclona* and *Reniera* sterol patterns established by Rovirosa et al. in 1990. These four species (2 *Haliclona flavescens* and 2 *Haliclona cinerea*) are namely characterized by the absence or, for one *H. flavescens*, a low percentage (1.8%) of Δ^5 sterols. Their main sterols appear to be stanols (60.0 - 75.0%) accompanied by a moderate quantities of Δ^7 sterols (22.0 - 47.0%). According to Elenkov et al., this difference in the sterol composition are an indication that *Haliclona flavescens* and *Haliclona cinerea* possess a sterol metabolism different from the other *Haliclona* species and form therefore a specific group (probably a subgenus).

Concerning the two *Haliclona* species investigated by us, none of them belong to this specific group suspected by Elenkov et al.. *Haliclona* sp. 1 in which we found exclusively Δ^5 nuclear unsaturation, confirms the chemotaxonomic study of Rovirosa et al.. *Haliclona* sp. 2 however with 76.6% of Δ^5 sterols and 20.8% of 5α-stanols seems to have a sterol composition similar to *Reniera* species. As it was suggested for *Haliclona rubens*, should we changed the name of *Haliclona* sp. 2 into *Reniera* sp. ? It is obvious that more chemical and taxonomic studies are necessary to elucidate this question. Taking into consideration the data we report, it appears of interest first to ascertain whether nuclear unsaturation of sterols is really sufficient to distinguish *Reniera* and *Haliclona* genera. A closer inspection, at once chemical and morphological, may also be required to support the recognition of a subgenus among the *Haliclona* species as suggested by Elenkov et al.

3. Material and Methods

General experimental procedures. Combined Gas Chromatography – Mass Spectrometry (GC-MS) analysis was performed on a Hewlett Packard 5971 mass spectrometer at 70 eV coupled to a Hewlett Packard 5890 gas chromatograph. An OV-1 column (0.25 μm phase thickness, 30 m × 0.32 m i.d.) programmed from 240 to 300°C (10°C/min) was used. Analytical Gas Chromatography (GC) of steryl acetates was carried out on a Delsi serie 30 gas chromatograph using an OV-1 column (0.15 μm phase thickness, 25 m × 0.32 m i.d.) at 260°C. The carrier gas was helium.

Sponge material. Sponges were collected by hand using scuba at depths between 10 and 30 m in Longogori Pass (or S pass), a strict fishing reserve located at the East of Mayotte

(Figure 2). They were frozen until extraction and identified by Dr. Nicole Boury-Esnault (Station Marine d'Endoume – Marseille – France) as two different species of Haliclona genus (class Demospongiae, order Haplosclerida, family Haliclonidae). Reference samples are deposited at the Station Marine d'Endoume.

Sterol extraction. The frozen sponges (81 g and 95 g respectively for *Haliclona* sp. 1 and *Haliclona* sp. 2, dry weight after extraction) were extracted at room temperature with chloroform-methanol 2:1 (v/v) in a suitable blender. After filtration, the solvent was evaporated in vacuum and the resulting crude lipid extracts of *H.* sp. 1 (5.52 g) and *H.* sp. 2 (3.24 g) were then partitioned between *n*-hexane and aqueous methanol 9:1 (V/V). The *n*-hexane residues (767 mg from *H.* sp. 1, 114 mg from *H.* sp. 2) were chromatographed on a silica gel column eluted with a gradient of *n*-hexane-ethyl acetate to provide the sterol mixtures (138 mg from *H.* sp. 1, 65 mg from *H.* sp. 2). Pure free sterols were afforded after recrystallization from methanol and then identified by gas chromatography-mass spectrometry (GC-MS). The sterols were also acetylated with acetic anhydride-pyridine 1:1 (v/v) for 12 hours at room temperature. The steryl acetates were analyzed by gas chromatography (GC) and identified by comparison of their relative retention time (RRT) with those given by the literature [Itoh et al. 1982].

Acknowledgements. The authors are grateful to Dr. N. Boury-Esnault (Station Marine d'Endoume, Marseille, France) for identification of the sponges. These studies were supported by the Regional council of La Réunion.

CONCLUSION

The marine environment of the Indian Ocean western part remains as one of the exceptional and largest untapped reservoirs for marine invertebrate taxonomy and the discovery of novel bioactive compounds. The Chemistry Laboratory of Natural Substances and Food Sciences (LCSNSA) located in the heart of the Indian Ocean western part, occupies clearly a strategic place for the marine natural products research in this area. The work presented above is just an illustration of the long-term project on marine natural products exploitation by the LCSNSA. The two main objectives of our research have been discussed. The first one is the use of marine invertebrates as a potential source for new drug discovery. This was illustrated by the case of the sponge *Luffariella variabilis* from which the sesterterpene manoalide, well-known for its potential anti-inflammatory properties, and four other related compounds have been isolated. The second one is the use of the chemical composition of the marine organisms as a tool for chemotaxonomic studies, in other words as a tool for a best knowledge of the marine biodiversity. This was discussed through our investigation on sterols from the sponges of the two closely related genera *Haliclona* and *Reniera* for which species identification mainly based on the skeletal structure may sometimes be confused. It has been shown that sponges belonging to these genera may be distinguished by the nuclear unsaturation of their sterols. A high amount of sterols with Δ^5 nuclear unsaturation has been attributed to species of the genus *Haliclona*. A mixture of sterols with both Δ^0 and Δ^5 nuclear unsaturations seems to be more characteristic of *Reniera* species. These findings should however be confirmed by further investigations.

The vast project launched by the LCSNSA is still at the very beginning dominated by the purely chemical exercise of isolating novel structures since 1990. However, a multidisciplinary team approach is being built up for collaboration with overseas scientists both in the area (Mayotte, Madagascar, Seychelles...) and in metropolitan France, which bring expertise of marine biology, microbiology and molecular biology, natural product chemistry and pharmacology. It is our hope that all these collaborations will help in transferring the chemistry-dominated research on marine invertebrates from Indian Ocean for the last two decades into bioactivity-directed pharmaceutical development and also chemosystematics in the coming decade. Our efforts are also focused on the way to rationalize our immense research program for discovering useful marine natural products by selecting a restrictive number of marine invertebrate groups limited to date at sponges, soft corals and tunicates.

REFERENCES

Aiello, A., Fattorusso, E. and Menna, M. (1999). Steroids from sponges: recent reports. *Steroids, 64*, 687-714.

Albizati, K. F., Holman, T., Faulkner, D. J., Glaser, K. B. and Jacobs, R. S. (1987). Luffariellolide, an anti-inflammatory sesterterpene from the marine sponge *Luffariella* sp. *Experientia, 43*, 949-950.

Bakus, G. J., Targett, N. M. and Schulte, B. (1986). Chemical ecology of marine organisms: an overview. *J. Chem. Ecol., 12*, 951-987.

Ballantine, J. A., Williams, K. and Burke, B.A. (1977). Marine sterols IV. C_{21} sterols from marine sources. Identification of pregnane derivatives in extracts of the sponge *Haliclona rubens*. *Tetrahedron Lett., 18*, 1547-1550.

Bennett, C. F., Mong, S., Clarke, M. A., Kruse, L. I. and Crooke, S. T. (1987). Differential effects of manoalide on secreted and intracellular phospholipases. *Biochem. Pharmacol., 36*, 733-740.

Bergmann, W. (1949). Comparative biochemical studies on the lipids of marine invertebrates, with special reference to the sterols. *J. Mar. Res., 8*, 137-176.

Bergmann, W. (1962). Sterols: their structure and distribution. In M. Florkin, and S. Mason (Eds.), *Comparative Biochemistry* (Academic Press, Vol. 3, pp. 103-162). New York.

Bergquist, P. R. (1980). A revision of the supraspecific classification of the orders Dictyoceratida, Dendroceratida, and Verongida (class Demospongiae). *N.Z. J. Zool., 7*, 443-503.

Bergquist, P. R., Hofheinz, W. and Oesterhelt, G. (1980). Sterol composition and the classification of the Demospongiae. *Biochem. Syst. Ecol., 8*, 423-435.

Bergquist, P. R., Lavis, A. and Cambie, C. (1986). Sterol composition and classification of the Porifera. *Biochem. Syst. Ecol., 14*, 105-112.

Bergquist, P. R., Karuso, P., Cambie, R. C., Smith D. J. (1991). Sterol composition and classification of the Porifera. *Biochem. Syst. Ecol, 19*, 17-24.

Bergquist, P. R. (1995). Dictyoceratida, Dendroceratida and Verongida from the New Caledonia lagoon (Porifera: Demospongiae). *Mem. Qld. Mus., 38*, 1-51.

Bloch, K. E. (1983) Sterol structure and membrane function. *CRC Crit. Rev. Biochem., 14*, 47-92.

Bourguet-Kondracki, M. L., Debitus, C. and Guyot, M. (1996). Biologically active sesterterpenes from a New Caledonian marine sponge *Hyrtios* sp.. *J. Chem. Research(S),* 192-193.

Bultel-Poncé, V., Berge, J. P., Debitus, C., Nicolas, J. L. and Guyot, M. (1999). Metabolites from the sponge-associated bacterium *Pseudomonas* species. *Mar. Biotechnol., 1*, 384-390.

Butler, M.S. and Capon, R.J. (1992). The luffarins (A-Z), novel terpenes from an Australian marine sponge, *Luffariella geometrica. Aust. J. Chem., 45*, 1705-1743.

Cambie, R. C., Craw, P. A., Bergquist, P. R. and Karuso, P. (1988). Chemistry of sponges, III. Manoalide monoacetate and thorectolide monoacetate, two new sesterterpenoids from *Thorectandra excavatus. J. Nat. Prod., 51*, 331-334.

Carballeira, N., Thomson, J. E., Ayanoglu, E. and Djerassi, C. (1986). Biosynthetic studies of marine lipids: the biosynthesis of long-chain branched fatty acids in marine sponges. *J. Org. Chem., 51*, 2751-2756.

Carballeira, N. M. and Shalabi, F. (1989). On the sterol composition of the marine sponge *Verongula gigantea*. A comparative study of the Verongidae - II. *Biochem. Syst. Ecol., 17*, 315-317.

Castellanos, L., Zea, S., Osorno, O. and Duque, C. (2003). Phylogenetic analysis of the order Halichondrida (Porifera, Demospongiae), using 3β-hydroxysterols as chemical characters. *Biochem. Syst. Ecol., 31*, 1163-1183.

Deems, R. A., Lombardo, D., Morgan, B. P., Mihelich, E. D. and Dennis, E. A. (1987). The inhibition of phospholipase A_2 by manoalide and manoalide analogues. *Biochem. Biophys. Acta, 917*, 258-268.

Dembitsky, V. M., Rezanka, T. and Srebnik, M. (2003) Lipid compounds of freshwater sponges: family Spongillidae, class: Demospongiae. *Chem. Phys. Lipids, 123*, 117-155.

Dini, A., Piccialli, V., Pronzato, R. and Sica, D. (1985). Sterol composition of marine sponges *Stryphnus mucronatus* and *Reniera sarai. Comp. Biochem. Physiol., 81B*, 111-114.

Djerassi, C. and Lam, W. K. (1991) Sponge phospholipids. *Acc. Chem. Res., 24*, 69-75.

Ekow Amoo, V., De Bernardo, S. and Weigele, M. (1988). Synthesis of (S)-manoalide diol and the absolute configuration of natural manoalide. *Tetrahedron Lett., 29*, 2401-2404.

Elenkov, I., Popov, S. and Andreev, S. (1999). Sterols from two Black sea sponges (*Haliclona* sp.). *Comp. Biochem. Physiol.,* Part B, *123*, 357-360.

Erpenbeck, D. and van Soest, R. W. M. (2005). A survey for biochemical synapomorphies to reveal phylogenetic relationships of halichondrid demosponges (Metazoa: Porifera). *Biochem. Syst. Ecol., 33*, 585-616.

Erpenbeck, D., van Soest, R. W. M. (2007). Status and perspective of sponge chemosystematics. *Marine biotechnology, 9*, 2-19.

Ettinger-Epstein, P., Motti, C. A., De Nys, R., Wright, A. D., Battershill, C. N. and Tapiolas, D. M. (2007). Acetylated sesterterpenes from the Great Barrier Reef sponge *Luffariella variabilis. J. Nat. Prod., 70*, 648-651.

Faulner, D. J. (1998). Marine natural products. *Nat. Prod. Rep., 15*, 113-158.

Faulkner, D. J., Newman, D. J. and Cragg, G. M. (2004). Investigations of the marine flora and fauna of the Islands of Palau. *Nat. Prod. Rep., 21*, 50-76.

Findlay, J.A. and Patil, A. D. (1985). Novel sterols from the finger sponge *Haliclona oculata*. *Can. J. Chem.*, *63*, 2406-2410.

de Freitas, J. C., Blankmeier, L. A. and Jacobs, R. S. (1984). In vitro inactivation of the neurotoxic action of β-bungarotoxin by the marine natural pr. and Murphy, P. (1994). Chemotaxonomic relationships within, and comparisons between, the orders Haplosclerida and Petrosida (Porifera : Demospongiae) using sterol complements. *Biochem. Syst. Ecol.*, *22*, 735-752.

Gauvin, A. (1998). Etude de cinq éponges de l'Océan Indien. Contribution chimiotaxonomique - Recherche de molécules bioactives. Ph.D. Thesis. University of Reunion Island, France.

Gauvin, A., Smadja, J., Aknin, M. and Gaydou, E. M. (1998). Cyclopropane-containing sterols in the marine sponge *Petrosia spheroïda*. *Comp. Biochem. Physiol., Part B, 121*, 451-456.

Gauvin, A., Smadja, J., Aknin, M., Faure, R. and Gaydou, E. M. (2000). Isolation of bioactive 5α,8α-epidioxy sterols from the marine sponge *Luffariella* cf. *variabilis*. *Can. J. Chem.*, *78*, 986-992.

Gauvin, A., Smadja, J., Aknin, M. and Gaydou, E. M. (2004). Sterol composition and chemotaxonomic considerations in relation to sponges of the genus *Xestospongia*. *Biochem. Syst. Ecol.*, *32*, 469-476.

Glaser, K. B. and Jacobs, R. S. (1986). Molecular pharmacology of manoalide. Inactivation of bee venom phospholipase A_2. *Biochem. Pharmacol.*, *35*, 449-453.

Glaser, K. B. and Jacobs, R. S. (1987). Inactivation of bee venom phospholipase A_2 by manoalide. A model based on the reactivity of manoalide with amino acids and peptide sequences. *Biochem. Pharmacol.*, *36*, 2079-2086.

Glaser, K. B., De Carvalho, M. S., Jacobs, R. S., Kernan, M. R. and Faulkner, D. J. (1989). Manoalide: structure-activity studies and definition of the pharmacophore for phospholipase A_2 inactivation. *Mol. Pharmacol.*, *36*, 782-788.

Griessinger, J. M. (1971). Etude des Réniérides de Méditerranée (Demosponges Haplosclérides). In Bulletin du Muséum National d'Histoire Naturelle, *Zoologie* (Vol. 3, pp. 97-181). Paris.

Hafner, D., Heiner, E. and Noack, E. (1977). Mathematical analysis of concentration-response relationships. *Arzneim.-Forsch.*, *27*, 1871-1873.

Hahn, S., Stoilov, I. L., Tam Ha, T. B., Raederstorff, D., Doss, G. A., Li, H. T. and Djerassi, C. (1988). Biosynthetic studies of marine lipids. 17. The course of chain elongation and desaturation in long-chain fatty acids of marine sponges. *J. Am. Chem. Soc.*, *110*, 8117-8124.

Itoh, T., Tani, H., Fukushima, K., Tamura, T. and Matsumoto, T. (1982). Structure-retention relationship of sterols and triterpene alcohols in gas chromatography on a glass capillary column. *J. Chromatogr.*, *234*, 65-76.

Jacobs, R. S., Culver, P., Langdon, R., O'Brien, T. and White, S. (1985). Some pharmacological observations on marine natural products. *Tetrahedron*, *41*, 981-984.

Jacobson, P. B., Marshall, L. A., Sung, A. and Jacobs, R. S. (1990). Inactivation of human synovial fluid phospholipase A_2 by the marine natural product, manoalide. *Biochem. Pharmacol.*, *39*, 1557-1564.

Katsumura, S., Fujiwara, S. and Isoe, S. (1985). Total synthesis of manoalide and seco-manoalide. *Tetrahedron Lett., 26*, 5827-5830.

Kernan, M. R., Faulkner, D. J. and Jacobs, R. S. (1987). The luffariellins, novel anti-inflammatory sesterterpenes of chemotaxonomic importance from the marine sponge *Luffariella variabilis. J. Org. Chem., 52*, 3081-3083.

Kernan, M. R., Faulkner, D. J., Parkanyi, L., Clardy, J., De Carvalho, M. S. and Jacobs, R. S. (1989). Luffolide, a novel anti-inflammatory terpene from the sponge *Luffariella sp.. Experientia, 45*, 388-390.

Kerr, R. G. and Baker, B. J. (1991). Marine sterols. *Nat. Prod. Rep., 8*, 465-497.

Kobayashi, J., Zeng, C. M., Ishibashi, M. and Sasaki, T. (1993a). Luffariolides F and G, new manoalide derivatives from the Okinawan marine sponge *Luffariella* sp.. *J. Nat. Prod., 56*, 436-439.

Kobayashi, J., Murata, O. and Shigemori, H. (1993b). Jaspisamides A-C, new cytotoxic macrolides from the Okinawan sponge *Jaspis* sp.. *J. Nat. Prod., 56*, 787-791.

Kobayashi, M., Okamoto, T., Hayashi, K., Yokoyama, N., Sasaki, T. and Kitagawa, I. (1994). Marine natural products. XXXII. Absolute configuration of C-4 of the manoalide family, biologically active sesterterpenes from the marine sponge *Hyrtios erecta. Chem. Pharm. Bull., 42*, 265-270.

König, G. M., Wright, A. D. and Sticher, O. (1992). Four new antibacterial sesterterpenes from a marine sponge of the genus *Luffariella. J. Nat. Prod., 55*, 174-178.

Kornprobst, J. M. (2005a). Les bases de la pharmacologie marine. In *Substances naturelles d'origine marine : Généralités, Micro-organismes, Algues* (Tec and Doc, Lavoisier, pp. 57-84). Paris.

Kornprobst, J. M. (2005b). Spongiaires. In *Substances naturelles d'origine marine : Invertébrés, Vertébrés* (Tec and Doc, Lavoisier, pp. 601-1035). Paris.

Lee, Y. K., Lee, J. H. and Lee, H. K. (2001). Microbial symbiosis in marine sponges. *J. Microbiol., 39*, 254-264.

Lam, W. K., Hahn, S., Ayanoglu, E. and Djerassi, C. (1989). Phospholipids studies of marine organisms. 22. structure and biosynthesis of novel brominated fatty acid from *Hymeniacidon* sponge. *J. Org. Chem., 54*, 3428-3432.

Lawson, M. P., Stoilov, I. L., Thompson, J. E. and Djerassi, C. (1988). Cell membrane localization of sterols with conventional and unusual side chains in two marine Demosponges. *Lipids, 23*, 750-754.

Lombardo, D. and Dennis, E. A. (1985). Cobra venom phopholipase A_2 inhibition by manoalide. A novel type of phospholipase A_2 inhibitor. *J. Biol. Chem., 260*, 7234-7240.

McConnell, O. J., Longley, R. E. and Koehn, F. E. (1994). The discovery of marine natural products with therapeutic potential. In V. P. Gullo (Ed.), *The discovery of natural products with therapeutic potential* (Butterworth-Henemann press, pp. 109-174). Boston.

Mclaughlin, J. C. (1991). Crown gall tumours on potato discs and brine shrimp lethality: two simple bioassays for higher plant screening. In K. Hostetmann (Ed.), *Methods in Plant Biochemistry: assays for Bioactivity* (Academic Press, Vol. 6, pp. 1-32). London.

Meyer, B. N., Ferrigni, N. R., Putnam, J. E., Jacobsen, L. B., Nichols, D. E. and McLaughlin, J. E. (1982). Brine Shrimp : a convenient general bioassay for active plant constituents. *Planta medica, 45*, 31-34.

Montagnac, A., Païs, M. and Debitus, C. (1994). Fasciospongides A, B and C, new manoalide derivatives from the sponge *Fasciospongia* sp.. *J. Nat. Prod., 57*, 186-190.

Namikoshi, M., Suzuki, S., Meguro, S., Nagai, H., Koike, Y., Kitazawa, A., Kobayashi, H., Oda, T. and Yamada, J. (2004). Manoalide derivatives from a marine sponge *Luffariella* sp. collected in Palau. *Fisheries Science, 70*, 152-158.

Ortiz, A. R., Pisabarro, M. T., Gago, F. (1993). Molecular model of the interaction of bee venom phospholipase A_2 with manoalide. *J. Med. Chem., 36*, 1866-1879.

Potts, B. C. M., Faulkner, D. J., De Carvalho, M. S. and Jacobs, R. S. (1992a). Chemical mechanism of inactivation of bee venom phospholipase A_2 by the marine natural products manoalide, luffariellolide, and scalaradial. *J. Am. Chem. Soc., 114*, 5093-5100.

Potts, B. C. M., Faulkner, D. J. and Jacobs, R. S. (1992b). Phospholipase A_2 inhibitors from marine organisms. *J. Nat. Prod., 55*, 1701-1717.

Potts B. C. M., Capon, R. J. and Faulkner, D. J. (1992c). Luffalactone and (4E,6E)-dehydromanoalide from the sponge *Luffariella variabilis*. *J. Org. Chem., 57*, 2965-2967.

Proksch, P. (1994). Defensive roles for secondary metabolites from marine sponges and sponge-feeding nudibranchs. *Toxicon, 32*, 639-655.

Reddy, M. V. R., Harper, M. K. and Faulkner, D. J. (1997). Luffasterols A-C, 9,11-secosterols from the Palauan sponge *Luffariella* sp.. *J. Nat. Prod., 60*, 41-43.

Reynolds, L. J., Morgan, B. P., Hite, G. A., Mihelich, E. D. and Dennis, E. A. (1988). Phospholipase A_2 inhibition and modification by manoalogue. *J. Am. Chem. Soc., 110*, 5172-5177.

Reynolds, L. J., Mihelich, E. D. and Dennis, E. A. (1991). Inhibition of venom phospholipase A_2 by manoalide and manoalogue. Stoichiometry of incorporation., *J. Biol. Chem., 266*, 16512-16517.

de Rosa, M., Minale, L. and Sodano, G. (1973). Metabolism in Porifera - II. Distribution of sterols. *Comp. Biochem. Physiol., 46B*, 823-837.

de Rosa, M., Giordano, S., Scettri, A., Sodano, G., Soriente, A., Pastor, P. G., Alcaraz, M. J., Payá, M. (1998). Synthesis and comparison of the antiinflammatory activity of manoalide and cacospongionolide B analogues. *J. Med. Chem., 41*, 3232-3238.

de Rosa, S., De Stefano, S. and Zavodnik, N. (1988). Cacospongionolide: a new antitumoral sesterterpene, from the marine sponge *Cacospongia mollior*. *J. Org. Chem.; 53*, 5020-5023.

de Rosa, S., Crispino, A., De Giulio, A., Iodice, C., Pronzato, R. and Zavodnik, N. (1995). Cacospongionolide B, a new sesterterpene from the sponge *Fasciospongia cavernosa*. *J. Nat. Prod., 58*, 1776-1780.

de Rosa, S., Crispino, A., de Giulio, A., Iodice, C. and Tommonaro, G. (1997). Cavernosolide, a new sesterterpene from a Thyrrhenian sponge. *J. Nat. Prod., 60*, 844-846.

de Rosa, S. and Carbonelli, S. (2006). The new luffarin derivatives from the Adriatic sea sponge *Fasciospongia cavernosa*. *Tetrahedron, 61*, 2845-2849.

Rovirosa, J., De la Luz Vasquez, M. and San-Martin, A. (1990). Chemotaxonomic considerations to sponges of the genera *Reniera* and *Haliclona*. *Biochem. Syst. Ecol., 18*, 53-55.

Santalova, E. A., Makarieva, T. N., Gorshkova, I. A., Dmitrenok, A. S., Krasokhin, V. B. and Stonik, V. A. (2004). Sterols from six marine sponges. *Biochem. Syst. Ecol., 32*, 153-167.

Seldes, A. M., Rovirosa, J., San Martin, A. and Gros, E. G. (1985a). Sterols from the sponge *Haliclona chilensis*. *Comp. Biochem. Physiol., 81B*, 119-122.

Seldes, A. M., Rovirosa, J., San Martin, A. and Gros, E. G. (1985b). A new sterol from the sponge *Haliclona chilensis*. *Experientia, 41*, 34-35.

Sheikh, Y. M. and Djerassi, C. (1974). Steroids from sponges. *Tetrahedron, 30*, 4095-4103.

de Silva, E. D. and Scheuer, P. J. (1980). Manoalide, an antibiotic sesterterpenoid from the marine sponge *Luffariella variabilis* (Polejaeff). *Tetrahedron Lett., 21*, 1611-1614.

Shigemori, H., Bae, M.A., Yazawa, K., Sasaki, T. and Kobayashi, J. (1992). Alteramide A, a new tetracyclic alkaloid from a bacterium *Alteromonas* sp. Associated with the marine sponge *Halichondria okadai*. *J. org. Chem., 57*, 4317-4320.

de Silva, E. D. and Scheuer, P. J. (1981). Three new sesterterpenoid antibiotics from the marine sponge *Luffariella variabilis* (Polejaeff). *Tetrahedron Lett., 22*, 3147-3150.

van Soest, R. W. M., Braekman, J. C. (1999). Chemosystematics of Porifera: a review. *Mem. Qld Mus., 44*, 569-589.

Soriente, A., De Rosa, M., Scettri, A., Sodano, G., Terencio, M. C., Payá, M. and Alcaraz, M. J. (1999). Manoalide. *Current Medicinal Chemistry, 6*, 415-431.

Tsuda, M., Shigemori, H., Ishibashi, M., Sasaki, T. and Kobayashi, J. (1992). Luffariolides A-E, new cytotoxic sesterterpenes from the Okinawan marine sponge *Luffariella* sp.. *J. Org. Chem., 57*, 3503-3507.

Tsuda, M., Endo, T., Mikami, Y., Fromont, J. and Kobayashi, J. (2002). Luffariolides H and J, new sesterterpenes from a marine sponge *Luffariella* species. *J. Nat. Prod., 65*, 1507-1508.

Unson, M. D., Holland, N. D. and Faulkner, D. J. (1994). A brominated secondary metabolite synthesized by the cyanobacterial symbiont of a marine sponge and accumulation of the crystalline metabolite in the sponge tissue. *Mar. biol., 119*, 1-11.

Wilkinson, C. R. (1978a). Microbial associations in sponges. I. Ecology, physiology and microbial populations of coral reef. *Mar. Biol., 49*, 161-167.

Wilkinson, C. R. (1978b). Microbial associations in sponges. II. Numerical analysis of sponge and water bacterial populations. *Mar. Biol., 49*, 169-176.

Wilkinson, C. R. (1978c). Microbial associations in sponges. III. Ultrastructure of the *in situ* associations in coral reef sponges. *Mar. Biol., 49*, 177-185.

Wilkinson, C. R. (1987). Significance of microbial symbionts in sponge evolution and ecology. *Symbiosis, 4*, 177-185.

Wilkinson C. R. and Fay, P. (1979). Nitrogen fixation in coral reef sponges with symbiotic cyanobacteria. *Nature, 279*, 527-529.

Wilkinson, C. R. and Garrone, R. (1980). Nutrition of marine sponges. Involvement of symbiotic bacteria in the uptake of dissolved carbon. In D. C. Smith, and Y. Tiffon (Eds.), *Nutrition in the lower Metazoa*, Pergamon Press, pp. 157-161). Oxford.

Wilkinson, C. R., Nowak, M., Austin, B. and Colwell, R. R. (1981). Specificity of bacterial symbionts in Mediterranean and Great Barrier Reef sponges. *Microb. Ecol., 7*, 13-21.

Wulff, J. L. (2006). Ecological interactions of marine sponges. *Can. J. Zool., 84*, 146-166.

Zhou, G. X. and Molinski, T. F (2006). Manoalide derivatives from a sponge, *Luffariella* sp.. *J. Asian Nat. Prod. Research, 8*, 15-20.

Zielinski, J., Milkova, T., Popov, S., Marekov, N. and Djerassi, C. (1982). Minor and trace sterols in marine invertebrates. 34 Isolation and structure elucidation of 24-methyl-5α-cholesta-7,9(11),24(28)-trien-3β-ol, the first naturally occurring $\Delta^{7,9(11)}$ unsaturated marine sterol. *Steroids, 39*, 675-680.

In: Atlantic and Indian Oceans ISBN 978-1-60692-475-4
Editors: E. S. Askew and J. P. Bromley, pp. 115-123

Chapter 4

TSUNAMI OCCURRENCE IN THE GULF OF BENGAL, INDIAN OCEAN

Mansour Ioualalen[1] and Ilayaraja Kathirkaman[2]
[1]Institut de Recherche pour le Développement, IRD
Géosciences Azur, Observatoire Océanologique, 2 Quai de la Darse, BP 48,
F-06235 Villefranche-sur-mer Cedex, France
[2]Department of Applied Geology, University of Madras,
Guindy Campus, Chennai – 600 025, India

ABSTRACT

A review of the tsunami hazard along the coastal area of the Gulf of Bengal, Indian Ocean, is provided. We consider here earthquakes as tsunami-triggering geophysical sources. For that purpose, the seismicity of the Indian Ocean area is described, i.e. the tectonics of the area, the earthquake recurrence, as well as historical tsunami events. In that context, recent events like the 26th December 2004 megathrust earthquake and associated tsunami are discussed. Finally, we describe the main physical processes that are responsible for wave amplification/attenuation and for the coastal vulnerability that may vary significantly along the Gulf of Bengal. Beside earthquake magnitude and modes of co-seismic rupture, we show in particular the importance of local effects for wave amplification (shape of the coast and local bathymetry), and of specific features, like underwater canyons, that are likely to inhibit wave amplification at the coast.

INTRODUCTION

A tsunami is a series of waves of possibly extreme long wavelength and long period caused by any fast sea floor deformation (earthquakes, submarine slides/slumps, underwater volcanic eruptions) or by an immersion of subaerial objects or flow (landslides, pyroclastic flow or extremely more occasional asteroids) (Bondevik *et al.*, 1997). Their main characteristics is a long wavelength (e.g. up to 200 km for the 26 December 2004 Indian Ocean tsunami) and may travel across the oceans at great velocity without encountering significant viscous damping because of its extremely low frequency. As a consequence they may travel over an entire ocean basin like the Pacific Ocean without significant loss of energy. Contrary to short wind-generated waves or the swell, the trajectories of the water particles describe very flat ellipses with a very weak stratification in horizontal induced current (long wave theory). As a result, bottom friction generates significant sediment arrachements increasing the water density and thus the water flooding impact.

The amplification of a tsunami is largely determined by three main processes: shoaling, focusing and bay resonance (more occasional). Shoaling occurs when a wave encounters a negative bathymetric gradient (water depth decreasing in the direction of propagation). The leading wave component slows down and the trailing water builds up behind, thereby increasing the wave height. This process takes place mainly in coastal areas or at the continental slope. When a tsunami encounters a concave-shaped (convergent) pattern in the bathymetry, focusing occurs, whereby refraction causes the wave front to concentrate. This phenomenon commonly arises near a cape and is enhanced when the region is surrounded by open bays of convex (divergent) shape. A recent example of study can be found in Ioualalen *et al.* (2009) who computed the vulnerability of the French Riviera coast to tsunamis generated by underwater landslides. The coastal area encounters numerous capes and open bays. They showed the efficiency of concave bathymetric shapes to amplify the wave through focusing (mostly around caps but not always). They also showed how the phenomenon can be enhanced when such bathymetric features are surrounded by open-convex type bays which force the wave to escape from its sides (divergence) and feed the wave amplitude growth at the capes. It is interesting to note that the occurrence of the focusing process does not depends on the initially triggered wave: Ioualalen *et al.* (2009) found a strong ratio between wave height at capes and elsewhere whatever is the absolute value of the wave amplitudes (cms in the case of deep and small volume landslides and ms for shallow large volumes). The general shape of isobaths is thus a robust criterion to identify the vulnerability of a coast to tsunamis. In that case the amplitude of the process depends on the isobaths rate of curvature, and on the position of a coastal location vs. the focalized area. Bay resonance generally takes place in closed bays where a trapped wave oscillates in a fashion similar to that of a standing wave. Resonance may then occur when a phase-locking begins between one particular mode of the tsunami and the bay frequency mode. The phase-locking is strictly linked to the time scales and wavelength of the tsunami and the size of the bay. The existence of a particular resonance mode associated with the phase-locking is linked to the nature of the dispersion of the incoming wave. All three main modes of amplification can be explained by linear theory. Contrary to wind-generated short wave, the nonlinear interactions of a tsunami are supposed to be weak offshore and are not proved to be determinant in coastal areas vs. site effects. This is partly because in the open ocean a tsunami does not encounter other waves with the same order wave spectrum.

Earthquake and Tsunami Occurrence in the Indian Ocean

As far as the Indian Ocean is concerned, the main tsunamigenic sources are past and potential earthquakes generated at its eastern boarder great subduction zone ranging from Myanmar to Indonesia through Nicobar and Andaman Islands (Figure 1).

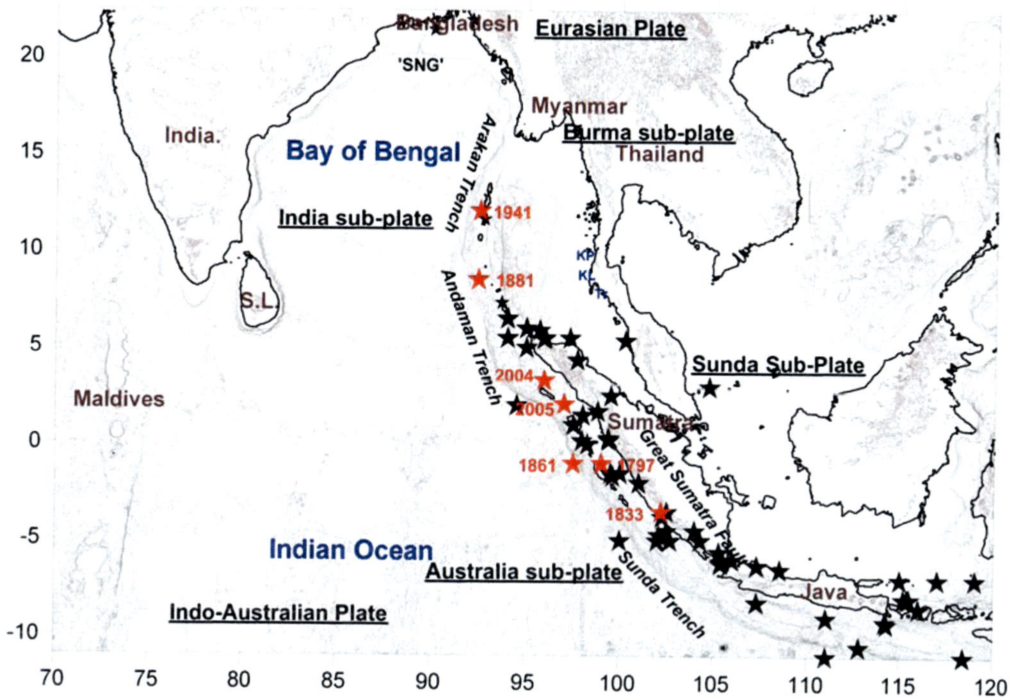

Figure 1. Morphology of the most seismic area of the Indian Ocean. The tsunamigenic earthquakes are reported (★) (Major event are in red (★). The events have been reported in Rastogi and Jaiswal (2006). S. L. is for Sri Lanka and 'SNG' is for 'Swatch of No Ground' Canyon. In blue: KL is for Khao Lak, KP is for Kho Pratong and Tr is for Trang. The bathymetry/topography is derived from 2' ETOPO-2 data set.

However, it has to be mentioned that one of the most devastating tsunamis ever recorded in the Indian Ocean (beside the 26th December 2004 Sumatra event) is the 27th August 1883 tsunami that has been triggered by the giant Krakatau volcanic eruption located between Java and Sumatra Islands. The event generated local waves of amplitude larger that 36 m (just as the 2004 Sumatra event near Banda Aceh, North Sumatra) and caused 36000 fatalities in Western Java and Southern Sumatra. Like the 2004 Sumatra tsunami also, it has also been recorded worldwide with tide gauges (Pelinovsky *et al.*, 2005a). The two extreme events (1883 and 2004) appear to be of same order of magnitude (in wave amplitude) despite the fact that remote locations amplitudes are twice more important for the 2004 probably because of the wave directivity vs. geographical locations of the tide gauges records (Pelinovsky *et al.*, 2005b). Up-to-date, the two events (of different origin) can be considered as "reference tsunamis" that one should refer to for any occurring tsunami in the Indian Ocean (Pelinovsky *et al.*, 2005b).

In the following however, we report only recurrent tsunami events (triggered by constant seismic activity and recurrent strain accumulation/rupture cycles) since the 1883 Krakatau tsunami is extremely rare. A compilation of the Indian Ocean tsunamis indicate that more than 80% of them are originated from the Java Sumatra and Adaman/Nicobar subduction zone (Rastogi and Jaiswal, 2006). Mega-earthquakes and associated tsunamis have approximately a two-century period of recurrence while the one of smaller but significant event is of the order of few decades (Rastogi and Jaiswal, 2006). The authors have updated former catalogs of tsunamis events in the Indian Ocean (about 90)(Newcomb and McCann, 1987; Heck, 1947; Berninghausen, 1966; Litzin, 1974, and USGS).

The Indian Ocean Western Boarder encounters a strong and recurrent seismicity because it hosts several sub-plates with an active tectonics (Figure 1). The Indo-Australia plate is subducting generally NNE (4-5 cm/year or more locally) under the southern part of the Eurasian Plate (Burma and Sunda sub-plates). Strong earthquakes are recorded in this area because as usual when large scale oceanic plate and a continental margin converge into each other (Lay *et al.*, 2005). More precisely, to the east, the Indo-Australia plate converges normal to the Java trench (Northward, 6.3 cm/year, Michel *et al.*, 2001; Bock *et al.*, 2003) and to the east along the Sunda then the Andaman trenches, the subduction is increasingly oblique (NE, 4.5 cm/year, Genrich *et al.*, 2000) (Fitch, 1972). In this area the amount of slip is of the order of 1 to 3 cm/year (Genrich *et al.*, 2000) which yields large amount of deformation and thus of strain. The region is also subject to a seismic segmentation in the presence of several sub-plates (Figure 1). Such segmentation might also be responsible for the occurrence of large earthquakes because it tends to inhibit stress accommodation at the sub-plates boundaries. The region encountered numerous tsunamigenic earthquakes in the past (Figure 1, Rastogi and Jaiswal, 2006). On the 10th February 1797 an estimated Mw=8.4 occurred on the Sumatra Java area (Berninghausen, 1966) followed by an estimated Mw=9 on the 24th November 1833 (Newcomb and McCann, 1987). The 1797 tsunami caused severe damages in Sumatra Island where Pandang city was inundated and caused more than 300 fatalities (Heck, 1947). The 1833 tsunami inundated over 550 km of South-Central Sumatra and caused severe amount of fatalities. To the north of this segment an estimated Mw=8.5 occurred on the 16th February 1861 (Newcomb and McCann, 1987) followed by the recent MW=8.6 on the 28th March 2005 (Briggs *et al.*, 2006). The 1861 tsunami caused thousands of fatalities at the west coast of Sumatra (7 m waves have been reported) and the 2005 tsunami generated 4-5 m high waves in Nias Island caused 665 fatalities (earthquake and tsunami). Further North is the location of the 26th December 2006 megathrust earthquake at the northern tip of Sumatra Island which involving a 1300 km rupture nearly Northward. The tsunami generated waves over 30 m (Banda Aceh, Northern Sumatra) and caused more than 300,000 fatalities in 13 countries. Northward Nicobar and Andaman islands area also encountered large earthquakes: an estimated Mw=7.9 on the 31st December 1881 off Car Nicobar Island (Berninghausen, 1966) and an estimated Mw=7.7 on the 26th June 1941 (Bilham *et al.*, 2005). Both events tsunamis caused minor damages with 1-2 m runup observed (highest inundated point). It is interesting to mention that only little seismicity has been observed in the last 40 years preceding the 2004 Sumatra megathrust in the region between the 2004 and the 1881 large earthquakes and in the region of the great 1861 earthquake until the 2005 event (Lay *et al.*, 2005). These large events and associated seismic gaps are consistent with long-term strain accumulation in the area (Lay *et al.*, 2005).

The 26TH December 2004 Indian Ocean Tsunami and Coastal Vulnerability

The dramatic tsunami of 26th December 2004 offshore Sumatra is in itself so extended (13 countries affected, 1300 km large axis rupture) that it released also the strain accumulated northward in the Andaman/Nicobar Islands (seismic gaps left after events of 1881 and 1941 (Figure 1). As a result, the large amount of coastal impact data that have been gathered is a strong basis for building a preventive vulnerability map of most of the Indian Ocean, at the exception of the Southern hemisphere east of 100°E (Java, Australia). The 2005 event caused severe damages but the impact has been quite confined (centre of Sumatra). Immediately after the 2004 event, international teams have sampled most of the damaged areas and provided a reasonably complete picture of the event, at least for damaged coasts: runup and inundation have been measured in most Indian Ocean countries: Fritz and Synolakis (2005) for the Maldives, Gusiakov (2005) for Sibolga and Nias Islands offshore Sumatra, Liu *et al.* (2005) and Synolakis *et al.* (2005) for Sri Lanka, Satake *et al.* (2006) for Myanmar, Sannasiraj and Sundar (2005) for Indian coast of Tamil Nadu, Yalciner *et al.* (2005a) for Malaysia and offshore islands, Yalciner *et al.* (2005b) for Northern Sumatra, Tsuji *et al.* (2006) for Thailand. To this (non-exhaustive) list must be added numerous national or Organizations reports and individual reports.

The compilation of runup and inundation data indicates a large variability of the tsunami coastal impact for the same event. Site effects, wave directivity and distance from the initial wave locations are supposed to be essential in the impact discrepancies. Choi *et al.* (2006) have analyzed a statistically representative ensemble of runup field surveys in several countries (nearly 1000 runup values). They found that the runup field follows a log-normal distribution of the form:

$$f(H) = \frac{\exp\left(-\frac{(\log H - a)^2}{2\sigma^2}\right)}{H\sigma\sqrt{2\pi}\ln 10},$$

where H is the runup height at any point of the coast (in meters), $a = \langle \log H \rangle$ is the average value of height logarithm and σ is standard deviation of the wave height logarithm (Choi *et al.* (2006). They explained that the main parameters controlling (non-explicitely) the distribution are the random repartition of bathymetry and topography (Choi *et al.*, 2001, 2002). The explanation is consistent with the numerical results of Ioualalen *et al.* (2007a) for the Thailand case study. The latter found that the simulated runups trend to a such theoretical distribution when the computational spatial grid spacing is reduced towards improved description of the bathymetry and topography field. As a result such distribution functions may be useful to recover missing data when facing such an extended event that impacted 1000s km of coast lines (Choi *et al.*, 2006). This aspect is important for mapping coastal vulnerability to tsunamis and the methodology applied in Choi *et al.* (2006). This event was the most sampled in the history: runup and inundation field surveys, numerous tide gauges records (Merrifield *et al.*, 2005), altimetric records (Gover, 2005). Nevertheless, still some areas have not been sufficiently sampled. This can be the case for example when an area is poorly dense in population; in such cases no survey has been operated because they were no reported fatalities. In such areas it is also necessary to estimate their vulnerability to tsunamis because the entire worldwide coastal area occupation is expanding ineluctably. However we believe that such statistical method can be inefficient in the case of strong coastal runup variability and or when having scarse runup observations (ensemble of observations not statistically representative). In that case only robust numerical simulations may provide coherent vulnerability mapping. Numerical simulations based on a satisfactory physical description of the propagation and inundation, a reliable numerical scheme, accurate boundary data (bathymetry and topography) and a reliable triggering seismic source may provide us with a synoptic picture of an event and, in particular, runup where no observations are available.

Numerous numerical models simulating propagation and inundation co-exist in the community. The models are all based on nonlinear shallow water or Boussinesq formulations (TUNAMI, GEOWAVE, MOST, CEA model, etc). Considering the sizes of the computational domains and the required time steps it is necessary to avoid the vertical stratification of the water column flow velocity. Once a robust numerical simulation is constructed (see above), the simulated runup are obtained in a prognostic way or at least partly: some kind of calibration is always needed to constrain the seismic source. The calibration may be performed with all available records: seismic rates of uplift/subsidence (distribution of the vertical movement of the sea floor that triggered the tsunami), tide gauge records and eventually sea level anomalies obtained with altimeters for the case of the 2004 event (inverse problem). In that case, if these data are sufficient, then there is no need to constrain a simulation with runup observations and thus the simulated runups are obtained in a prognostic mode. However generally only estimates of the seismic rupture parameters and runup data are available (or exploitable) and thus the simulation provides with partial prognostic results only. In all cases the simulation is validated with available runups and the non-sampled coastal runups are estimated with the simulation. For the 2004 event numerous numerical reliable simulations have been performed dealing with case studies, process studies, inverse problems, studies to sensitivity to parameters.

Chronologically, the first issue after the 2004 event was to provide coherent explanation on the earthquake mode of rupture. An immense efforts have been made in that direction. Several inversion studies, using all available seismic records (Ammon *et al.*, 2005; Lay *et al.*, 2005), GPS ground displacements records (Chlieh *et al.*, 2005; Vigny *et al.*, 2005), the tide gauge records evaluating local rate of uplift/subsidence being used systematically, have been successfully realized. Then numerous coherent sources have been calibrated with available hydrographic records (JASON altimeter, tide gauges) based on the first guess solutions delivered from seismic and GPS records (Grilli *et al.*, 2007; Sladen and Hebert, 2007; Fujii and Satake, 2007; Piatanesi and Lorito, 2007). Then case studies were simulated along with process studies.

We briefly provide here a review of two studies that deal with coastal vulnerability of the Indian Ocean based on numerical studies. The two studies are chosen because we want to show how numerical studies may complement missing observations and thus increase our knowledge on the repartition of the vulnerability over an area. These two studies are also mentioned because their aim was to provide some vulnerability criteria that stand for any tsunami propagating in the Indian Ocean.The first example is The Andaman coast of Thailand which encountered severe runup around Khao Lak (more than 10 m runups) and non-significant wave heights north of Kho Pratong or south of Trang region (1 m) (Figure 1 for location) (Tsuji *et al.*, 2006). Another example is the weak impact on the Bangladesh coast and the presence of significant waves for the North-East Indian coast which received the tsunami at nearly equivalent incidence.

The six provinces of the Andaman coast of Thailand that have been exposed to the 2004 tsunami have been impacted irregularly: Khao Lak (Figure 1), Phang Nga province, was the most affected with runups ranging from 4 to 15 m (tsujii *et al.*, 2006). Within the same province, observations reveal a strong variability of the wave impact. The international survey teams sampled the segment of the province where human damages have been particularly severe. Numerical simulations performed by Ioualalen *et al.* (2007a) matched the ensemble of observations with a 0.94 correlation coefficient and a 17% standard deviation. The simulations were not constrained with the observed runup values so that they can be qualified as prognostic simulations. About 70 km south of Khao Lak The Island of Phi Phi also experienced severe damages and numerical simulations fit reasonably observations. Between the two areas, no observations have been operated probably because no damage has been recorded, the region being weakly dense in population. It happens that the prognostic numerical results within this area (approx. 40 km south of Khao Lak) indicate 10 m values for the runup. Here is an example showing how numerical experiments may improve drastically the identification of vulnerable areas to tsunami in the absence of observations. For information, the authors found that Khao Lak region encountered large runup because of wave focusing in that specific area.

The coast of Bangladesh has been very weakly impacted by the 2004 tsunami. This is something unexpected because of the tsunami source neighbourhood. Besides, we may expect that the wave directivity (nearly zonal) might have limited the impact. In that case how can we explain that the wave has reached Northern India? The explanation might come from the presence of a very extend continental shelf (up to 200 km of extension). Ioualalen *et al.* (2007b) made some sensitivity tests to estimate the effects of dispersion, nonlineariy and bottom friction within the shelf. It resulted that these physical effects are not crucial in the 2004 wave damping. The simulations reveal that the maxima of wave steepness occur within the continental shelf (and not at the coastline). Later Ioualalen (2008) suggested that the presence of the impressive 'Swatch of No Ground' canyon (Figure 1) might have been crucial in the wave propagation. They suggested in particular that when the tsunami wave reaches the continental shelf, it experiences a huge lateral gradient of ocean depth. Such gradient may re-orient the wave to the sides of the canyon and thus defocus within the shelf. As a result the maxima of wave steepness do not reach the coast. They also proposed that at the canyon termination (near the coast) , the guided wave along the canyon encounters a convex shape bathymetry (divergence) and thus defocuses. The proof of these suggestions has not been established yet but, clearly, the analysis of the numerical simulations encourages to go further in the study of the impact of underwater canyons in the coastal vulnerability to tsunamis. Further theoretical studies or the use of more conceptual models are also another issue for evaluating coastal vulnerability to tsunamis.

ACKNOWLEDGEMENTS

Tha authors would like to thank the French *Agence Nationale Pour la Recherche*, ANR, for funding this work through the grant *TSUMOD* ANR-05-CATT-016-02 M. Ioualalen would addresses his undebtful thanks to the Department *Milieu et Environement* of the *Institut de Recherche pour le Developpement*, IRD, for having granted him a long stay at Madras University, especially R. Louat and J. Boulegue and P. Charvis (Géosciences Azur) for their constant support. They are also grateful to Prof. Efim Pelinovsky for having graciously accepted to review this paper.

References

Ammon, C. J.; Ji, C.; Thio, H. K.; Robinson, D.; Ni, S. D.; Hjorleifsdottir, V.; Kanamori, H.; Lay, T.; Das, S.; Helmberger, D.; Ichinose, G.; Polet, J.; Wald, D. *Science*. 2005, *Vol.* 208, 1133-1139.

Berninghausen, W. H. *Bull. Seism. Soc. Am.* 1966, *Vol.* 56(1), 69-74.

Bock, Y.; Prawirodirgo, L.; Genrich, J. F.; Stevens, C. W.; McCaffrey, R.; Subarya, C.; Puntodewo, S. S. O.; Calais, E. *J. Geophys. Res.* 2003, *Vol.* 109(B8), 2367, doi:10.1029/2001JB00324.

Bondevik, S.; Svenden, J. L.; Johsen, G.; Mangerud, J.; Kalan, P. E. *Boreas*. 1997, Vol. 261, 29-54.

Briggs, R. W.; Sieh, K.; Meltzner, A. J.; Natawidjaja, D.; Galetzka, J.; Suwargadi, B.; Hsu, Y. J.; Simons, M.; Hananto, N.; Suprihanto, I.; Prayudi, D.; Avouac, J. P.; Parawirodirdjo, L.; Bock, Y. *Science*. 2006, *Vol.* 311, 1897-1901.

Chlieh, M.; Avouac, J.-P.; Hjorleifsdottir, V.; Song, T.-R., A.; Ji, C.; Sieh, K.; Sladen, A.; Hebert, H.; Prawirodirdjo, L.; Bock, Y.; Galettzka, J. *Bull. Seism. Soc. Am.* 2007, *Vol.* 97(1), S152-S173.

Choi, B. H.; Hong, S. J.; Pelinovsky, E. *Geophys. Res. Letters*. 2001, Vol. 28, 2013-2016.

Choi, B. H.; Hong, S. J.; Pelinovsky, E. *Geophys. Res. Letters*. 2006, Vol. 33, doi: 0094-8276/06/2006GL025867$05.00.

Choi, B. H.; Pelinovsky, E.; Riabov, I.; Hong, S. J. *Nat. Hazards*. 2002, *Vol.* 25, 1-21.

Fitch, T. J. *J. Geophys. Res.* 1972, *Vol.* 77, 4432-4460.

Fritz, H. M.; Synolakis, C. E. *Proc. 5th Int. Conf. on Ocean Wave Meas. And Analysis, WAVES 2005,* Madrid, Spain. 2005, Paper number 219.

Fujii, Y.; Satake, K. *Bull. Seism. Soc. Am. 2007*, Vol. 97, S174-S207.

Genrich, J. F.; Bock, Y.; McCaffrey, R.; Prawirodirgo, L.; Stevens, C. W.; Puntodewo, S. S. O.; Subarya, C.; Wdowinsky, S. *J. Geophys. Res.* 2003, *Vol.* 105, 722-743.

Gover, J. *EOS* (Wash. D. C.). 2005, Vol. 86(3), 37-38.

Grilli, S. T.; Ioualalen, M.; Asavanant, J.; Shi, F.; Kirby, J. T.; Watts, P. *J. Wtrwy, Port, Coast, and Oc. Engrg.* ASCE. 2007, *Vol.* 133 (6), 414-428.

Gusiakov, V. K. (2005). Basic list of measurements made in Sibolga and Nias Islands. http://www.pmel.noaa.gov/tsunami/indo20041226/sibloga_nias.htm.

Heck, N. H. *Bull. Seis. Soc. Am.* 1947, *Vol.* 37, 269-286.

Ioualalen, M. *Environ. Model. and Softw.* 2008a, doi :10.1016/j.envsoft.2007.07.007.

Ioualalen, M. *La houille Blanche, La Revue Internationale de l'Eau.* 2008b, *Vol.* 1, 27-32, doi :10.1051/lhb:2008002.

Ioualalen, M.; Asavanant, J.; Kaewbanjak, N.; Grilli, S. T.; Kirby, J. T.; Watts, P. *J. Geophys. Res./Oceans*. 2007a, *Vol.* 112, C07024, doi:10.1029/2006JC003850.

Ioualalen, M.; Migeon, S.; Sardou, O. *Geophys. J. Int.* 2009, submitted.

Ioualalen, M.; Pelinovsky, E.; Asavanant, J.; Lipikorn, R.; Deschamps, A. *Natural Hazards and Earth System Sciences*. 2007b, *Vol.* 7, 141-147.

Lay, T.; Kanamori, H.; Ammon, C. J.; Nettles, M.; Ward, S. N.; Aster, R.; Beck, S. L.; Bilek, B. L.; Brudzinski, M. R.; Butler, R.; DeShon, H. R.; Ekstrom, G.; Satake, K.; Sipkin, S. *Science*. 2005, *Vol.* 308, 1127–1132.

Litzin, E. *Elsevier Oceanographic Series*. 1974, *Vol.* 8, New York, 273 pp.

Liu, P. L.-F.; Lynett, P.; Fernando, H.; Jaffe, B. E.; Higman, B.; Morton, R.; Goff, J.; Synolakis, C. S. *Science*. 205, *Vol.* 308, 1595.

Merrifield, M. A.; Firing, Y. L.; Aarup, T.W.; Agricole, W.; Brundrit, G.; Chang-Seng, D.; Farre, R.; Kilonsky, B.; Knight, W.; Kong, L.; Magori, C.; Manurung, P.; McCreery, C.; Mitchell, W.; Pillay, S.; Schindele, F.; Shillington, F.; Testut, L.; Wijeratne, E. M. S.; Caldwell, P.; Jardin, J.; Nakahara, S.; Porter, F.-Y.; Turestky, N. *Geophys. Res. Letters*. 2005, Vol. 32, L09603, doi: 10.1029/2005GL022610.

Michel, G. W.; Yu, Y. Q.; Zhu, S. Y.; Reigber, C.; Becker, M.; Reinhart, E.; Simons, W.; Ambrosius, B.; Vigny, C.; Chamot-Rooke, N.; Le Pichon, .; Morgan, P.; Matheussen, S. *Earth, Planet, Sci. Lett.*. 2001, *Vol.* 187, 239–244.

Newcomb, K. R.; McCann, W. R. *J. Geophys. Res.* 1987, *Vol.* 92(B1), 421-439.

Pelinovsky, E.; Choi, B. H.; Stromkov, A.; Didenkulova, I.; Kim, H. S. *Advances in Natural and Technological Hazards Research*. Springer. 2005a, *Vol.* 23, 57-77.

Pelinovsky, E.; Choi, B. H.; Zaitsev, A.; Didenkulova, I. *Proc. 5th Int. Conf. on Ocean Wave Meas. And Analysis, WAVES 2005,* Madrid, Spain. 2005b.

Piatanesi, L.; Lorito, S. *Bull. Seism. Soc. Am.* 2007, Vol. 97, S223-S231.

Rastogi, B. K.; Jaiswal, R. K. *Science of Tsunami Hazard*. 2006, *Vol.* 25, 128-143.

Sannasiraj, S. A.; Sundar, V. *Proc. 5th Int. Conf. on Ocean Wave Meas. And Analysis, WAVES 2005,* Madrid, Spain. 2005, Paper number 214.

Satake, K.; Aung, T. T.; Sawai, Y.; Okamura, Yukinobu; Win, K. S.; Swe, W.; Swe, C.; Swe, T. L.; Tun, S. T.; Soe, M. M.; Oe, T. Z.; Zaw, S. H. *Earth, Planets Space*. 2006, *Vol.* 58, 243-252.

Sladen, A.; Hebert, H. *Eos Trans. AGU*. 2007, Vol. 86(52), Abstract U22A-07.

Synolakis, C. E.; Fritz, H. M.; Titov, V. V. *Proc. 5th Int. Conf. on Ocean Wave Meas. And Analysis, WAVES 2005,* Madrid, Spain. 2005, Paper number 219.

Tsuji, Y.; Namegaya, Y.; Matsumoto, H.; Iwasaki, S.; Kanbua, W.; Sriwichai, M.; Meesuk, V. *Earth, Planets Space*. 2006, *Vol.* 58, 223-232.

Vigny, C.; Simons, W. J. F.; Abu, S.; Bamphenyu, R.; Satirapod, C.; Choosakul, N.; Subarya, C.; Socquet, A.; Omar, K.; Abidin, H. Z.; Ambrosius, B. A. C. *Nature*. 2005, Vol. *436*, 201-206.`

Yalciner, A. C.; Ghazali, N. H.; Wahab, A. K. A. (2005a). December 26, 2004 Indian Ocean tsunami field survey (July 09-10, 2005) at North west peninsular Malaysia coast, Penang and Langkawi Islands. http://Yalciner.ce.metu.edu.tr/malaysia-survey.

Yalciner, A. C.; Peirncek, D.; Ersoy, S.; Presateya, G.; Hidayat, R.; McAdoo, B.. (2005b). Report on December 26, 2004 Indian Ocean tsunami, field survey on Jan. 21-31 at north of Sumatra. http://Yalciner.ce.metu.edu.tr/sumatra/survey.

Reviewed by

Prof. Efim Pelinovsky,
Department of Nonlinear Geophysical Processes,
Institute of Applied Physics, Russian Academy of Sciences, 46 Uljanov Street,
Nizhny Novgorod, 603950 Russia.

In: Atlantic and Indian Oceans
Editors: E. S. Askew and J. P. Bromley, pp. 125-140 ISBN 978-1-60692-475-4

Chapter 5

ADAPTIVE AND GROWTH OUTCOMES FOLLOWING TSUNAMI: THE EXPERIENCE OF THAI COMMUNITIES FOLLOWING THE 2004 INDIAN OCEAN TSUNAMI

Douglas Paton and Catherine So-kum Tang
University of Tasmania, Tasmania, Australia
National University of Singapore, Singapore

ABSTRACT

Drawing on research conducted in coastal communities in Thailand following the 26th December Indian Ocean tsunami, this chapter discusses how mainstream, pre-existing community characteristics derived from dealing with the challenges posed by daily life predict community members' resilience and their capcity to experience posttraumatic growth in the aftermath of the tsunami impact. The chapter also discusses the implications of these findings for developing cross-cultural understanding of natural hazard resilience. Specifically, it explores the procedures that need to be adopted to assess the cross-cultural validity of resilience and growth (collective efficacy, response efficacy and posttraumatic growth) measures developed in Western, individualistic cultures with members of collectivistic communities. Having established the cross-cultural validity of the measures, the authors proceed to identify the predictors of each outcome. The chapter concludes with a discussion of the need to develop a better understanding of indigenous and culture-specific resilience.

Keywords: Tsunami, Adaptive Capacity, Resilience, Posttraumatic Growth, Cross-cultural.

INTRODUCTION

On December 26 2004, an earthquake measuring 9.3 on the Richter Scale triggered a powerful tsunami along coastal regions of countries bordering the Indian Ocean. This tsunami resulted in some 300,000 deaths, displaced more than 1,600,000 people, and resulted in more than

500,000 people becoming homeless. Overall, this tsunami affected some five million people in Indonesia, Sri Lanka, Maldives, India, Thailand, Seychelles, and Myanmar. Thailand was one of the hardest hit countries. The tsunami resulted in some 5,395 deaths and affected over 60,000 Thai residents.

It is commonly assumed that when community members are exposed to catastrophic disaster events, the experience of substantial levels of posttraumatic pathology is inevitable and affected communities are suddenly rendered completely reliant on external aid for their physicl and psychological recovery. However, such assumptions are not, however, be representative of all possible outcomes for people and communities. A growing body of research attests to the fact that not only can communities adapt to, and cope with, the consequences of challenging traumatic events using their own resources, they may also experience enduring psychological and social growth from their encounter with adversity.

While adaptive and growth outcomes have been found repeatedly in research involving Western populations, this chapter examines whether similar phenomena exist in a non-Western population. The scale of the 26th December tsunami exceeded, by a considerable margin, the capacity of the formal response to meet all the needs of those affected. It thus created a context in which it was possible to examine whether, and to what extent, members of Thai communities affected by the tsunami demonstrated a capacity to adapt to, and experience a sense of social and psychological growth from, the challenging circumstances in which they found themselves (Paton et al., 2008; Tang, 2007).

Adaptive Capacity and Collective Efficacy

Adaptive capacity describes people's ability to draw upon resources and competences that already exist within their community to confront, cope with, and adapt to, the losses and disruption created by hazard activity (Klein, Nicholls, and Thomalla, 2003; Paton, 2006). When disaster strikes, community members are compelled to confront demands that differ radically in scale, complexity and duration from anything they would encounter under normal circumstances. It also requires people to function in the absence of the community and societal resources normally available to them. Given the typical suddenness with which disasters such as the December 26th tsunami strike, community members' capacity to effectively confront these consequences will reflect, to some extent at least, the social competencies that developed from dealing with the everyday challenges that characterize community life.

The community characteristics discussed in this chapter reflect competencies derived from peoples' everyday experiences (e.g., participation in activities with other members of their community) and thus allow assessment of the degree to which mainstream aspects of community life influence whether or not people chose to develop their disaster resilience. This approach is based on the premise that mainstreaming the development of risk management strategies with other community development activities increases the likelihood that a sustained capacity to adapt to infrequent hazard events will develop.

One such social competence is collective efficacy. Collective Efficacy is the shared belief that a group can effectively meet environmental demands and improve their lives through concerted effort (Bandura, 1997). As a result of its potential to transcend cultural boundaries (see below), this construct is capable of informing our understanding of resilience, if it can predict the ability of tsunami-affected populations to adapt to, and cope with, the consequences of this disaster.

Posttraumatic Growth

Posttraumatic growth (PTG) refers to the outcome of experiencing positive changes following challenging life events (Linley and Joseph, 2004; Tedeschi and Calhoun, 2004). PTG represents a movement beyond pre-event levels of adaptation, and is typically manifest in an increased appreciation of life, a sense of greater personal strength, improved relationships with others, and a more integrated philosophy of life (Tedeschi and Calhoun, 1996). A wide variety of events have been found to trigger PTG (Linley and Joseph, 2004; Tedeschi and Calhoun, 2004). For example, one US study found that 90-95% of surveyed survivors indicated positive benefits at 4-6 weeks, as well as at 3 years, following a tornado impacting on their community (McMillen, Smith, and Fisher, 1997). Prominent predictors of PTG are survivor characteristics (e.g., gender, age, personality, coping style), cognitive-emotional appraisal of traumatic events, pattern of traumatic stress responses, and characteristics of the socio-cultural (e.g., support from family and community sources) environment (Tang, 2007). Traumatic stress responses such as intrusive, hyper-arousal, and avoidant symptoms also show positive associations with PTG in trauma survivors. Despite an accumulation of literature on PTG, little is known about positive changes in the aftermath of natural disaster across cultural groups. To date, no research has been undertaken to investigate PTG in Asian survivors of natural disaster. This chapter examines whether a measure (collective efficacy) known to predict disaster resilience and an established measure of posttraumatic growth provide insights into how Thai citizens' experience of disaster can be resolved in positive ways in a manner comparable to those recorded in their Western counterparts.

The Challenges of Researching Resilience in Cross-cultural Contexts

When researching resilience and posttraumatic growth in cross-cultural contexts, the research process must accommodate several issues not encountered when conducting comparable research in Western communities. One important question concerns whether any proposed predictor variables have universal (i.e., cross-cultural) applicability. That is, whether existing measures remain valid and reliable when used to assess resilience in populations whose cultural (e.g., collectivist versus individualist) and belief (e.g., Buddhist) systems differ from those of the Western populations in which the measures were developed. Careful attention must first be paid to the equivalence of constructs before using them in research (Brislin, 2000; Norenzayan and Heine, 2005). If these measures can demonstrate cross-cultural validity, it opens the way to using existing models and measures of resilience and growth in cross-cultural contexts, providing a cost-effective approach for assessing these phenomena in all cultures (Brislin, 2000; Dasen and Mishra, 2000).

Cultural Influences on Collective Efficacy and Growth

The Collective Efficacy measure (Zaccaro, Blair, Peterson, and Zazanis, 1995) discussed here had proven its utility as a predictor of adaptive behaviour in disaster studies undertaken with western populations (Benight, 2004) characterised by an individualistic cultural orientation. Within individualist cultures, collective efficacy reflects the degree to which people choose the nature and level of collaboration and cooperation they have with others. In contrast, in collectivist cultures (e.g., Thailand), cooperation is motivated by culturally-embedded beliefs that emphasize social relations and engaging in activities related to achieving collective goals (Diener and Suh, 2000; Triandis, 1995; 1996).

Cross-cultural differences in social motivations for collaboration and cooperation (i.e., personal choice versus cultural predisposition) mean that the cultural equivalence of any scales used for assessment cannot be assumed (Brislin, 2000; Diener and Suh, 2000; Norenzayan and Heine, 2005; Poortinga, 1997). In other words, it cannot be assumed that knowledge of factors that contribute to resilience in individualistic cultures, in which people act consistently across situations in accordance with a self-concept that is relatively independent of social situation and in which achieving personal goals is a prominent objective, will automatically be applicable to members of collectivist societies. The foundation for assuming that this collective efficacy measure could be used for conducting exploratory research into resilience with members of a collectivist culture was the fact that it was designed to reflect community members' perceptions of their collective capabilities (Benight, 2004). Similar issues confront the researcher interested in the cross-cultural investigation of PTG.

Three major explanatory models have been proposed to explain how and why individuals experience PTG following the experience of significant adverse life events; the biopsychosocial-evolutionary model (Christopher, 2004), the organismic valuing model (Joseph and Linley, 2005), and the process model (Tedeschi and Calhoun, 2004). When natural disaster occurs, it not only generates massive destruction and casualties within a short period of time, it also greatly challenges individuals' perceptions of both their own abilities and their assumptions about the benevolence, predictability, and controllability of the world. According to these models, the experience of a disaster can provide a unique opportunity for growth as individuals rebuild their assumptive world with new meanings. However, the bulk of PTG research has been conducted with trauma survivors residing in Western countries. As specific dynamics of PTG may vary with the degree of disparity between pre-existing culturally shaped beliefs and trauma-related information, there is a need to examine PTG in more diverse ethnic and cultural groups following severe natural disaster to determine whether PTG represents a universal psychological construct.

All three models of PTG argue that people's intrinsic motivation toward growth has a strong socio-cultural foundation. However, as with the collective efficacy measure, the cross-cultural equivalence of the measure cannot be assumed. There exist several reasons why the equivalence of the PTG construct must be assessed. For example, Western conceptualisations of the relationship between traumatic events and growth tend to underestimate cross-cultural differences in the role of suffering in human development and growth and differences in the role of the social context in which events are rendered meaningful.

The Buddhist perspective, the major religious force in Asian cultures, denies the reality of a permanent self and views all things pertaining to the phenomenal world as undergoing constant transformation (Daya, 2005). It views human growth as remaining present with one's suffering without trying to flee from it. Thus, depending on one's culture, traumatic events may actually validate rather than violate one's assumptive worldviews, and human suffering may be viewed as inevitable and essential to achieving self and spiritual transcendence/ growth.

Cultures can also be differentiated with regard to their dominant ideas about the ontology of self as well as about relationships between self and others, between self and the universe, and between life and death (Markus and Kitayama, 1991; Sampson, 2000). For example, the inter-dependent concept of self common in many Asian cultures results in their conceptualizing themselves living in a network of interpersonal relationships, with family relations being the most dominant influence. When traumatic events happen, they affect the whole network rather than just the individual. Culture may thus influence any predisposition for growth.

Consequently, while sound theoretical reasons for assuming cross-cultural equivalence of both measures exists, this must be established before they could be used. The widespread impact of the tsunami in coastal Thai communities provided the context to test the cross-cultural equivalence of resilience and PTG.

The Impact of the Tsunami on Thai Communities

The investigation of resilience and growth took place in communities in six provinces (Ranong, Phang Nga, Phuket, Krabi, Trang and Satun) located along Thailand's Andaman coast that borders the Indian Ocean between April and June 2005 when recovery activities were at their height (Paton et al., 2008; Tang, 2007). In addition to their having had direct experience of the tsunami impact (e.g., some 90% of the residents had to evacuate their homes, and 27% lost at least one family member), participants also had to confront the recovery issues associated with the widespread damage to their home province. All respondents were affected by substantial loss to both physical infrastructure and disruption to sources of economic livelihood (e.g., fishing, agricultural production, tourism). Conducting the research during the period of recovery, when residents were actively confronting the substantive consequences of the tsunami impact, provided an opportunity to gain valuable insights into how community members were adapting to this circumstance and to assess whether the consequences of this disruption acted as a catalyst for psychological and social growth. As mentioned earlier, conducting cross-cultural research into resilience and PTG poses challenges that are not encountered when conducting research with Western populations. These challenges concern the preparation of the scale for data collection and assessing the cross-cultural equivalence of the scales.

EVALUATING THE CROSS-CULTURAL EQUIVALENCE OF THE MEASURES

The first step in cross cultural work involves preparing scales for data collection in a language and cultural context that differs from that in which the scale was originally developed. The procedure (Brislin, 1986) involves first translating the English version into Thai. Next, after translating it back into English, the original and back-translated versions are compared, examined for meaning errors, and corrections made as required to provide a Thai language-adapted measure that is equivalent to the original version. Only after this procedure has been followed is it possible to use these measures to conduct cross-cultural research.

Collective Efficacy

To determine the cross-cultural equivalence of the collective efficacy scale, its face validity (the degree to which Thai participants perceived the items to be appropriate for them) was first assessed. Positive feedback from participants regarding its face validity and the results of a principal components analysis demonstrated that the scale comprised a coherent set of items. The collective efficacy scale had sufficient cross-cultural equivalence to warrant its use as a potential predictor of resilience (adaptive capacity) with a Thai population. The collective efficacy scale comprised three components; organization (e.g., ability to coordinate community activities, resolve conflicts), cooperation (e.g., ability to work well with others), and responsibility (e.g., the community is able to take responsibility to deal with a crisis).

Posttraumatic Growth

Participants' perceived positive changes resulting from the earthquake-tsunami were assessed using a version of the Posttraumatic Growth Inventory (Tedeschi and Calhoun, 1996) adapted for Chinese samples (Ho, Chan, and Ho, 2004). In this version, items of the inventory clustered into four dimensions: (1) change in interpersonal relationship; (2) change in self, including willingness to express emotions, acceptance of how things work out in life, and developing new interests and opportunities; (3) change in life orientation and priority; and (4) change in religious beliefs and spiritual life. These four dimensions could in turn be broadly dichotomized into an interpersonal and an intrapersonal dimension. Confirmatory factor analysis revealed a moderately acceptable fit for this measure with a Thai population (Tang, 2007). Having established the equivalence of the measures, they could then be used to examine the resilience/adaptive and growth experiences of Thai citizens in the aftermath of the tsunami. Discussion commences with an examination of resilience.

Resilience in Tsunami Affected Communities

Operationalizing and Predicting Resilience

The next question is whether Collective Efficacy predicts the resilience of Thai citizens. To do so, it is first necessary to operationalise resilience (measured here as their capacity to adapt to the tsunami consequences they encountered).

To be resilient is to be able to adapt to whatever atypical demands a community encounters. Before it is possible to test predictors of resilience, it is necessary to identify what adapting to disaster demands would look like (Klein, Nicholls, and Thomalla, 2003). The circumstances that community members are compelled to confront differ radically from those they have to deal with under normal circumstances. Operationalising what people must adapt to is complicated by the diverse range of specific issues that people could have to contend with. For example, a common challenge for disaster-affected communities is dealing with the economic disruption wrought by the disaster. However, assessing how well people can adapt to, or cope with, this consequence is complicated by the fact that, even within a given area, some people may have to adapt to disruption of tourism, while others would have to contend with the loss of a fishing fleet or agricultural production. This diversity raises a new challenge. If people's resilience is to be assessed, any measure must be able to transcend the range of specific circumstances that comprise this diversity and so facilitate comparability between groups. One approach is to develop more generic measures of what people have to adapt to. For example, Paton, Millar and Johnston (2001) found that resilience could be assessed in ways that accommodated community diversity using questions that tapped into community member's ability to, for example, reconstruct or develop alterative sources of livelihood or being able to secure the resources necessary to meet community needs. A similar approach was adopted here.

For example, in the aforementioned example of disruption to livelihood, it would be possible to compare groups experiencing diverse economic impacts (e.g., fishing versus tourism) by developing a generic measure that included questions to assess, for example, how effective community members were in 'reconstructing livelihood' or 'developing alternative sources of livelihood' in the aftermath of a disaster. Generic measures framed in this way facilitate comparison across groups, irrespective of the specific economic disruption encountered. Similarly, a community capability to 'secure the resources they require to meet their unique needs' (irrespective of whether these resources are secured from other communities, government or relief agencies) would represent another activity that could be included in a scale to measure how effectively people were adapting irrespective of the specific circumstances that communities had to contend with. These generic measures provide a basis for systematically examining the utility of personal and community variables to predict whether, how and to what extent people can adapt by, for example, 'reconstructing livelihood.' However, only when resilience is operationalised can predictors be tested.

Community members were first invited to describe the activities they believed allowed them to deal effectively with the demands they faced during the recovery period. This process identified three factors consistently (Paton et al., 2008): the ability to actually help those affected by the tsunami; being able to secure the resources required to meet their specific needs from outside the community when they were actually required; and the degree to which they could utilise support from government institutions to allow them to meet their own needs. These items provide a measure of resilience that describes what was involved in adapting to the challenging circumstances encountered during the disaster. The fact that these activities were identified while participants were actively involves in their own disaster recovery conferred upon these items considerable face validity. Factors analysis confirmed the utility of this measure (Paton et al., 2008).

The resilience measure is assessing the degree to which community members can adapt to the practical recovery demands posed by the consequences of the tsunami. The question is then, does collective efficacy, which taps into an enduring aspect of the fabric of community life that reflects peoples' beliefs regarding their collective ability to confront challenging circumstances under normal circumstances, predict resilience? The results are summarised in Figure 1 (full details of the analyses can be found in Paton et al., 2008). Multiple regression analyses revealed that organization (Beta = .506) accounted for 31.5% of the variance in resilience. While, responsibility (Beta = .191) accounted for an additional 3.3% of the variance, the addition of cooperation did not account for any additional variance in response capability. Together, organization and responsibility accounted for 35% of the variance in resilience and supports the argument that this Collective Efficacy measure can be used to predict resilience in Thai communities.

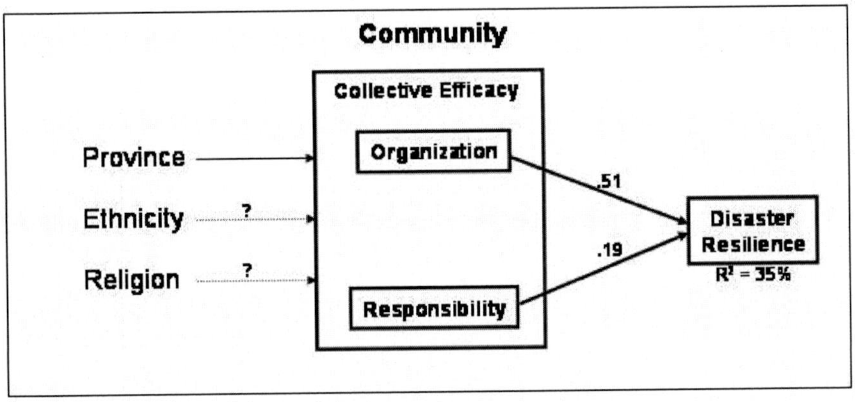

Figure 1. The role of collective efficacy as a predictor of resilience in Thai survivors of the 2004 Indian Ocean tsunami. Adapted from Paton et al. (2008).

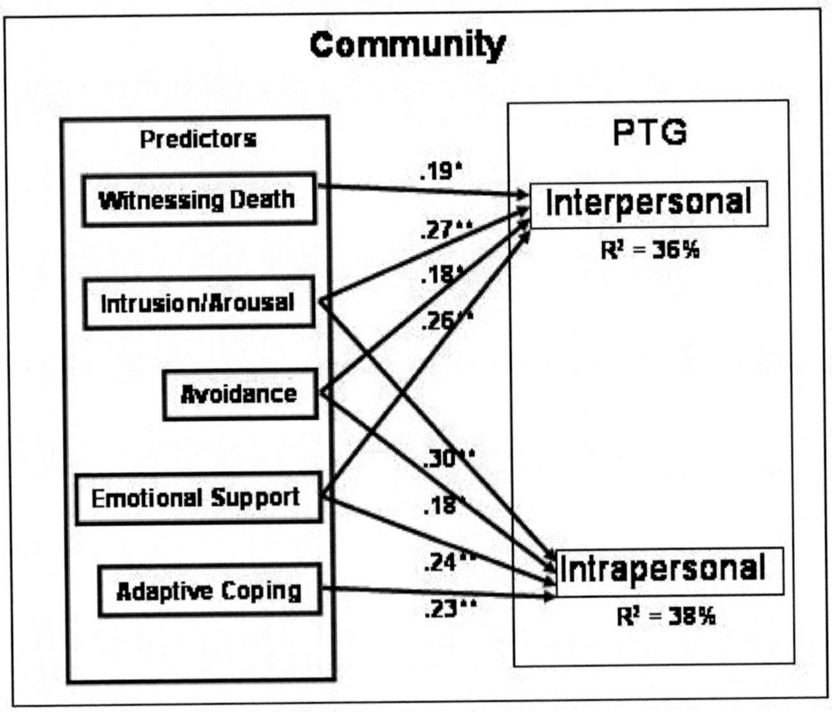

Figure 2. Predictors of interpersonal and intrapersonal aspects of posttraumatic growth (Chinese version) in Thai survivors of the 2004 Indian Ocean tsunami. Adapted from Tang (2007).

The fact that collective efficacy reflects levels of a pre-existing community competence, derived from dealing with day-to-day challenges encountered within a community begs the question of whether it is possible to identify the community activities that represent its source. Two possibilities were examined. One source could be the characteristics of community life. For example, in communities in which collective involvement in traditional forms of livelihood (e.g., fishing or agriculture) prevail, involvement in the collective aspects of these activities could, it can be hypothesised, afford different opportunities for confronting problems or accessing resources that are required to deal with the challenges that community members encounter in everyday life. In contrast, in communities in which new forms of livelihood (e.g., tourism), that rely less on traditional collectivist social structures, could lessen collective efficacy. Collective efficacy can also be sustained by social institutions that sustain traditional ways of life and beliefs.

In collectivist societies, prominent influences on people's beliefs regarding their ability to deal with everyday challenges are the socio-cultural institutions that structure life in ways that facilitate and sustain cooperative action in community life (Brislin, 2000; Jang, 2005). Prominent sources of this predisposition are ethnic and religious affiliation (Jang, 2005). Data were collected on ethnicity (Thai and Chao Lay) and religious affiliation (Muslim and Buddhist). A three-way (province, ethnicity, religious affiliation) between-groups analysis of variance found that only the main effect for province could account for differences in prevailing levels of collective efficacy (Figure 1). Full details of the analyses can be found in Paton et al. (2008).

Predicting Posttraumatic Growth in Thai Communities

Prevalence rates for PTG were assessed at six months following the Indian Ocean tsunami. When the least stringent criteria were used (mean score $\geqq$ 1 on the PTG Inventory), some 88% of participants reported at least some degree of positive change in their PTG scores. When more stringent criteria were adopted (mean score $\geqq$ 3), some 32% and 37% of participants reported at least moderate degrees of positive changes in interpersonal and intrapersonal dimensions of the PTG scale respectively. Among those reporting at least moderate levels of PTG (n = 59), 51% also met the full criteria for PTSD according to the DSM-IV (APA, 1994).

Hierarchical multiple regression analyses were conducted to determine the most salient predictors of PTG. Results (Figure 2) showed that all predictors together explained 36% and 38% of variances in interpersonal and intrapersonal PTG, respectively (Tang, 2007).

Resilience and PTG in Thai Communities

Resilience and adversarial growth are realities for Thai community members who found themselves having to confront the consequences of a disaster of epic proportions. In this section, the implications of the findings of the research are discussed.

Collective efficacy is a valid predictor of resilience. This finding has several important implications. First, the fact that this collective efficacy scale has predictive utility within a population with cultural characteristics that differ substantially from those on whom the scale was developed fuels the argument that some predictors of community resilience represent universal aspects of the human spirit. Secondly, since no participant had had any disaster readiness training, the levels of collective efficacy brought to bear on the circumstances that these Thai residents had to contend with reflect experiences accumulated in the course of the day-to-day community life. This view is tentatively reinforced by the finding of provincial differences in levels of collective efficacy. It was proposed above that observed levels of collective efficacy could reflect involvement in traditional, collective ways of living (e.g., fishing). While additional work is needed to test this contention, support for this possibility comes from the observation that differences in levels of collective efficacy were associated with different sources of livelihood. For example, collective efficacy scores were significantly higher in provinces (e.g., Phang Na, Ranong, Trang) in which there were more traditional sources of livelihood (e.g., fishing) than there were in areas (e.g., Krabi, Phuket) in which tourism has developed as a significant contributor to the provincial economy. This possibility remains tentative, until more searching analysis of the predictors of resilience is undertaken.

Even though not developed or sustained for this express purpose, collective efficacy can be pressed into service to deal with extraordinary circumstances, such as those that prevailed during the aftermath of the disaster. If this position is accepted, it means that people's day-to-day experience of confronting issues within normal community contexts (e.g., dealing with the challenges posed by changes in fishing or agriculture) over time can translate into a capacity to effectively confront more significant disaster consequences, even in the absence of any specific disaster preparedness programs because they increase community members' shared belief that their community can effectively meet environmental demands and improve their lives through concerted effort (Bandura, 1997).

The fact that a level of disaster resilience can reflect the operation of a pre-existing community resource provides some valuable insights into how risk management strategies can be developed and implemented. For example, it provides opportunities to increase people's potential to adapt by focusing on community development activities (Anckermann, et al., 2005; Paton, 2000, 2005). Adopting this approach can have other benefits. For example, strategies based on community development have the advantage of being more likely to be perceived, by community members and civic authorities alike, as offering a solution that has immediate benefits, by facilitating the development of social capital that will show a return on investment in everyday life, and not just in the event of the occurrence of a disaster at some indeterminate time in the future (Paton, 2005). Furthermore, adopting this approach will ensure that community capacity can be sustained in ways that accommodate community characteristics (e.g., livelihood, size, administrative arrangements etc.,), community needs and expectations, and, in particular, the in-group/out-group differences (e.g., livelihood, ethnic, religious) that make important contributions to reconciling the influence of these important community dynamics with the process of sustaining adaptive capacity in collectivist societies.

Knowledge that collective efficacy can predict resilience provides emergency management, community development, humanitarian aid and other agencies with a role to play in disaster risk management, with one means of assessing community resilience to disasters. Scale content can be used to identify the content (e.g., developing a capacity for self-organization within a community) of the risk management programs that could complement community development initiatives (Paton, 2005). Furthermore, if the contention that levels of collective efficacy reflect the outcomes of ways of life that reflect traditional, collective activity (e.g., fishing, agriculture), audits of livelihood activities throughout a region can inform the development of peoples' adaptive capacity. It follows from this line of argument that intervention can also focus on increasing or maintaining levels of collective problem solving through the establishment of local industry groups (e.g., encouraging membership of tourism associations). Because it is possible to identify how livelihood activities are distributed throughout at-risk regions, this knowledge could be used to assist decision making about the allocation of recovery resources (e.g., to areas with lowest collective efficacy and, therefore, with relatively greater need for assistance).

Not only was a capacity to effectively confront recovery issues a reality for many Thai communities, a sizable proportion also experienced positive, beneficial changes in spite of the devastating disaster. This finding has implications for mental health delivery, particularly with regard to developing strategies that can promote and sustain beneficial outcomes.

Consistent with the findings of other research (see review by Linley and Joseph, 2004), intrusive and hyper-aroused states, more commonly assumed to be automatic predictors of posttraumatic pathology, were salient predictors of PTG. Clearly, the automatic assumption of a link between intrusive and hyper-aroused states and posttraumatic pathology must be challenged and alternative explanations sought. It may be that repetitive ruminations and intrusive images enable the storage of disaster-related information in active memory for reappraisal and revision of schema that manifest as a growth experience (Creamer, Burgess, and Pattison, 1992; Linley and Joseph, 2004).

An interesting finding was the fact that PTG and PTSD were not mutually exclusive (see also Salter and Stallard, 2004). Some 50% of the tsunami survivors who reported PTG also fulfilled the criteria for DSM-IV PTSD. This reinforces other work indicating the potential for positive (PTG) and negative (PTSD) states to co-exist (Linley and Joseph, 2004). This is an important issue. For example, it means that PTG strategies will be required to increase the likelihood of PTG outcomes, but these strategies may have little impact on managing PTSD and vice versa. Strategies to facilitate PTG outcomes and minimise PTSD must be developed and be capable of implementation. Of particular importance in this context is that aid and mental health agencies are aware of the potential for psychological and social growth to occur in the first place. If not, intervention, offered under the assumption of universal vulnerability to PTSD, may undermine natural coping strategies and, ironically, create vulnerability in those who may otherwise have experienced more salutary outcomes from naturally-occurring recovery processes (as happened in Thailand).

Thus, the clinical goal of post-disaster mental health management should also include the facilitation of PTG in addition to minimizing distressing symptoms. The predictors of PTG (e.g., social and family support) reflect the action of social and psychological processes developed and sustained from experience of everyday community and cultural life. PTG can be influenced by the availability of social support networks in the aftermath of trauma and adversity (Christopher, 2004; Joseph and Linley, 2005; Tedeschi and Calhoun, 2004). This is of particular relevance to Asian collectivist cultures that emphasis mutual support among family members and kinships at all times (Markus and Kitayama, 1991; Sampson, 2000). Indeed, perceived family emotional support and being married were salient predictors of both inter- and intrapersonal PTG. However, instrumental support from the family and the government was found to be unrelated to PTG in this study. It may be that Asian cultural and religious thinking tends to view human suffering as inevitable, and individuals should endure hardship with dignity rather than trying to flee from it with external assistance (Daya, 2005; Jang, 2005). In fact, researchers have also suggested that PTG may be impeded by overly supportive social environment in which survivors are precluded from developing their autonomy and competence to integrate new trauma-related information and to devise their own adaptive coping strategies (Linley and Joseph, 2004). This is consistent with the findings from the adaptive capacity study in Thailand. People identified that being empowered by external sources, rather than having aid thrust upon them, was a crucial element of their resilience.

Adaptive coping was a specific predictor of intrapersonal PTG, especially with regard to participants reporting having a sense of greater personal competency and a more spiritual life. Coping strategies that emphasize active re-interpretation and problem solving are positively related to PTG (Manne et al., 2004; Sears, Stanton, and Danoff-Burg, 2003).

Survivors should be educated about the role of intrusive disaster-related images and thoughts in growth experience, and ways to channel aroused symptoms toward constructive activities and strengthening of support networks. In large-scale, severe natural disasters such as the 2004 earthquake-tsunami, it may affect the entire family and community and deplete its ability to provide support to affected individuals. Post-disaster programs should thus make use of the influx of funds to assist the strengthening of family relations and community social support network in affected regions. Future study should also pay specific attention to the relationship between Buddhist doctrines on suffering and existence and resilience and growth outcomes. The latter view highlights a need for cultural, as well as cross-cultural, understanding of resilience.

Cultural Influences on Resilience

There are two general ways to investigate cultural phenomena (Brislin, 2000). The first, and the one that has been the focus of this chapter, is to use measures whose cross-cultural equivalence can be demonstrated. This approach provides the most cost-effective approach to investigating the phenomena of interest. In the present case, the work discussed in this chapter provides community development, emergency management, aid and mental health agencies with a cost-effective means of advancing both their understanding of resilience and posttraumatic growth and their ability to implement strategies to develop the potential to realise these outcomes in communities susceptible to experiencing natural hazard impacts.

The contents of this chapter represent a foundation for future work on identifying other cross cultural predictors of resilience. There is, however, a second approach that can be used to inform understanding of the relationship between culture, resilience and posttraumatic growth. This involves investigating whether culturally-specific predictors of resilience exist. While the cost of this approach tends to mitigate against its widespread use, there are good grounds for giving this approach a more prominent position in future disaster research agenda. The benefits that can accrue from this approach can be illustrated with reference to some recent work in Taiwan (Jang, 2005).

Jang discussed how an indigenous cultural predisposition in Tung Shih in Taiwan, the Hakka Spirit, represented a significant cultural influence on how people respond to disaster. The *Hakka Spirit* (the *spirit of sturdy neck*) means to hold on firmly, despite extreme adversity. The Hakka Spirit derives from a unique set of cultural beliefs and practices that has its foundation in Taiwanese peoples' belief that natural disasters reflect the will of gods. They do not try to control nature, but strive to live with it in a harmonious way (Jang, 2005). The beliefs and practices that reflect the co-existence with nature are incorporated into a unique set of cultural beliefs - the Hakka Spirit.

Following an earthquake, some 25% of the survivors in Tung Shih demonstrated high levels of posttraumatic growth, with the Hakka Spirit being a significant predictor of this outcome (Jang, 2005). The Hakka Spirit represents a set of social beliefs that increase the capacity of people to thrive in adversarial situations, and, when enacted, encourages performance in a manner that promote resilience and growth (Jang, 2005). It consists of qualities such as frugality, diligence, self-reliance, responsibilities, persistence and other qualities. It is many things: a lifestyle, a belief, a history of their ancestors, and the essence of their unique culture that has been developed over time. In addition to providing valuable insights into specific facets of cultural life, the investigation of culturally-specific phenomena may also inform the wider understanding of resilience and posttraumatic growth in Western populations.

Parallels can be drawn between the elements of the Hakka Spirit and variables implicated as predictors of resilience in the literature (Paton, 2006). For example, perseverance despite adversity bears some relationship to the constructs of collective efficacy and self efficacy. Likewise, some similarity between the Hakka Spirit's focus on responsibilities and issues rather than on emotions and the constructs of problem-focused/adaptive coping is evident, as is a possible link between optimism, an underlying sense of self-reliance, social responsibility and commitment to family and neighbours and social/family support. The difference is that for those for whom the Hakka Spirit underpins a way of life, the beliefs and competencies that facilitate their resilience is transmitted and sustained through cultural mechanisms.

Conclusion

The work cited in this chapter demonstrated that a capacity to adapt and experience a sense of psychological and social growth from large disaster impacts was a reality for many Thai citizens. These findings have several implications. At a general level, these findings demonstrate the cross-cultural validity of several predictors of salutary outcomes following highly challenging experiences. At a specific level, the findings identify how adaptive capacity can be assessed (e.g., using the measure of collective efficacy), provide insights into how collective efficacy is sustained over time (e.g., through patterns of economic livelihood), and provide a foundation for the cost-effective development of resilience by building on, or complementing, pre-existing activities through community engagement and development programs.

The approach adopted in this chapter focused on understanding how a mainstream competency (collective efficacy) could influence resilience to tsunami consequences. The fact that a level of resilience in the face of the devastation created by the tsunami can reflect the influence of pre-existing community characteristics and competencies supports the view that the effectiveness of risk management strategies can be enhanced by integrating them with mainstream community development activities (Anckermann, et al., 2005; Paton & Bishop, 1996; Rich et al., 1995). Risk management strategies that dovetail with community development activities are more likely to be perceived, by community members and civic authorities alike, as offering a solution that has immediate benefits, by facilitating the development of social capital that will show a return on investment in everyday life, and not just in the event of the occurrence of a disaster at some indeterminate time in the future.

Building on capacities developed through mainstream activities increases the likelihood that some level of resilience will be sustained over time. That is, resilience can be forged and sustained through community engagement in activities concerned with identifying and dealing with local issues even if they have little or nothing to do with hazard readiness per se (e.g., farming in the case of the Hakka). Participation in identifying shared problems and collaborating with others to develop and implement solutions to resolve them engenders the development of resilience competencies like collective efficacy that, in turn, will facilitate the development of a capacity to sustain societal functioning during a disaster.

In addition to providing insights into the benefits that can accrue from the cross cultural application of Western findings, future study should also pay specific attention to how knowledge of the role of culture-specific factors (e.g., the Hakka Spirit, the relationship between Buddhist doctrines on suffering and existence and resilience and growth outcomes) can inform understanding of resilience and how it is sustained. As the world becomes a more dynamic place, questions of sustainability becoming increasingly more important. In order to emulate the levels of sustainability found when investigating resilience in indigenous context, Western populations will require more active community engagement and development, if they are to achieve the same capacity for adaptation and growth. The investigation of cross-cultural similarities and differences remains an area for future research. The contents of this chapter suggest that such inquiry will have highly positive implications for understanding and promoting adaptation and growth in all cultures.

REFERENCES

American Psychiatric Association (1994) *Diagnostic and Statistical manual for Mental Disorders*, 4th Ed. Washington D.C. American Psychiatric Association.

Anckermann, S., M. Dominguez, N., Sotol, F. Kjaerulf, P. Berliner and Mikkelsen, E.N. (2005). Psycho-social Support to Large Numbers of Traumatized People in Post-conflict Societies: An Approach to Community Development in Guatemala. *Journal of Community and Applied Social Psychology, 15*, 136–152.

Bandura, A. (1997). *Self-efficacy and agency of change*. New York: Raven Press.

Benight, C. (2004). Collective Efficacy following a series of natural disasters. *Anxiety, Stress, and Coping,* 17, 401-420.

Brislin, R. (2000). *Understanding culture's influence on behaviour*. Melbourne: Wadsworth.

Brislin, R. (1986). The wording and translation of research instruments. In W. Lonner and J. Berry (Eds.) *Field methods in cross-cultural research*. Newbury Park, CA.: Sage.

Christopher, M. (2004). A broader view of trauma: A biopsychosocial-evolutionary view of the role of the traumatic stress response in the emergence of pathology and/or growth. *Clinical Psychology Review, 24,* 75-98.

Creamer, M., Burgess, P., and Pattison, P. (1992). Reactions to trauma: A cognitive processing model. *Journal of Abnormal Psychology, 101,* 452-459.

Davis, C.G., Nolen-Hoeksema, S., and Larson, J. (1998). Making sense of loss and benefiting from the experience: Two construals of meaning. *Journal of Personality and Social Psychology, 75,* 561-574.

Dasen, P.R. and Mishra, R.C. (2000). Cross-cultural views on human development in the third millennium. *International Journal of Behavioural Development, 24,* 428-434.

Daya, R. (2005). Buddhist moments in psychotherapy. In R. Moodley and W. West (Eds.), *Integrating traditional healing practices into counseling and psychotherapy* (pp.182-193). Thousand Oaks, CA: Sage Publications, Inc.

Diener, E. and E.M. Suh (2000). *Culture and subjective well-being.* Cambridge, MA.: MIT Press.

Frazier, P., Conlon, A., and Glasser, T. (2001). Positive and negative life changes following sexual assault. *Journal of Consulting and Clinical Psychology, 69,* 1048-1055.

Ho, S.M., Chan, C., and Ho, R. (2004). Posttraumatic growth in Chinese cancer survivors. *Psycho-Oncology, 13,* 377-389.

Jang, L. J. (2005). *The 921 Earthquake: A study of the effects of Taiwanese cultural factors on resilience.* Unpublished doctoral dissertation, University of Denver, Denver.

Joseph, S., and Linley, P.A. (2005). Positive change to threatening events: An organismic valuing theory of growth through adversity. *Review of General Psychology, 9,* 262-280.

Klein, R., Nicholls, R. and Thomalla, F. (2003). Resilience to natural hazards: How useful is this concept? *Environmental Hazards* 5, 35-45.

Linley, P.A., and Joseph, S. (2004). Positive change following trauma and adversity: A review. *Journal of Traumatic Stress, 17,* 11-21.

Manne, S., Ostroff, J., Winkel, G., Goldstein, L., Fox, K., and Grana, G. (2004). Posttraumatic growth after breast cancer: Patient, partner, and couple perspectives. *Psychosomatic Medicine, 66,* 442-454.

Markus, H.R., and Kitayama, S. (1991). Culture and the self: Implications for cognition, emotion, and motivation. *Psychological Review, 98,* 224-253.

McMillen, J.C., Smith, E.M., and Fisher, R.H. (1997). Perceived benefit and mental health after three types of disaster. *Journal of Consulting and Clinical Psychology, 63,*733-739.

Norenzayan, A. and Heine, S.J. (2005). Psychological universals: What are they and how can we know? *Psychological Bulletin,* 131, 763-784.

Paton, D. (2000). Emergency Planning: Integrating community development, community resilience and hazard mitigation. *Journal of the American Society of Professional Emergency Managers* 7, 109-118.

Paton, D. (2005). Community Resilience: Integrating Hazard Management and Community Engagement. Proceedings of the International Conference on Engaging Communities. Brisbane: Queensland Government/UNESCO.

Paton, D. (2006). Disaster Resilience: Integrating individual, community, institutional and environmental perspectives. In D. Paton and D. Johnston (Eds.), *Disaster Resilience: An integrated approach* (pp. 305-318). Springfield, Ill., Charles C. Thomas.

Paton, D., Gregg, C.E., Houghton, B.F., Lachman, R., Lachman, J., Johnston, D.M., and Wongbusarakum, S. (2008). The impact of the December 26th 2004 tsunami on coastal Thai communities: Assessing adaptive capacity. *Disasters.* In press.

Paton, D., Millar, M., and Johnston, D. (2001). Community Resilience to Volcanic Hazard Consequences. *Natural Hazards, 24,* 157-169.

Poortinga, Y. (1997). Towards convergence? In J. Berry, Y. Poortinga, and J. Pandey (Eds.), *Theory and method: Vol. 1. Handbook of Cross Cultural Psychology* (2nd ed.) (pp. 347-387). Boston, MA: Allyn and Bacon.

Rich, R.C., Edelstein, M., Hallman, W.K. & Wandersman, A.H. (1995). Citizen participation and empowerment: the case of local environmental hazards, *American Journal of Community Psychology, 23*, 657-677.

Salter, E., and Stallard, P. (2004). Posttraumatic growth in child survivors of a road traffic accident. *Journal of Traumatic Stress, 17,* 335-340.

Sampson, E.E. (2000). Reinterpreting individualism and collectivism: Their religious roots and monologic versus dialogic person-other relationship. *American Psychologist, 55,* 1425-1432.

Sears, S.R., Stanton, A.L., and Danoff-Burg, S. (2003). The yellow brick road and the emerald city: Benefit finding, positive reappraisal coping and posttraumatic growth in women with early-stage breast cancer. *Health Psychology, 22,* 487-497.

Sheikh, A. (2004). Posttraumatic growth in the context of heart disease. *Journal of Clinical Psychology in Medical Settings, 11,* 265-273.

Sinclair, V.G., and Wallston, K.A. (2004). The development and psychometric evaluation of the Brief Adaptive Coping Scale. *Assessment, 11,* 94-101.

Tang, C. S. (2007). Posttraumatic Growth of Southeast Asian Survivors with Physical Injuries: Six Months after the 2004 Southeast Asian Earthquake-Tsunami. *Australasian Journal of Disaster and Trauma Studies*, 2007-2. http://www.massey.ac.nz/%7Etrauma/issues/2007-1/tang.htm

Tedeschi, R.G., and Calhoun, L.G. (1996). The Posttraumatic Growth Inventory: Measuring the positive legacy of trauma. *Journal of Traumatic Stress, 9,* 455-471.

Tedeschi, R.G., and Calhoun, L.G. (2004). Posttraumatic growth: Conceptual foundations and empirical evidence. *Psychological Inquiry, 15,* 1-18.

Triandis, H.C. (1995). *Individualism and collectivism.* Boulder, CO: Westview.

Triandis, H.C. (1996). The psychological measurement of cultural syndromes. *American Psychologist, 51,* 407-415.

Weiss, T. (2002). Posttraumatic growth in women with breast cancer and their husbands: An intersubjective validation study. *Journal of Psychosocial Oncology, 20,* 65-80.

Zaccaro, S.J., Blair, V., Peterson, C. and Zazanis, M. (1995). Collective Efficacy. In J.E. Maddux (Ed.), *Self efficacy, adaptation, and adjustment: Theory, research, and application* (pp. 305-328). New York: Plenum Press.

In: Atlantic and Indian Oceans
Editors: E. S. Askew and J. P. Bromley, pp. 141-158
ISBN 978-1-60692-475-4

Chapter 6

TIME SERIES ANALYSIS OF THE INDIAN OCEAN DIPOLE INDEX

Elizabeth Ann Maharaj
Department of Econometrics and Business Statistics,
Monash University, Australia

ABSTRACT

The Indian Ocean Dipole (IOD) (Saji et al., 1999) is an ocean-atmosphere phenomenon in the Indian Ocean. It is normally characterized by a positive phase which sees greater-than-average sea-surface temperatures (SST) leading to greater precipitation in the western equatorial Indian Ocean, and a negative phase which sees a corresponding cooling of waters in the south eastern Indian Ocean which tends to cause droughts in adjacent land areas of Indonesia and Australia. The IOD also affects the strength of monsoons over the Indian subcontinent. The variability of the IOD causes climatic extremes and socio-economic hardship throughout the tropical Indian Ocean region (Birkett et al., 1999). The ability to predict the strength of the IOD two to three seasons ahead will help communities prepare, if need be for adverse climatic events. This study considers a selection of statistical time series models to analyze and predict the IOD. These models are based only on the IOD index time series and do not require the input of any other climate time series variables. We will show that one of these time series models produces credible forecasts that are comparable to those produced by a forecasting system developed by Luo et al, 2007, and which incorporates several time series variables.

1. INTRODUCTION

The IOD is one aspect of the general cycle of global climate, interacting with similar phenomena such as the El Nino-Southern Oscillation (ENSO) in the Pacific Ocean. The IOD phenomenon was first identified by climate researchers in 1999. However, evidence from fossil coral reefs demonstrates that the IOD has functioned since at least the middle of the Holocene period, about 6500 years ago (Abram et al., 2007).

The variability of the IOD causes climatic extremes and socio-economic hardship throughout the tropical Indian Ocean region (Birkett et al., 1999). The ability to predict the strength of the IOD two to three seasons ahead will help communities prepare, if need be for adverse climatic events. A significant positive IOD occurred in 1994 and 1997, with another occurring in 2006. Luo et al. (2007) developed a forecasting system based on a relatively high resolution coupled ocean-atmosphere global climate model (GCM). This system produced mixed prediction results of past IOD events.

This chapter considers statistical time series models to analyze and predict the IOD. In particular, we will consider an exponential smoothing method, an autoregressive model, a harmonic model and a dynamic harmonic model. These models are based only on the IOD index time series and do not require the input of any other climate time series variables. These time series models are discussed in Section 2.

The fit of each of the models in the estimation period (in-sample forecasts), and forecasts in the validation period (out-of-sample forecasts) consisting of a time horizon of one to twelve months are given and discussed in Section 3. One to twelve step-ahead out-of-sample forecasts for each month for the last two years of the available IOD index record obtained by the best performing model, and its skills measures are discussed in Section 4, while in Section 5, we examine the prediction results of past IOD events using this selected method. In Section 6 we summarize the results and conclude.

2. Time Series Models

We are interested in time series models that are applicable to climate time series. These type of time series are generally non linear in structure, may have one or more cyclical components and generally have changing variance over time.

Many different time series methods and models exist to predict different types of time series. The simple exponential smoothing method generally does quite well in capturing the dynamics of most time series types, in-sample, i.e., during the estimation period, but is only useful for very short term forecasting. This is an iterative method where forecasts are obtained by the following equation.

$$F_{t+k} = \alpha Y_t + (1-\alpha)F_t \quad . \tag{1}$$

Y_t is the time series variable, α is the smoothing constant which takes on a value between 0 and 1, and F_{t+k} is a forecast made of Y_{t+k} at time t, $k \geq 1$.

To initialize the simple exponential smoothing process, $F_1 = Y_1$. The smoothing constant is optimized to give the minimum root mean square error for the in-sample data. For further details, see Chatfield (2004).

The simple exponential smoothing method can be extended to capture trend and seasonal components in a time series, if they exist. These methods require distinct special specification of the trend and seasonal components to enable useful predictions.

Autoregressive integrated moving average models (ARIMA) which are a class of linear models (see Chatfield, 2004) suitable for modeling both stationary and non-stationary time series are defined by

$$\left(1-\sum_{i=1}^{p}\phi_i B^i\right)(1-B)^d\ Y_t = \left(1-\sum_{i=1}^{q}\theta_i B^i\right)e_t \qquad (2)$$

In general the model is described as ARIMA(p,d,q), where Y_t is the time series variable, B is a backshift operator such that $BY_t = Y_{t-1}$, d is the order of differencing required to make the series stationary, ϕ_i, $i = 1, 2,\ldots, p$ are the autoregressive (AR) parameters, θ_i, $i = 1, 2,\ldots, q$ are the moving average (MA) parameters, and ε_t is the random error which is white noise. The orders of the AR and MA parameters are determined from an examination of the autocorrelation and partial autocorrelation functions of the time series. The ARIMA class of models can also be extended to incorporate seasonal structure in a time series if it exists. ARIMA models are generally not very useful for forecasting a time series with non-linear structure.

Many time-varying parametric models (see e.g. Box et al., 1994; Young, 1993; Priestley, 1988; Nicholls and Pagan, 1985) have been proposed to handle non-linear time series. Many of these models are based on piecewise linear models that require the specification of the points of structural changes in the time series. In finance, autoregressive conditionally hetroscedastic (ARCH) and generalised autoregressive conditionally hetroscedastic (GARCH) models are used to model time series with changing variances (see Bollerslev et al.,1992, 1994). However these models are regarded as being more useful in understanding the changing structure of time series and the assessment of risk rather than their ability to make forecasts (Chatfield, 2004).

Harmonic regression models may be useful in describing and forecasting time series with one or more cycles. A harmonic regression model (Bloomfield, 2000) with a single cycle is defined as

$$Y_t = \mu + A\cos(\omega t) + B\sin(\omega t) + \varepsilon_t,$$

where Y_t is the time series, μ is the mean of the underlying time series, $A = R\,cos\,\phi$, $B = -R\,sin\,\phi$, R the amplitude or height of the cycle peaks, ϕ is the phase, or location of peaks relative to time zero and ε_t is the random error which is white noise. $\omega = \frac{2\pi}{\tau}$ is the frequency, and τ is, the cycle length, i.e., the distance from one peak to the next. The errors ε_t are uncorrelated with the cosine and sine terms. By varying the A and B parameters that indicate how much weight is given to the sine and cosine components, it is possible to generate a sinusoid waveform of period τ with any particular amplitude and phase. Sine and cosine functions have the same waveform shape but differ in phase, that is, the cosine function was a peak value of one at $t = 0$, whereas the sine function has a peak value of one at a quarter of a cycle later. The sine and cosine functions of period τ are orthogonal, i.e., they are uncorrelated. Therefore by varying the relative size of the A and B coefficients, it is possible to generate all possible sinusoids of period τ; sinusoids with any mean μ, amplitude R, and phase ϕ. Given an assumed value of τ, the period or cycle length, the remaining three parameters, i.e., the mean, amplitude and the phase of the cycle can be estimated using least squares regression to obtain the best possible fit to an observed time series.

Now if a series displays multiple cycle, these can be incorporated into the harmonic model as follows:

$$Y_t = \mu + \sum_{i=i}^{k} \left(A_i \cos(\omega_i t) + B_i \sin(\omega_i t) \right) + \varepsilon_t , \qquad (3)$$

where k is the number of cycles.

While exponential smoothing and ARIMA models can also be used to model time series with multiple cycles, the estimation of the harmonic regression model with multiple cycles is fairly straight forward compared to these other models.

While harmonic regression models can capture multiple cycles that may be present in a time series, they unable to capture the time varying nature of the cycles.

The dynamic harmonic regression (DHR) model as developed by Young et al., (1999) is better able to do so. The DHR can be considered as an extension of the harmonic regression model defined in Equation (3) in which the amplitude and phase of the harmonic components can vary as a result of estimated temporal changes in the parameters A_i and B_i which now become A_{it} and B_{it}.

The estimation of the time varying parameters is obtained through state space and frequency domain formulation. For details of this estimation process, see Young et al., (1999) who state that their DHR model is asymptotically equivalent to other time varying approaches, but it is more computationally attractive. Hence, the DHR model has been selected for use here over the other time-varying models.

3. Predicting the IOD Index

3.1. Assessing the Accuracy of the Model Fits and Forecasts

In the first instance, using all four methods, with January 1982 to July 2006 taken as the estimation period, we forecast the IOD[1] index for this period and then obtain forecasts for the validation period for twelve months, August 2006 to July 2007. We compare the performance of these methods in both the estimation and validation periods in terms of their root mean squared errors (RMSE).

Figure 1 shows the IOD index from January 1982 to July 2007. The series is stationary about the mean but does not appear to have a seasonal component, i.e., while there do appear to be cycles, they are not regular.

[1] The IOD index time series used in this study was obtained from http://www.jamstec.go.jp/frsgc/research/d1/iod/

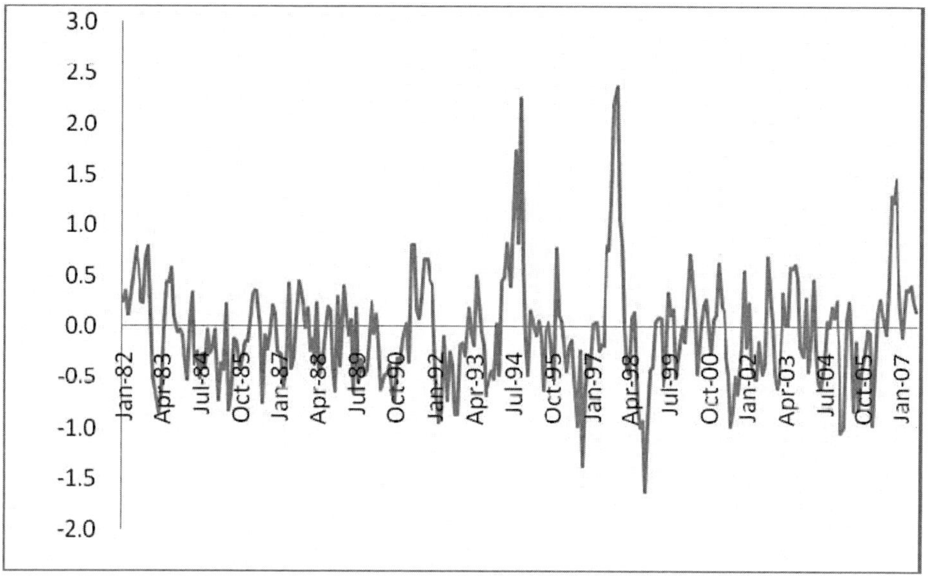

Figure 1. IOD Index: January 1982 to July 2007.

3.1.2. Exponential Smoothing

Figure 2 shows the IOD index series with the series of forecasts generated from exponential smoothing for the estimation period. It can be seen that the exponential smoothing forecasts track the IOD index time series s quite well, albeit, a step ahead of it, often reaching its peaks and troughs.

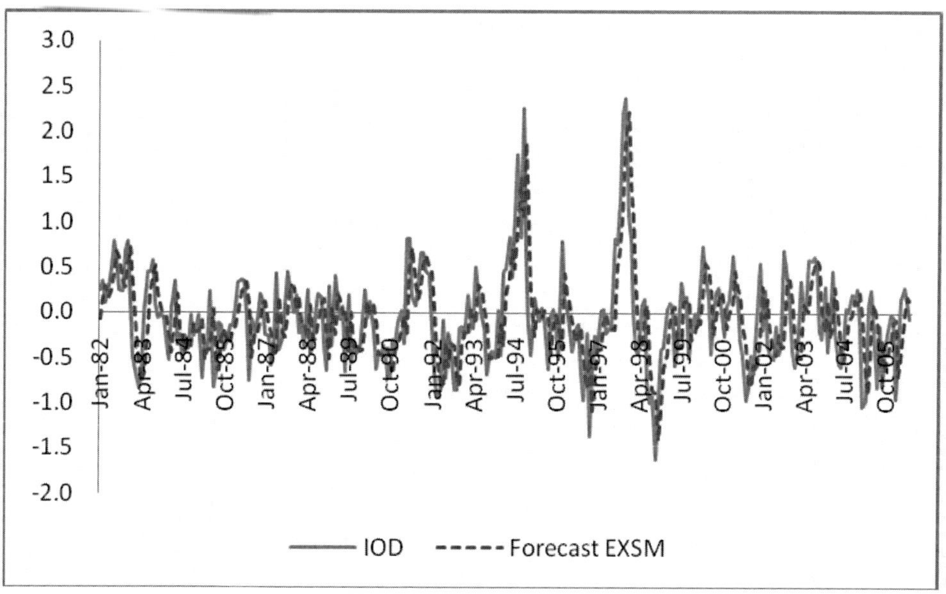

Figure 2. IOD Index and Exponential Smoothing Forecasts for the Estimation Period.

3.1.2. ARIMA Modeling

In order to fit an ARIMA model, we must first ensure that the original time series is transformed in order to make it stationary. Since the IOD index time series has no trend, no differencing is required to make it stationary. A highly significant partial autocorrelation function at lag one as shown in Figure 3 indicates that an ARIMA(1,0,0), i.e., an AR(1) model which is a stationary model is appropriate.

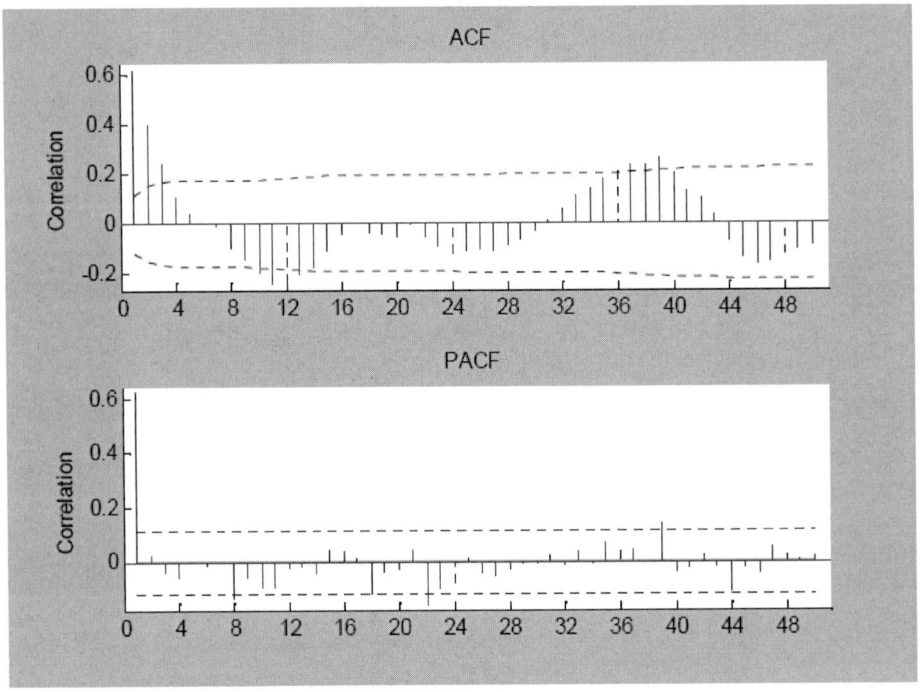

Figure 3. Autocorrelations and Partial Autocorrelations of the IOD Index.

Figure 4 shows the IOD with the series of forecasts generated from the estimated AR(1) model for the estimation period. While the forecasts do to some extent track the IOD index time series in the estimation period, albeit one step ahead of the original time series values, they are often unable to reach its peaks and trough.

3.1.3. Harmonic Regression and Dynamic Harmonic Regression Models

Prior to fitting either the harmonic regression or dynamic harmonic regression models, the length(s) of the cycles(s) have to be determined. This is done by examining the periodogram of the time series which indicates the contribution to the variance of the time series at a particular frequency and hence at a particular cycle length (see Chatfield, 2004 for more details). The total area under the periodogram is equal to the variance of the time series. Figure 5 shows the periodogram of the IOD index time series. Because the periodogram is fairly noisy, a smoothed version of it usually gives a better indication of where the actual peaks might arise. Figure 6 shows the smoothed periodogram using Welsh's method (see Percival, 1995).

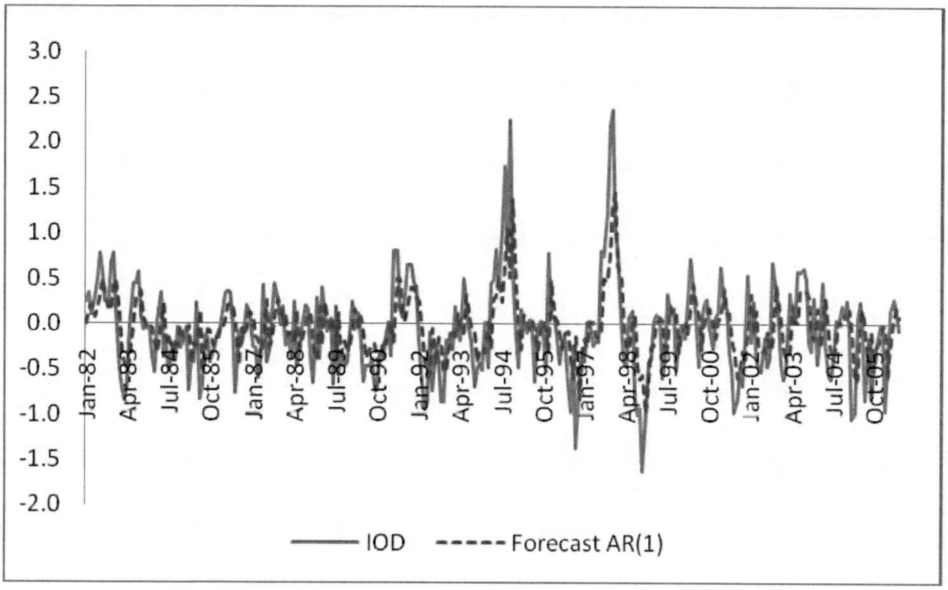

Figure 4. IOD Index and AR(1) Forecasts for the Estimation Period.

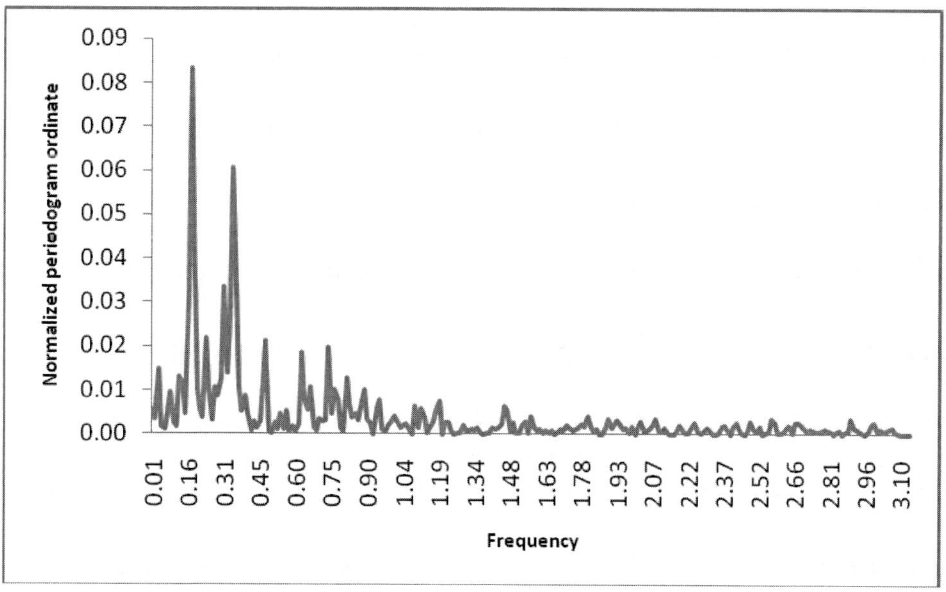

Figure 5. Periodogram of the IOD Index.

In both cases, distinct peaks occur around frequency bands that are equivalent to cycle lengths of approximately 36 and 18 months. A third smoothed-out peak occurs in a frequency band corresponding to a cycle length of about 8 to 9 months. Both the harmonic regression model and dynamic harmonic regression model were estimated using the cycle lengths of 36 and 18 months. The inclusion of the third cycle length did not produce better out-of-sample forecasts.

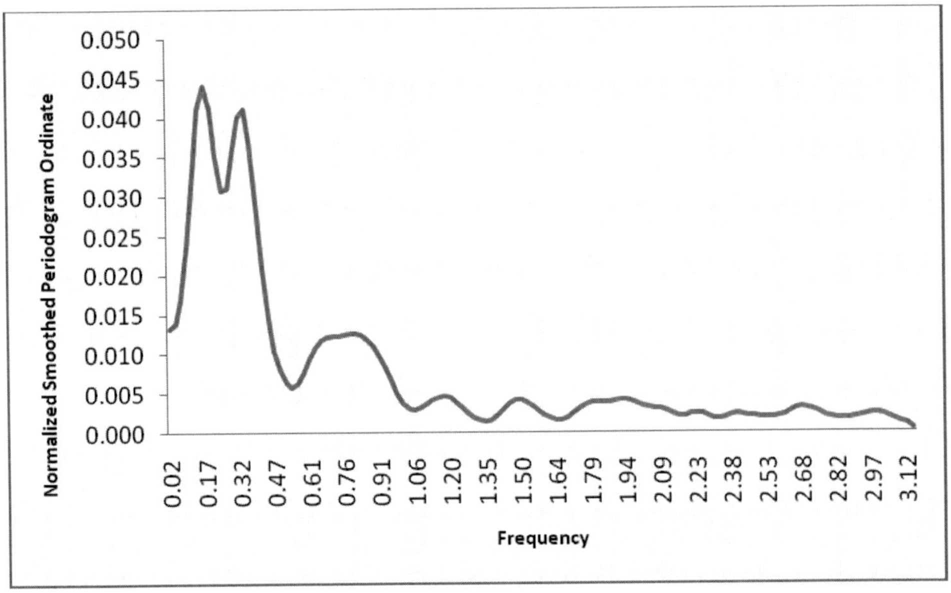

Figure 6. Smoothed Periodogram of the IOD Index.

Figure 7 shows the IOD Index with the series of forecasts generated from harmonic regression model for the estimation period. The peaks and troughs of the forecast series are regular, i.e., they have the same amplitude and while they do correspond to those of the IOD index time series in some cases, they fail to reach its heights and depths.

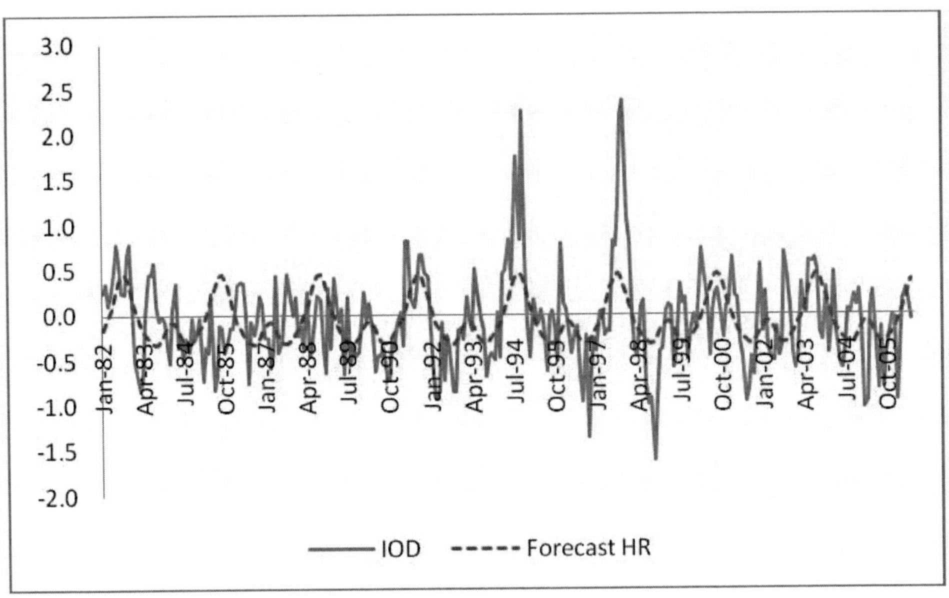

Figure 7. IOD Index and Harmonic Regression Forecasts for the Estimation Period.

The IOD Index series together with the forecasts generated from the dynamic harmonic regression (DHR) model for the estimation period are given in Figure 8. These forecasts track the IOD index time series very well with no lag shift and get fairly close to reaching the heights and depths of the peaks and troughs respectively. The superior fit of the dynamic harmonic regression model over the other models is confirmed by the fact that it has the smallest root mean squared error in the estimation period as shown in Table 1.

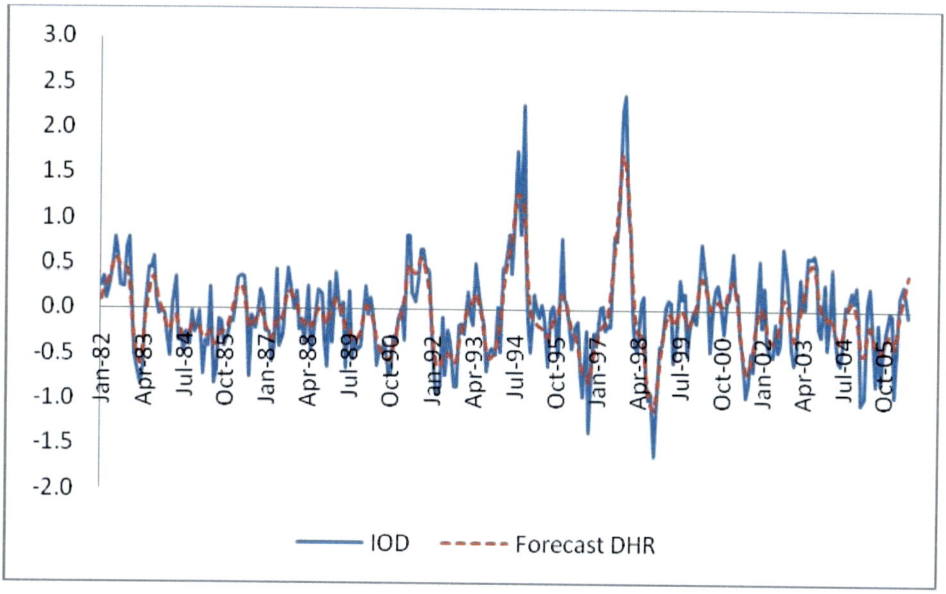

Figure 8. IOD Index and Dynamic Harmonic Regression Forecasts for the Estimation Period.

Table 1. Root Mean Squared Errors in the Estimation Period: January 1982 to July 2006

	RMSE
Exponential Smoothing (EXP)	0.4138
Autoregression Model (AR(1))	0.4446
Harmonic Regression Model (HR)	0.4617
Dynamic Harmonic Regression Model (DHR)	0.2503

3.1.4. Validation Period Forecasts

Figure 9 shows forecasts for the validation period from the four methods. The root mean squared errors from one to twelve forecast horizons for each method are given in Table 2. The validation period forecasts for the exponential smoothing method and the AR(1) model are both flat and hence are incapable of capturing the evolution of the IOD index into the real future. However, the harmonic regression and dynamic harmonic regression models are much more credible in that they do capture the evolution of the IOD index over the last twelve months to some extent. It can be seen from Table 2 that with the exception of the first forecast horizon, the dynamic harmonic regression method performs must better than the other methods in predicting the true future.

Since the dynamic harmonic regression model has the best fit in-sample and produces the better out-of-sample forecasts, all further analysis will be conducted on the IOD index time series using this model.

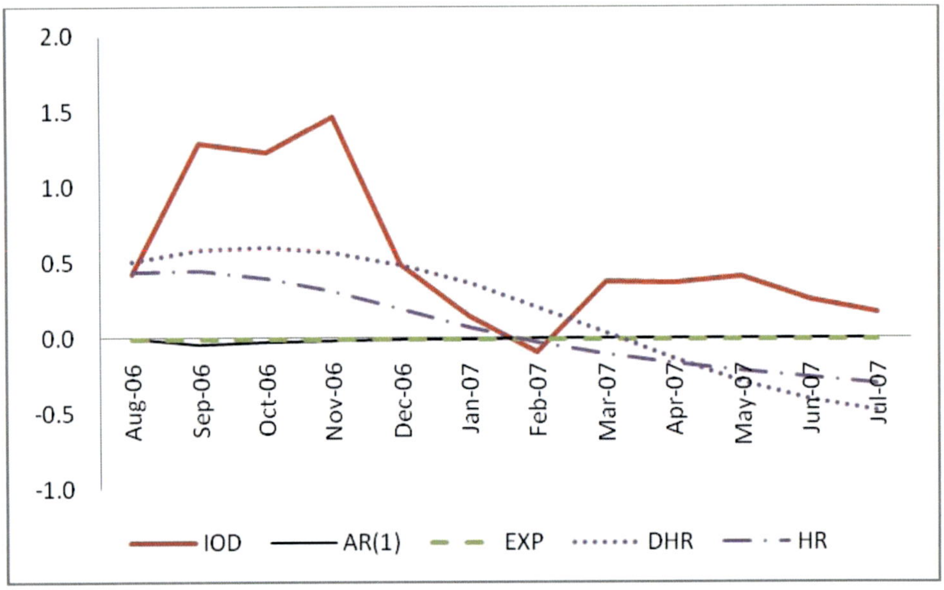

Figure 9. IOD Index and Forecasts for the Validation Period August 2006 to July 2007.

Table 2. Root Mean Squared Errors in the Validation Period, August 2006 to July 2007

Validation Period	Forecast Horizon	EXP	AR(1)	HMR	DHR
Aug-06	1	0.4302	0.4641	0.0161	0.0895
Sep-06	2	0.9697	0.9887	0.6001	0.5063
Oct-06	3	1.0690	1.0830	0.6883	0.5511
Nov-06	4	1.1846	1.1941	0.8323	0.6551
Dec-06	5	1.0819	1.0899	0.7554	0.5860
Jan-07	6	0.9896	0.9966	0.6900	0.5424
Feb-07	7	0.9167	0.9234	0.6394	0.5148
Mar-07	8	0.8681	0.8738	0.6215	0.4964
Apr-07	9	0.8278	0.8327	0.6119	0.4970
May-07	10	0.7962	0.8002	0.6126	0.5202
Jun-07	11	0.7631	0.7666	0.6039	0.5343
Jul-07	12	0.7323	0.7355	0.5933	0.5450

3.2. Assessing the Accuracy and Skill of the Dynamic Regression Model

Using the dynamic harmonic regression model, for the validation period of the last two years (August 2005 to July 2007), we determined one to twelve step-ahead forecasts for each month. The accuracy and skill of these forecasts were determined by root mean squared error and the correlation coefficients. Figure 10 shows the root means squared errors and the correlation coefficients for the one to twelve step-ahead forecasts in the validation period. Figure 11 shows the IOD Index with the one to twelve step-ahead forecasts in this validation period.

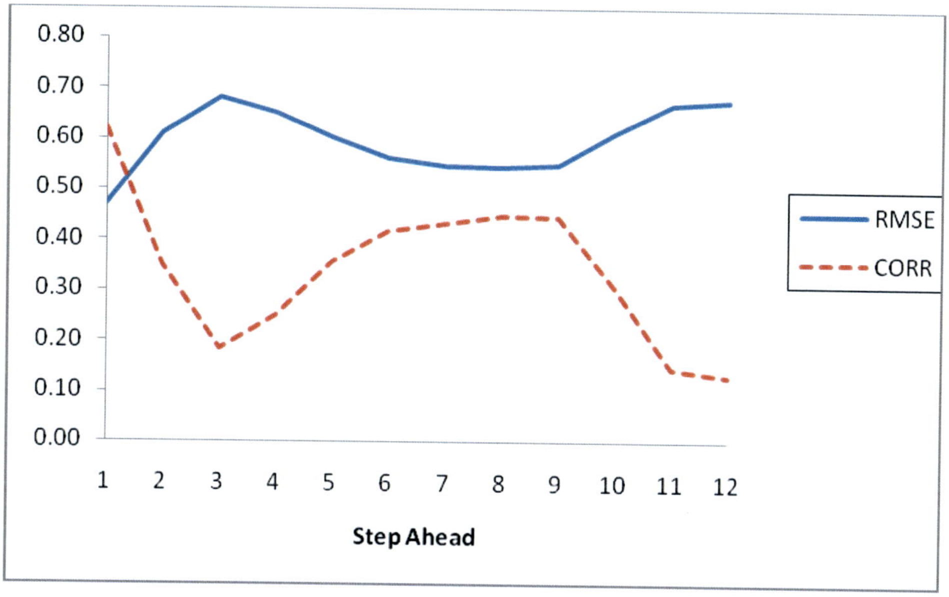

Figure 10. Root mean squared errors and correlation coefficients of the one to twelve step-ahead forecasts in the validation period.

The results are mixed for the different step-ahead forecasts, with the one step-ahead forecasts obviously being the most accurate (root mean squared error ≈ 0.47) and skillful (correlation coefficient ≈ 0.62). As seen from Figure 10, apart from the first step-ahead forecasts, the six, seven, eight and nine step-ahead forecasts perform better than the other step-ahead forecasts in terms of both root mean squared errors and correlations.

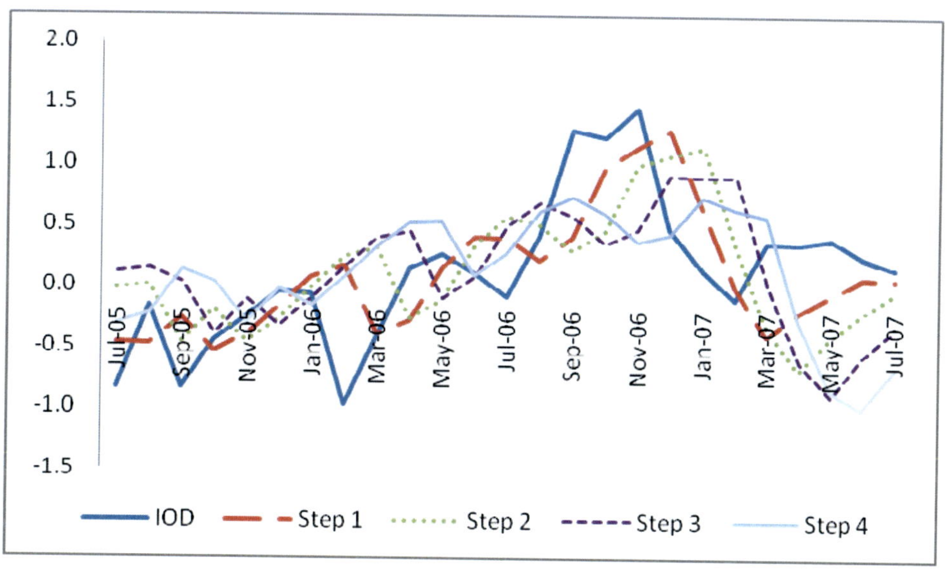

Figure 11. (Continued)

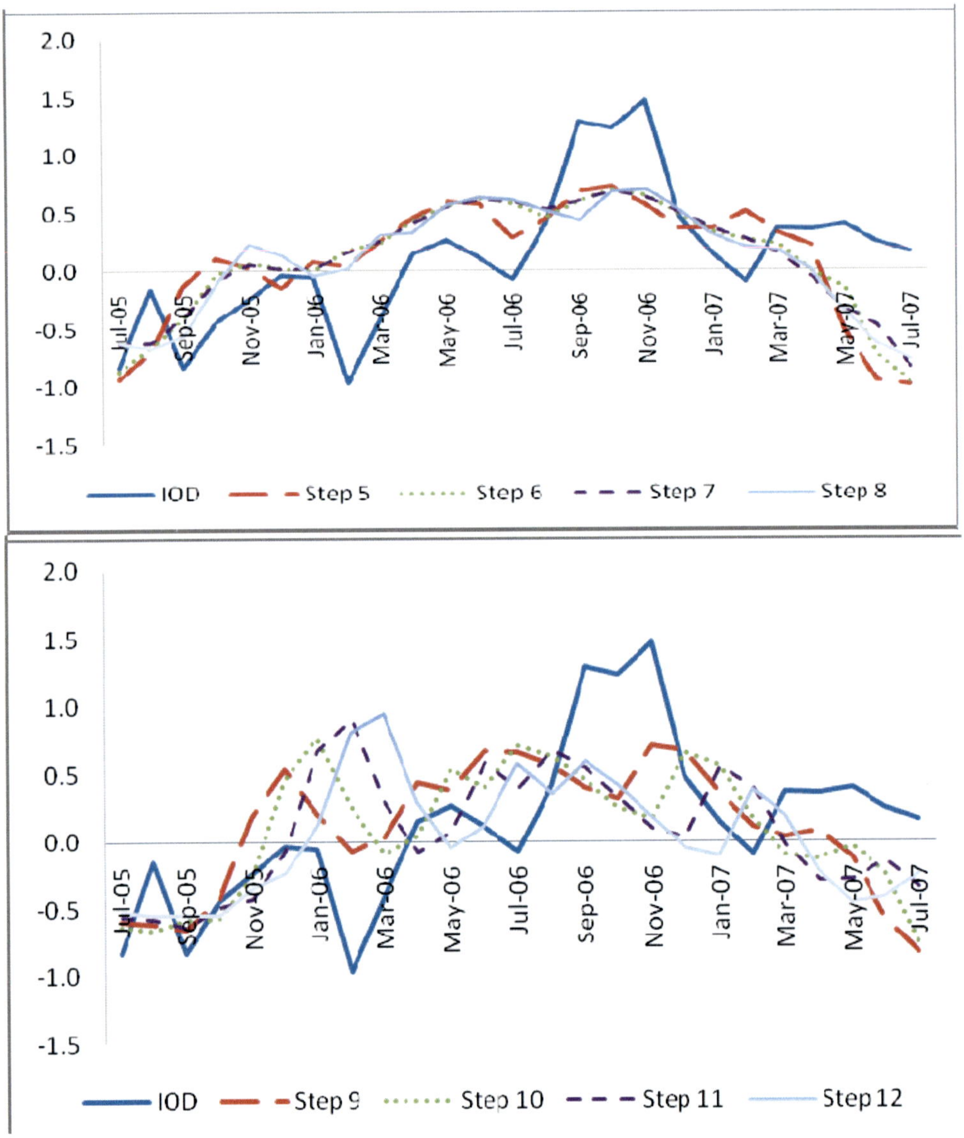

Figure 11. IOD Index with the one to twelve steps ahead forecasts in the validation period.

3.3. Assessing the Ability the Dynamic Regression Model to Capture Past IOD Events

Forecasts were generated for the IOD events in 1994, 1997 and 2006 to assess to what extent the DHR model forecasts are able capture these events. Table 3 shows the months at which the procedure was initiated to obtain one to twelve step-ahead forecasts. Figures 12 to 14 show the IOD and forecasts for the 1994, 1997 and the 2006 events, respectively.

For the all three IOD events, most of forecasts, regardless of which month they were initiated from, do to some extent capture the peak, either a few months before the event or around it.

Table 3. Months at which forecasts were initiated

	IOD Event		
Step-ahead	1994	1997	2006
12	Oct-93	Nov-96	Nov-05
11	Nov-93	Dec-96	Dec-05
10	Dec-93	Jan-97	Jan-06
9	Jan-94	Feb-97	Feb-06
8	Feb-94	Mar-97	Mar-06
7	Mar-94	Apr-97	Apr-06
6	Apr-94	May-97	May-06
5	May-94	Jun-97	Jun-06
4	Jun-94	Jul-97	Jul-06
3	Jul-94	Aug-97	Aug-06
2	Aug-94	Sep-97	Sep-06
1	Sep-94	Oct-97	Oct-06

However, for the 1994 IOD event, the amplitude of the forecasted peak decreased when forecasts were initiated from between one to eight months ahead; then increased slightly when forecasts where initiated from between nine to twelve months ahead. The forecasts initiated from one month ahead, and from nine to twelve months ahead appear to do a reasonably good job in capturing the actual location of the highest peak of the IOD index in 1994.

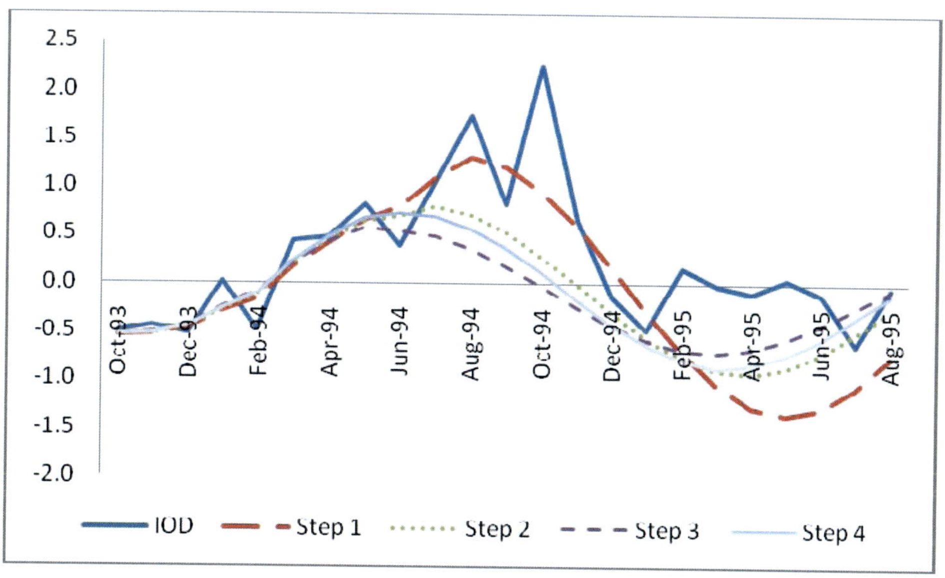

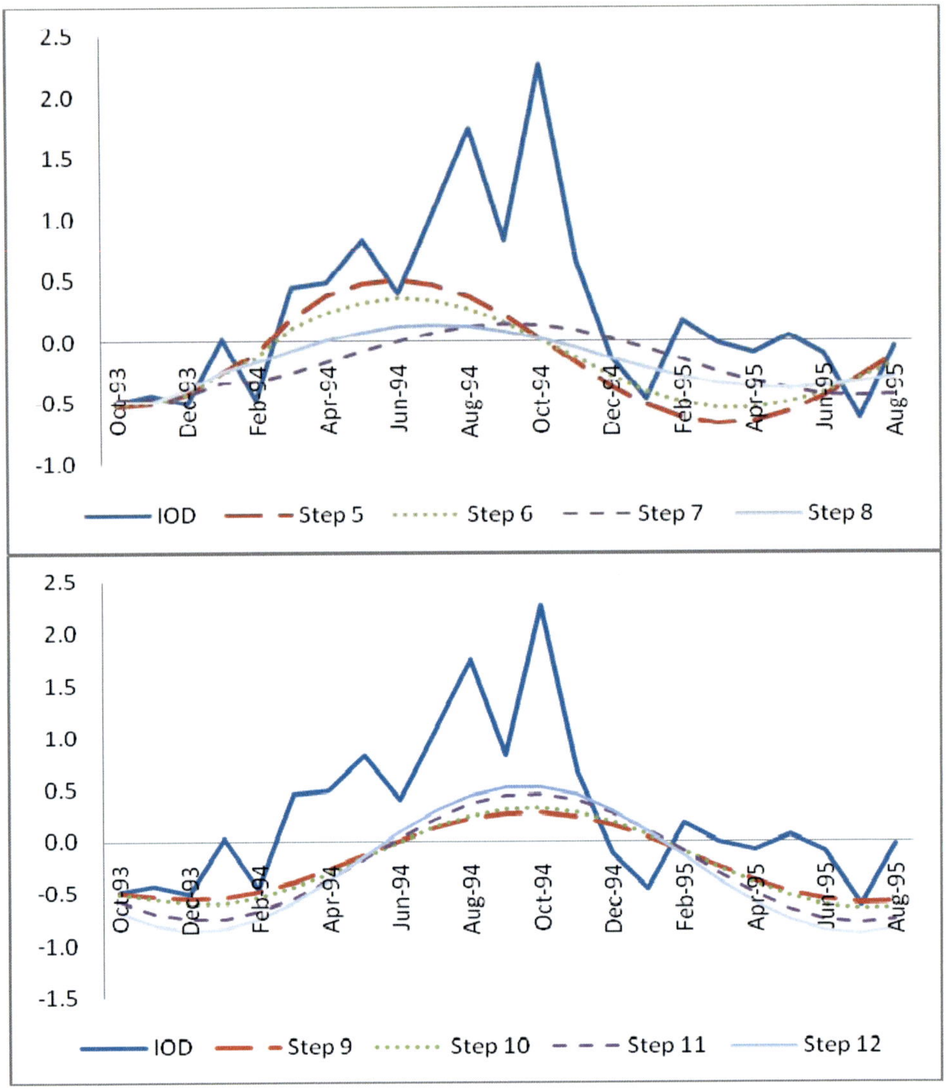

Figure 12. IOD Index with Forecasts for the 1994 IOD Event.

For the 1997 IOD event, the amplitude of the forecasted peak decreased when forecasts were initiated from between one and six months ahead, then increased when the forecasts were initiated from seven months ahead. When forecasts were initiated from between eight to twelve months ahead, the forecasted peak remained at the same height. However, when forecasts were initiated from between nine and twelve months ahead, the forecasted peak shifted by quite a few months to the left of the actual location of the peak of the IOD index. The forecasts initiated from one to three months ahead appear to do a reasonably good job in capturing the actual location of the highest peak of the IOD index in 1997.

For the 2006 IOD event, the amplitude of the forecasted peak decreased when forecasts were initiated from between one and three months ahead; then increased when the forecasts were initiated from between four and eight months ahead. When forecasts were initiated from between six to eight months ahead, the forecasted peak remained at the same height.

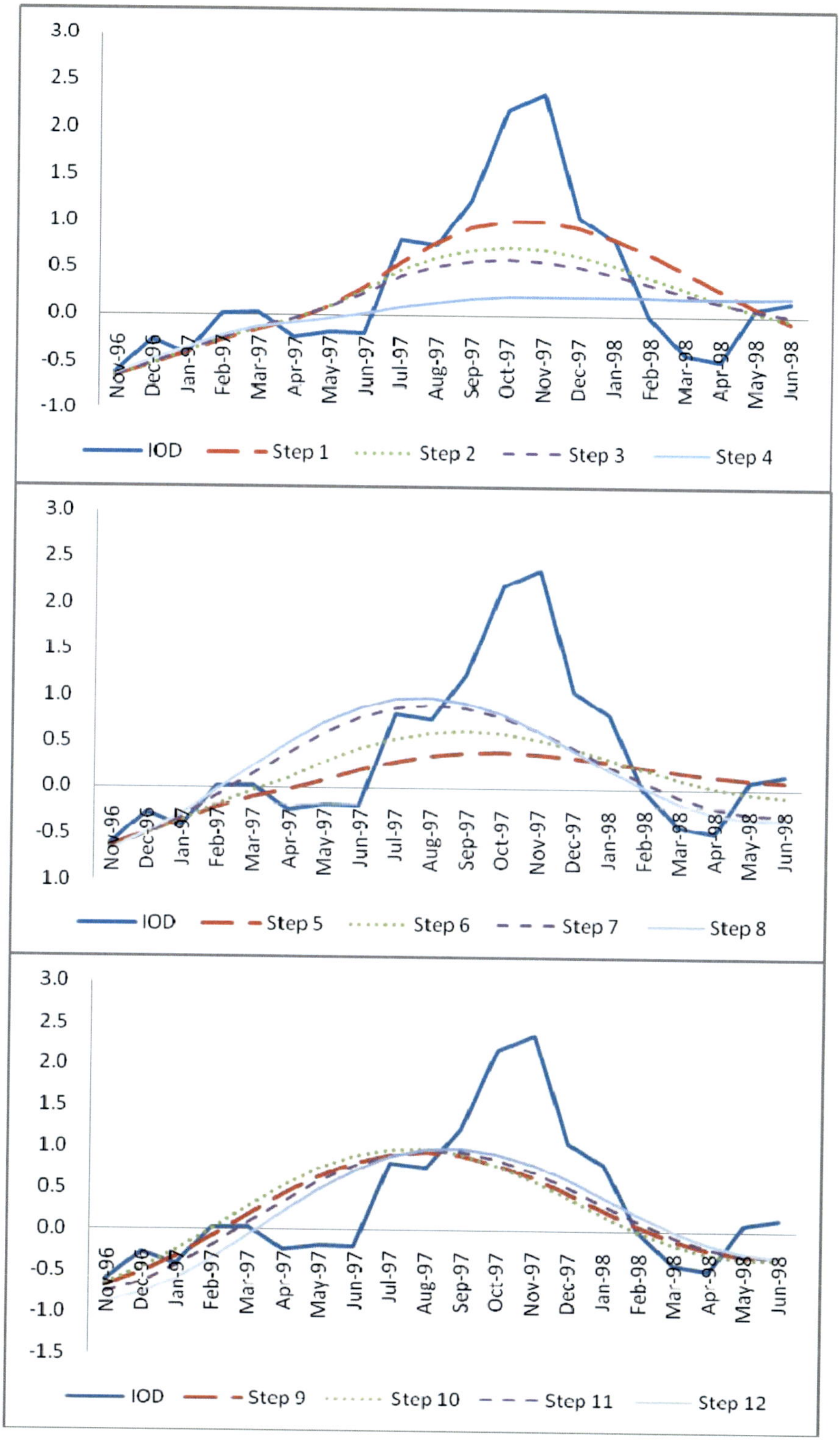

Figure 13. IOD Index with Forecasts for the 1997 IOD Event.

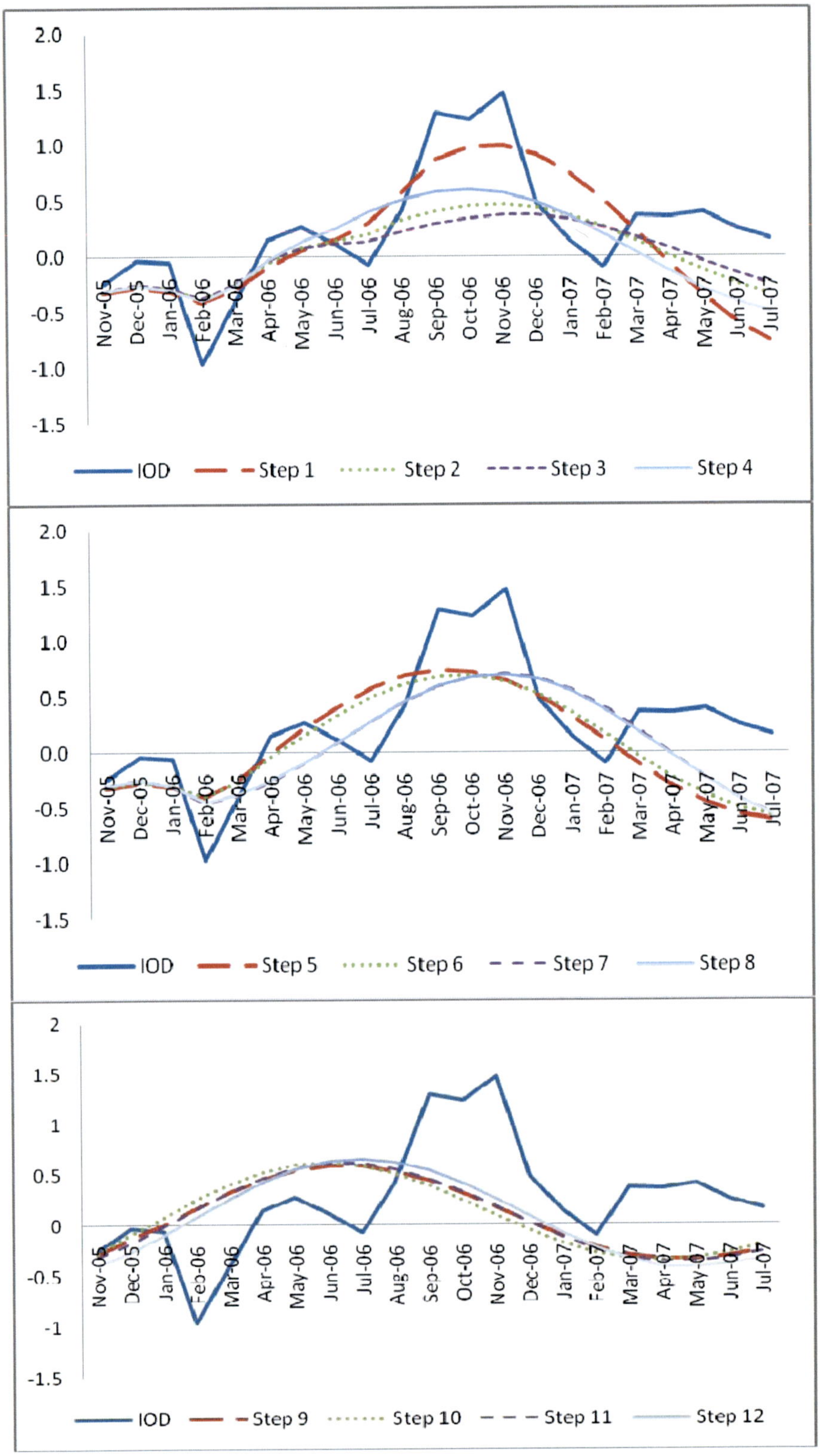

Figure 14. IOD Index with Forecasts for the 2006 IOD Event.

However, while the amplitude of the forecasted peak was almost at the same height when forecasts were initiated from between nine and twelve months ahead, it shifted by quite a few months to the left of the location of the peak of actual IOD index. The forecasts initiated from one month ahead, and from four to eight months ahead appear to do a reasonably good job and capturing the actual location of the highest peak of the IOD index in 2006.

CONCLUSION

In this study we first investigated the performance of several time series models in forecasting the IOD index using a validation period of the last twelve months of the available IOD Index record where single one to twelve step-ahead forecasts were generated. We concluded that the dynamic harmonic regression (DHR) model produced the most credible forecasts.

We then investigated the performance of the DHR model by generating multiple one to twelve step-ahead forecasts using a validation period of the last twenty four months of the available IOD index record. It was found that this model produced forecasts with reasonable accuracy and skill when initiated from the various steps ahead.

Finally, we investigated the performance of the DHR model in capturing the 1994, 1997 and 2006 IOD events. We found that in many cases, while the amplitude for peaks the actual IOD index were not reached, the location of the peaks were reasonably captured even though in some case the forecasted peaks were shifted a few months to the left.

This study investigated the predictability of the DHR model of the IOD index time series using only the observations of this time series. No other influencing time series variables were taken into account. This methodology is in contrast to that of the forecasting system employed by Luo et al. (2007) which took into account several other influencing factors. Whereas the DHR model in no way outperforms the forecasting system of Luo et al, it does tend to produce comparable predictive results. Thus the main advantage in using the DHR model to forecast the IOD would be its simplicity in implementation, i.e., applying a model to the single time series without the need to input any other time series variables.

ACKNOWLEDGEMENTS

Results for the dynamic harmonic regression model have been generated using a suite of MATLAB programs developed by Pedregal et.al., (2007).

REFERENCES

Abram N.J., Gangan M.K., Zhenhyu L., Hantoro W.S., McCulloch M.T. and Suwargadi B.W. (2007), Seasonal characteristics of the Indian Ocean Dipole during the Holocene epoch, *Nature*, DOI:10.1038/nature05477.

Birkett C., Murutgudde R. and Allan T., (1999) Indian Ocean climate events brings floods to East Africa's lakes and the Sudd Marsh, *Geophysical Research Letters,* 26, 1031-1034.

Bloomfield P. (2000), *Fourier Analysis of Time Series: An Introduction*, Wiley, New York.

Bollerslev T., Chou Y. and Kroner K.F. (1992), ARCH models in finance, *Journal of Econometrics,* 52, 5-59.

Bollerslev T., Engle R.F. and Nelson D.B. (1994) ARCH models. In *Handbook of Econometrics*, Vol 4 (eds. R.F. Engle and D.L. Mcfadden), 2959-3038, Amsterdam, Elsevier.

Box G.E.P., Jenkins G.M. and Reinsel, G.C. (1994), *Time Series Analysis: Forecasting and control, 3rd edition*, Prenctice-Hall Inc.

Chatfield C. (2004), *The Analysis of Time Series, 6th edition*, London, Chapman and Hall.

Luo J., Masson S., Behera S. and Yamagata T. (2007), Experimental forecasts of the Indian Ocean Dipole using a Coupled OAGCM, *Journal of Climate,* 20, 2178-2190.

Nicholls D.F. and Pagan A.R. (1995), Varying Coefficients Regression. In *Handbook of Statistics,* Vol 5 (eds. E.J. Hannan, P.R.Krishnaiah and M.M. Rao), 413-449, Amsterdam, North-Holland.

Pedregal D.J., Taylor C.J. and Young P.C. (2007), System identification, time series analysis and forecasting, the Captain Toolbox, in *Handbook, Version 2.0, Centre for Research on Environmental Systems and Statistics (CRES),* Lancaster University, Lancaster, LAI 4YQ, United Kingdom.

Percival D.B. and Walden A.T. (1993), *Spectral Analysis for Physical Applications,* Cambridge, Cambridge University Press.

Priestley, (1988), *Non-linear and non-stationary time series analysis*, London, Academic Press.

Saji H. H., Goswami B.N., Vinayachandran P.H. and Yamagata T. (1999), A dipole mode in the tropical Indian Ocean, *Nature,* 401, 360-363.

Young P.C. (1993), Time variable and state dependent modelling of nonstationary and nonlinear time series, in Subba Rao, T.(ed), *Developments in Time Series Analysis,* London: Chapman and Hall, 374-413.

Young P.C., Pedregal D.J. and Tych W. (1999), Dynamic harmonic regression, *Journal of Forecasting*, 18, 369-394.

In: Atlantic and Indian Oceans
Editors: E. S. Askew and J. P. Bromley, pp. 159-167
ISBN 978-1-60692-475-4

Chapter 7

HISTORY OF OCEANOGRAPHIC RESEARCH IN THE INDIAN OCEAN FROM THE EARLY YEARS UNTIL NOW

Rolf Koppelmann* and Bettina Martin
Universität Hamburg, Institut für Hydrobiologie und Fischereiwissenschaft,
Große Elbstraße 133, 22767 Hamburg, Germany

ABSTRACT

Investigations in the Indian Ocean with respect to sea routes, harbor locations and favorable winds became increasingly important when Vasco da Gama rounded the Cape of Good Hope and discovered the sea route to India at the end of the 15th century. However, it took more than two centuries before the Danish Arabia Felix Expedition with the Danish Navy ship Grønland investigated the Indian Ocean between 1761 and 1767. Since then, several worldwide expeditions were performed which visited also the Indian Ocean. The first voyage into this area with a focus on biology was the British John Murray Expedition in the years 1933 to 1934.

After World War II, the numbers of expeditions to the Indian Ocean increased. Based on fundamental scientific interest, the International Indian Ocean Expeditions (IIOE) took place between 1959 and 1965. Twenty-three countries were involved in 180 research cruises on 40 different research vessels in this venture. After this period only occasional cruises were obtained by several countries until, between 1990 and 1998, participants from nine countries investigated the northeastern part of the Indian Ocean, the Arabian Sea, on 80 cruises with 18 ships in the course of the international Joint Global Ocean Flux Study (JGOFS).

INTRODUCTION

The Indian Ocean covers approximately 20 % of the total area of the worlds oceans. As the youngest of the three major oceans (Atlantic, Pacific, Indian Oceans) its formation set in

* Phone: +49 40 42838 6679; Fax: +49 40 42828 6618; e-mail: koppelmann@uni-hamburg.de, bmartin@uni-hamburg.de

about 150 million years ago by the dispartment of the ancient continent Gondwana. The ocean first opened approximately 125 millions years ago, most of its basin being less than 80 million years old. The Indian Ocean persists in its present configuration since approximately 36 million years.

The northern part is a tropical region influenced by seasonal changing monsoon conditions which lead to a total reversal of the surface current system twice a year. During November to February, high atmospheric pressure over Asia and low pressure over an area from northern Australia to approximately 10°S generate the NE monsoon winds and a wet season for southern Indonesia and northern Australia. Between May to September the conditions change, low pressure over Asia and high pressure over Australia and the ocean cause a strong SW monsoon, with wind speeds up to 24 knots, and a wet season in South Asia.

EARLY INVESTIGATIONS

The northern Indian Ocean was explored by Greek, Egyptian and Arabian seafarers with respect to sea routes, harbor locations and favorable winds since 2500 BC, accumulating in the 15th century when Vasco da Gama rounded the Cape of Good Hope and discovered the sea route to India (1497-1498). An overview on these early and later investigations is given by Rao and Griffiths (1998). A major step in worldwide oceanographic research took place in the late 18th and the early 19th century by the expeditions of James Cook (H.M.S. *Resolution*) and Charles Darwin (H.M.S. *Beagle*). Deacon (1971) compiled these and earlier investigations of marine science for the period between 1650 and 1900. One of the first expeditions to the Indian Ocean (Table 1) was the Danish Arabia Felix Expedition with the Danish Navy ship Grønland. The expedition took place between 1761 and 1767, and a large number of flora and fauna was collected during the voyage (see Wolff 1968). Nearly one century later, the Danish *Galathea* (1845-1847) and the Austrian frigate *Novara* (1857-1859; Scherzer 1864-65) circumnavigated the world and visited the Indian Ocean. The famous British expedition with H.M.S. *Challenger* (1873-1876) under the scientific leadership of Sir Wyville Thomson crossed the Atlantic, Pacific and the southern Indian Ocean (Thomson and Murray 1885-95). During the same period (1874-1876), the German S.M.S. *Gazelle* visited also the southern part of the Indian Ocean for mainly physical research. Between 1884 and 1894, the Indian R.I.M.S. *Investigator* undertook oceanographic research in the northern part of the Indian Ocean under the leadership of A.W. Alcock. Some of the zooplankton data collected during these cruises were added to the description of the free-swimming planktonic Copepoda from the John Murray Expedition by Sewell (1947 a,b). During a cruise of the german steamer *Valdivia* in 1898-1899, the central part of the Indian Ocean and the southern Arabian Sea were investigated (Chun 1902 ff). The main focus of this expedition was the biology of the great depths in order to give detailed descriptions on the taxonomy and distribution of its faunal elements.

Table 1. Major research activities in the Indian Ocean. Historical data sets (before 1990)

Expedition/Campaign	Country	Data
Arabia Felix-Expedition	Denmark	1761-1767
Galathea-Expedtion	Denmark	1845-1847
Novara-Expedition	Austria	1857-1859
Challenger-Expedition	Great Britain	1872-1876
Gazelle-Expedition	Germany	1874-1876
Investigator-Expedition	India	1884-1894
Valdivia-Expedition	Germany	1898-1899
Sealark- Expedition	Great Britain	1905
Planet- Expedition	Germany	1906-1907
Möwe- Expedition	Germany	1912-1913
Dana- Expedition	Denmark	1928-1930
Willebrod Snellius- Expedition	Netherlands	1929-1931
Discovery II- Expeditions	Great Britain	1930,32,35,38
John Murray/Mabahiss- Expedition	Great Britain/Egypt	1933-1934
William Scoresby- Expedition	Great Britain	1935,36,50
Albatross- Expedition	Sweden	1947-1948
Commander Charcot- Expedition	France	1948-1949
Galathea- Expedition	Denmark	1950-1952
Umitaka Maru- Expedition	Japan	1956,57
Vityaz-Expeditons	Russia	1957-1962
International Indian Ocean Expedition	see Text	1959-1965
Indian Expeditions	India	since 1976
GEOSECS	USA	1977-1978
INDEX	USA	1979
French Expeditions	France	since 1980
Charles Darwin-Expedition	Great Britain	1986,87
Meteor	Germany	1987

THE 20TH CENTURY

At the beginning of the 20th century, several ships from Great Britain (*Sealark*, *Discovery II*), Denmark (*Dana*), Germany (*Planet*, *Möwe*) and the Netherlands (*Willebrord Snellius*) visited mainly the southern and central part of the Indian Ocean to carry out biological, geological and physical research. The first step forward to a better understanding of the zooplankton community in the Arabian Sea was done by the John Murray Expedition (JME) onboard the Egyptian research vessel *Mabahiss* (Sewell 1935/36) to the Red Sea and the Indian Ocean in the years 1933 and 1934 (see Rice 1986 for a review). The main objective of the JME was the study of the biology of the Arabian Sea, particularly of the bottom living animals which could be collected in trawls and dredges (Deacon and Rice 1984). Nevertheless, water column research was carried out as well during the cruise. Gilson (1937) described the nitrogen cycle of the Arabian Sea and gave a summary of the state of knowledge of phytoplankton ecology. Sewell (1947 a,b) gave an account of planktonic

Copepoda in Indian waters and a description of their geographical distribution. He discussed his results from the Indian Ocean with other published data from the world oceans with respect to the pelagic provinces published by Steuer (1933). Furthermore, he gave some insights in the vertical distribution of Copepoda down into meso- and bathypelagic depths. In the years 1935 and 1936, the British expeditions on the R.R.S. *William Scoresby* were the last scientific cruises to the southern Indian Ocean before the Second World War started.

After World War II, the number of expeditions increased. Early cruises were undertaken by the Swedish research vessel *Albatross*, the French research vessel *Commander Charcot* and the British R.R.S. *William Scoresby* in 1947 to 1950. Between 1950 and 1952, the great Danish *Galathea* Expedition took place which circumnavigated the world and the Japanese *Umitaka Maru* Expeditions collected samples in the southern part of the Indian Ocean during December 1956 and March 1957. The composition and distribution of the Copepoda sampled during the latter expeditions was published by Seno et al. (1963 a,b, 1966). Between 1957 and 1962, the Russian research vessels *Vityaz* and *Ob* conducted a vast amount of full-depth samples in the Indian Ocean including some stations in the Arabian Sea.

The zooplankton data collected during these cruises were compiled by Vinogradov (1968). Several years later, Tseitlin and Rudyakov (1999) used most of the Russian data to calculate the seasonal variation of vertical distribution of zooplankton in the mesopelagic zone, down to 1000 m depth.

In August 1957, the International Indian Ocean Expedition (IIOE) was first proposed. The expedition was organized and coordinated by the Scientific Committee on Oceanic Research (SCOR) and the Intergovernmental Oceanographic Commission (IOC). The work was undertaken by scientists from 23 countries. Forty research vessels were involved in approximately 180 research cruises (see Wooster 1984). The original proposal for an Indian Ocean expedition was justified on the basis of its fundamental scientific interest. Marine physicists wished to examine the response of density layers and currents to changing monsoon winds, while biologists wanted to elucidate the causes of the highly varying productivity of different parts of the Arabian Sea; geologists and geophysicists emphasized how little was known of submarine topography, sediments, and their structure. The important scientific questions as then perceived were set out by SCOR in early 1959 (Deacon 1959, cited in Wooster 1984). Many of the biological results of the IIOE can be found in the book "The Biology of the Indian Ocean" edited by Zeitzschel (1973); Rao and Griffiths (1998) give an overview on the investigations in the Indian Ocean until the IIOE. A retrospective view on both expeditions, the JME and the IIOE, is given by Aleem and Morcos (1984).

Since 1976 several expeditions took place by Indian research vessels (*Gaveshani, Sagar Kanya, Sagar Sampada, Samudra, Manthan*) and French research vessels (*Jean Charcot, Atalante, Marion Dufrense*), respectively, to collect data on physical oceanography, meteorology, marine chemistry, biology, geology, and geophysics. The book "Oceanography of the Indian Ocean" edited by Desai (1992) compiles data sets collected by the Indian research vessels and gives an excellent overview on various parts of the Indian Ocean sciences. As part of the Geochemical Ocean Sections Study (GEOSECS), the RV *Melville* from the USA visited the Arabian Sea. During the Indian Ocean Experiment (INDEX), samples were collected in the Somali Current during the NE and SW monsoons of 1979 by the U.S. research vessel *Columbus Iselin* (Smith 1982).

The German research vessel *Meteor* visited the Arabian Sea in 1987 during an expedition to the Eastern Mediterranean Sea, the Red Sea and the Indian Ocean (Nellen et al. 1996) to investigate spatial and temporal variability in biological structures and pelagic processes. Burkill et al. (1993) outline the results of two expeditions of the British R.R.S. *Charles Darwin* in the Arabian Sea (September and December 1986) and investigations obtained by the Indian O.R.V. *Sagar Kanya.* This list may be not complete, but to summarize, the biology of the Arabian Sea was primarily known from the *Valdiavia*-, the John Murray-, the *Vityaz*-, and the International Indian Ocean-Expedition, several cruises of the O.R.V. *Sagar Kanya*, the INDEX-Expedition and cruises by the R.V.'s *Charles Darwin* and *Meteor* before 1990. In the 90th, an international program took place to investigate especially the northern part of the Indian Ocean - the JGOFS (Joint Global Ocean Flux Study) Arabian Sea Process Study (SCOR 1995).

THE JOINT GLOBAL OCEAN FLUX STUDY

Under the auspices of the Scientific Committee on Oceanic Research (SCOR) and as a core project of the International Geosphere-Biosphere Program (IGBP), the Joint Global Ocean Flux Study (JGOFS) has been established in order to determine and understand processes controlling time-varying fluxes of carbon and associated biogenic elements in the ocean; and to predict the response of marine biogeochemical processes to climate change (SCOR 1992).

The Arabian Sea in the northwestern Indian Ocean was one of four ocean regions identified by JGOFS for process studies. In comparison with the North Atlantic, the Equatorial Pacific and the Southern Ocean, the Arabian Sea is important for its large seasonal signal of biogenic processes in the euphotic zone and for its oxygen minimum zone over vast areas between 100 and 1200 m depth (SCOR 1995). The aims of the Arabian Sea process study as identified by SCOR (1995) were:

- to investigate the seasonally varying biogeochemical responses to intense and regularly oscillating physical forcing in the Arabian Sea;
- to quantify the role of the Arabian Sea to global biogeochemical cycling of carbon and associated biogenic elements;
- to quantify the role of the Arabian Sea as a source or sink of atmospheric CO_2 and other gases relevant to climate change;
- to quantify primary production, transformation and fate of phytogenic carbon in the Arabian Sea;
- to quantify the role of the oxygen minimum zone in biogeochemical cycling;
- to quantify monsoonal variability in relation to fluctuations in ocean palaeo-circulation.

Biological processes are controlling (1) the uptake rate of inorganic carbon and its modification into organic matter, (2) the amount of particulate organic matter (POM) reaching the deep sea, and (3) the burial rate of carbon into deep-sea sediments. On the way through the water column, POM is utilized by the marine biota to fulfill its energetic and somatic requirements. In this context, the Arabian Sea recently gained in interest in order to examine the role of the oceans and its biota in global carbon cycling under changing monsoon conditions. Banse (1994) gives an excellent overview on the coupling of hydrography, phytoplankton, zooplankton, and settling organic particulate matter offshore in the Arabian Sea. In the course of JGOFS and JGOFS-related studies like the Pakistan-US program NASEER (North Arabian Sea Environment and Ecosystem Research), the Dutch Indian Ocean Program (NIOP), the GLOBEC project (Global Ocean Ecosystem Dynamics), the WOCE program (World Ocean Circulation Experiment) and the German BIGSET project (Biogeochemical transports of energy and matter in the deep-sea), the Arabian was visited between 1990 and 1998 by 18 ships from 9 countries, resulting in a total of 80 cruises (see Table 2).

Table 2. JGOFS and JGOFS-related cruises to the Arabian Sea since 1990. NEM = Northeast monsoon, SWM = southwest monsoon, SI = spring intermonsoon, FI = fall intermonsoon

Year	Country	Ship	Cruises	Season
1990	South Africa	RV Benguela	1	NEM
1991	South Africa	RFS Africana	1	NEM
		SA Agulhas	1	SI
1992	Pakistan	MV Aghyar-S	2	NEM, SW
	South Africa	RFS Africana	1	NEM
	Australia	RV Franklin	1	SI
	Netherlands	RV Tyro	9	SWM, FI, NEM
	India	ORV Sagar Kanya	1	SWM
1993	Netherlands	RV Tyro	2	NEM
	Pakistan	MV Mangan	1	NEM
	South Africa	SA Agulhas	1	SWM
	Australia	RV Lady Basten	2	SWM, FI
	India	ORV Sagar Kanya	1	SWM
1994	India	ORV Sagar Kanya	1	SI
	Pakistan	SV Behr Paima	2	SWM, NEM
	Great Britain	RRS Discovery	4	SWM, FI, NEM
	USA	RV Thomas Thompson	4	SWM, FI, NEM
		RV Knorr	1	NEM
1995	USA	RV Thomas Thompson	12	NEM, SI, SWM, FI
		RV Knorr	8	NEM, SI, SWM, FI
		RV Malcom Baldrige	3	NEM, SI, SWM
	India	ORV Sagar Kanya	2	NEM, SWM
	Germany	RV Meteor	8	NEM, SI, SWM, FI
	Australia	RV Franklin	1	SWM
1996	USA	RV Thomas Thompson	1	NEM
	India	ORV Sagar Kanya	1	SWM
1996/ 1997	Japan	Hakuho Maru	1	NEM
		Shoyo Maru	1	NEM
1997	India	ORV Sagar Kanya	1	NEM
	Germany	RV Sonne	4	NEM, SI, SWM
1998	Germany	RV Sonne	1	NEM

A vast amount of results obtained by these cruises was published in the book "Biogeochemistry of the Arabian Sea" edited by Lal (1994), in a special section of *Current Science* "JGOFS (India)" (Krishnaswami and Nair 1996), in several special issues of *Deep-Sea Research* (Weering et al. 1997, Smith 1998, 1999, 2000, 2001, 2002, Burkill 1999, Gage et al. 2000, Pfannkuche and Lochte 2000) and in *Progress of Oceanography* (Smith 2005) and as articles in regular scientific journals.

CONCLUSION

In old days seagoing expeditions to the Indian Ocean were mainly undertaken to map foreign and unknown countries and to explore new trade routes to Asia. In the 18th century the voyages of James Cook and Charles Darwin started a new era of investigative cruises that were mainly arranged out of scientific interest. Not only abutting nations, above all India, but the USA, Russia, Japan and several european countries sent out research vessels to study the biology, physical oceanography and chemistry of this part of the worlds ocean. Whereas in the first half of the 20th century these studies were carried out on a national basis, after World War II it came to several international cooperations to investigate different topics to extend the scientific knowledge of the Indian Ocean. Some of the leading international projects were the International Indian Ocean Expedition (IIOE), the Indian Ocean Experiment (INDEX), and the parts concerning the Indian Ocean or, particularly, the Arabian Sea, of GEOSECS (Geochemical Ocean Sections Study), JGOFS (Joint Global Ocean Flux Study) and WOCE (World Ocean Circulation Experiment).

REFERENCES

Aleem, A.A. and Morcos, S.A. (1984). John Murray/Mabahiss expedition versus the International Indian Ocean Expedition (IIOE) in retrospect. *Deep-Sea Research I, 31*, 583-588.

Banse, K. (1994). On the coupling of hydrography, phytoplankton, zooplankton, and settling organic particles offshore in the Arabian Sea. *Indian Academy of Science. Earth and Planetary Sciences, 103*, 125–161.

Burkill, P.H., Mantoura, R.F.C. and Owens, N.J.P. (1993). Biogeochemical cycling in the northwestern Indian Ocean: a brief overview. *Deep-Sea Research II, 40*, 643-649.

Burkill, P.H. (1999). ARABESQUE: An overview. *Deep-Sea Research II, 46*, 529-547.

Chun, C. (1902ff). Wissenschaftliche Ergebnisse der Deutschen Tiefsee-Expedition auf dem Dampfer □Valdivia“ 1898-1899. Verlag von Gustav Fischer, Jena, 24 Bände.

Deacon, G.E.R. and Rice, A.L. (1984). The significance of the John Murray / Mabahiss expedition to the Arabian Sea. *Deep-Sea Research I, 31,* 573-581.

Deacon, M. (1971). *Scientists and the Sea 1650 – 1900. A study of marine Science.* London/New York: Academic Press.

Desai, B.N. (1992). *Oceanography of the Indian Ocean.* New Delhi: Oxford and IBH Publishing Co. PVT. LTD.

Gage, J.D., Levin, L.A. and Wolff, G.A. (2000). Benthic processes in the deep Arabian Sea: introduction and overview. *Deep-Sea Research II, 47*, 1-7.

Gilson, H.C. (1937). The nitrogen cycle. *The John Murray Expedition, Sci. Rep., 2.*

Krishnaswami, S. and Nai, R.R. (1996). JGOFS (India) – Introduction. *Current Science, 71*, 831-833.

Lal, D. (1994). Biogeochemistry of the Arabian Sea. *Indian Academy of Science. Earth and Planetary Sciences, 103.*

Nellen, W., Bettac, W., Roether, W., Schnack, D., Thiel, H., Weikert, H. and Zeitzschel, B. (1996). MINDIK – Meteor Reise Nr. 5 vom 2. Januar 1987 – 24. September 1987. *Meteor-Berichte, Universität Hamburg, 96-1.* 2 Volumes.

Pfannkuche, O. and Lochte, K. (2000). The biogeochemistry of the deep Arabian Sea: overview. *Deep-Sea Research II, 47*, 2615-2628.

Rao, T.S.S. and Griffiths, R.C. (1998). *Understanding the Indian Ocean. Perspectives on Oceanography*. Paris: UNESCO press.

Rice, A.L. (1986). *Deep-sea challenge: the John Murray/Mabahiss expedition to the Indian Ocean, 1933-34.* Bungay: The Chaucer Press Ltd.

Scherzer, K. (1864-1865): *Reise der Österreichischen Fregatte Novara um die Erde in den Jahren 1857, 1858, 1859 unter den Befehlen des Commodore B. von Müllerstorf-Urbair / beschreibender Teil von Karl von Scherzer.* Wien: Gerold.

SCOR (1992) Joint Global Ocean Flux Study. Implementation Plan. Scientific Commitee on Oceanic Research. *JGOFS Report, 9.*

SCOR (1995). The Joint Global Ocean Flux Study. Arabian Sea Process Study. Scientific Committee on oceanographic Research. *JGOFS Report, 17.*

Seno, J., Komaki, Y. and Takeda, A. (1963a). Reports on the biology of the Umitaka-Maru expedition. Plankton collected by the Umitaka-Maru in the Antarctic and adjacent waters by larva net, with special references to Copepoda. *Journal of the Tokyo University of Fisheries, 50*, 1-10.

Seno, J., Komaki, Y. and Takeda, A. (1963b). Reports on the biology of the Umitaka-Maru expedition. Plankton collected by the Umitaka-Maru in the Antarctic and adjacent waters, with special reference to Copepoda. *Journal of the Tokyo University of Fisheries, 49*, 53-62.

Seno, J., Komaki, Y. and Takeda, A. (1966). Report on the biology of the Umitaka-Maru expedition. Plankton collected in the Antarctic Ocean and adjacent waters by the closing net, with special reference to copepods. *Journal of the Tokyo University of Fisheries, 52*, 1-16.

Sewell, R.B.S. (1935/36): *The John Murray Expedition 1933-34. Introduction and list of stations.* London: British Museum (Natural History).

Sewell, R.B.S. (1947a): The free-swimming planktonic Copepoda. *Scientific Reports. The John Murray expedition 1933-34, 8,* 1-303.

Sewell, R.B.S. (1947b): The free-swimming planktonic Copepoda. Geographical distribution. *Scientific Reports. The John Murray expedition 1933-34, 8,* 317-592.

Smith, S.L. (1982). The northwestern Indian Ocean during the monsoons of 1979: distribution, abundance, and feeding of zooplankton. *Deep-Sea Research I, 29*, 1331-1353.

Smith, S.L. (1998). The 1994-1996 Arabian Sea Expedition: Oceanic Response to Monsoonal Forcing, Part 1. *Deep Sea Research II, 45*, Nos. 10-11

Smith, S.L. (1999). The 1994-1996 Arabian Sea Expedition: Oceanic Response to Monsoonal Forcing, Part 2. *Deep Sea Research II, 46*, Nos. 8-9.

Smith, S.L. (2000). The 1994-1996 Arabian Sea Expedition: Oceanic Response to Monsoonal Forcing, Part 3. *Deep Sea Research II, 47*, Nos. 7-8.

Smith, S.L. (2001). The 1994-1996 Arabian Sea Expedition: Oceanic Response to Monsoonal Forcing, Part 4. *Deep Sea Research II, 48*, Nos. 6-7.

Smith, S.L. (2002). The 1994-1996 Arabian Sea Expedition: Oceanic Response to Monsoonal Forcing, Part 5. *Deep Sea Research II, 49*, No. 12.

Smith, S.L. (2005). The Arabian Sea of the 1990s: New biogeochemical understanding. *Progress in Oceanography, 65*, 113-115.

Steuer, A. (1933). Zur planmässigen Erforschung der geographischen Verbreitung des Halipanktons besonders der Copepoden. *Zoogeographica I, Pt. 3.*

Thomson, C.W. and Murray, J. (1885-1895). *Reports on the scientific results of the voyage of H.M.S. Challenger during the years 1873 - 1876*. London: Her Majesty's Government.

Tseitlin, V.B. and Rudyakov, Y.A. (1999). Seasonal variations of vertical distribution of zooplankton in the mesopelagic zone of the Indian Ocean. *Oceanology, 39*, 521-527.

Vinogradov, M.E. (1968). *Vertical distribution of the Oceanic zooplankton.* Academy of Science of the U.S.S.R., Institut of Oceanography. In Russian, translated by Israel Program for Scientific Translation Ltd. Jerusalem: Keter Press.

Weering, T.C.E.v., Helder, W. and Schalk, P. (1997). The Netherlands Indian Ocean Expedition 1992-1993, first results and an introduction. *Deep-Sea Research II, 44*, 1177-1193.

Wolff, T. (1968). The Danish Expedition to "Arabia Felix" (1761-1767). *Bulletin de l'Institut Océanographique de Monaco, Numero Special 2*, 581-601.

Wooster, W.S. (1984). International studies of the Indian Ocean, 1959-1965. *Deep-Sea Research I, 31*, 589-597.

Zeitzschel, B. (1973). *The biology of the Indian Ocean.* Berlin, Heidelberg, New York: Springer.

In: Atlantic and Indian Oceans
Editors: E. S. Askew and J. P. Bromley, pp. 169-178
ISBN 978-1-60692-475-4

Chapter 8

MERIDIONAL HEAT FLUXES IN THE NORTH INDIAN OCEAN

***T. Rojsiraphisal*[a] *and L. Kantha*[b]**

[a]Department of Mathematics, Burapha University, Thailand
[b]Department of Aerospace Engineering Sciences,
University of Colorado, Boulder, Colorado, USA

ABSTRACT

The seasonal cycle of the cross-equatorial heat transport is of great importance to South Asian monsoons since it is a component of the heat budget and hence determines the SST of the North Indian Ocean. In this chapter, we use a data-assimilative model of the North Indian Ocean assimilating altimetry and MCSST data and driven by ECMWF and QuikSCAT wind stresses, to explore the seasonal and inter-annual variability of the meridional heat fluxes during the years 1993 to 2005. In addition to a strong seasonal variability, the cross-equatorial heat transport also exhibits considerable inter-annual variability. There are also strong, short period intra-seasonal fluctuations on time scales of 20-50 days. The meridional heat flux values are 20-30% higher when the hindcast is driven by observed QuikSCAT wind stresses than when it is driven by ECMWF winds.

1. INTRODUCTION

The overwhelming determinant of the circulation in the North Indian Ocean are the seasonally reversing monsoon winds. During the boreal summer, the winds blow from the southwest. The Southeast Trade Winds south of the equator cross the equator and continue into the Arabian Sea in the form of a strong, southwesterly low-level jet along the Somali and Oman coasts. The currents in the surface layers across the equator are northward. During the boreal winter, the winds are exactly in the opposite direction to summer monsoon winds, northeasterly. This gives raise to southward cross-equatorial currents in the upper layers. Thus, there is a strong seasonal component to the meridional heat, mass and salt fluxes across the equator in the upper layers. However, the net cross-equatorial transport of heat depends on the currents in the entire water column. A numerical model can shed light on the seasonal and inter-annual variability of the meridional fluxes. We have used a data-assimilative model of the North Indian Ocean assimilating altimetry and MCSST data and driven by ECMWF (EC henceforth) and QuikSCAT (QS henceforth) wind stresses, to explore this variability during the years 1993 to 2005.

The numerical model used for the hindcast is the University of Colorado version of the Princeton Ocean Model (CUPOM). It is a primitive equation model using topographically conformal coordinate in the vertical and orthogonal curvilinear coordinates in the horizontal. The sea surface height is calculated explicitly using the split-mode technique. CUPOM includes a second moment closure-based model (Kantha and Clayson, 1994) of turbulent mixing in the upper and bottom layers. More details about the basic features of CUPOM can be found in Kantha and Clayson (2000b).

The model has 38 sigma levels in the vertical, with the levels closely spaced in the upper 300 m. The sigma levels are 0, 1, 2, 5, 10, 20, 30, 40, 50, 60, 70, 80, 90, 100, 110, 120, 130, 140, 150, 160, 170, 180, 190, 200, 210, 220, 230, 240, 250, 270, 300, 400, 700, 1000, 2000, 3000, 4000, 5000 m in a water column of 5000 m depth. The high vertical resolution in the upper 300 m is deliberately chosen to better simulate the near-surface circulation.

Figure 1 shows the model domain, along with bottom topography. The model is forced by 6-hourly EC winds. It assimilates altimetric sea surface height anomalies and weekly composite MCSST using a simple optimal interpolation-based assimilation technique. The model was run at two different values of horizontal resolution: 1/2° and 1/4° to explore the sensitivity of the results to model resolution.

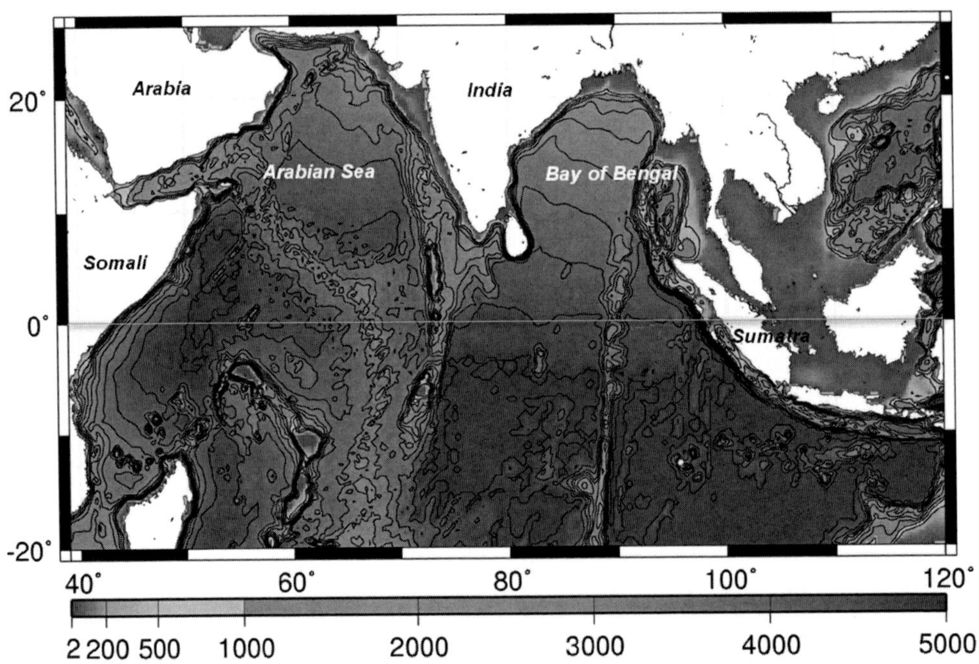

Figure 1. The North Indian Ocean showing the bottom topography and the model domain.

The model, and the assimilation methodology based on conversion of SSH anomalies into pseudo-BT anomalies for adjusting the model temperature field via optimal interpolation, have been described in detail by Lopez (1998), Lopez and Kantha (2000a and b) and Kantha et al. (2008), and will not be repeated here. For details of the model and the hindcast, including comparisons of model results with observational data, see Kantha et al. (2008).

2. Meridional Heat Fluxes

The surface transport is generally southward on both sides of the equator during summer, but northward in winter. The net annual heat flux across the equator is southward and occurs predominantly via the cross-equatorial cell (Schott and McCreary, 2001). Figure 2 shows the meridional heat fluxes in PW (1 PW = 10^{15} W) across various zonal transects (5 °S, 0 °N, 5 °N, 10 °N, 15 °N and 20 °N) averaged over the entire hindcast period, compared with results of other studies in the past. Both 1/2° and 1/4° results are shown. The hindcast results are in good agreement with the values of Hastenrath and Greischar (1993) derived from air-sea flux calculations, and Lee and Marotzke (1998) derived from a numerical model, except that the zero heat flux occurs at about 12-13 °N instead of 18°N. The observed QS wind stresses produce a higher value of meridional heat fluxes than the EC winds obtained from the coarse resolution Numerical Weather Prediction (NWP) model in general. The net southward meridional heat flux across the equator varies between 0.18 and 0.55 PW.

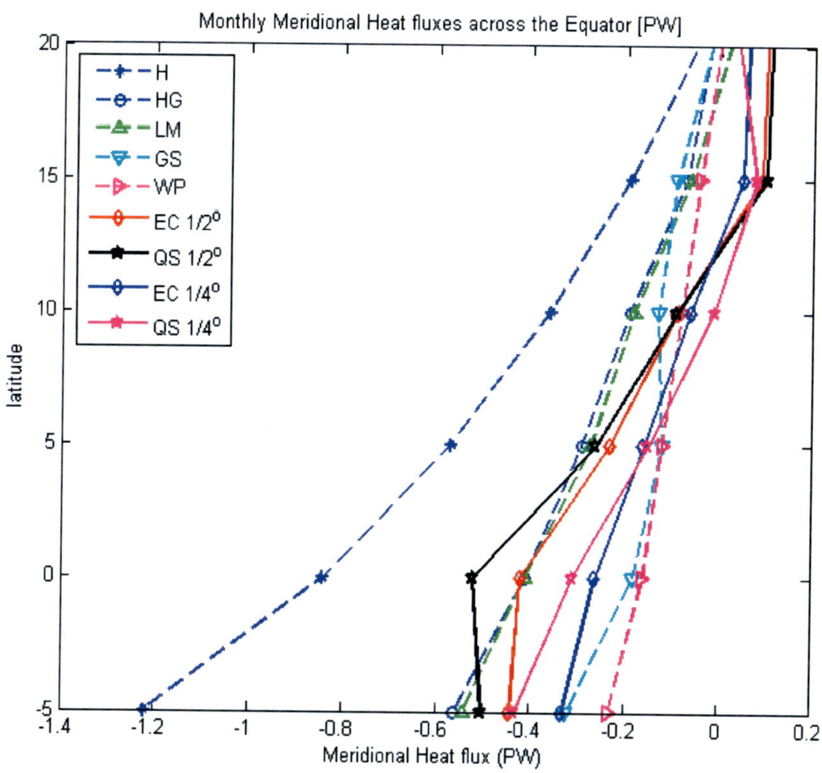

Figure 2. Meridional heat flux across zonal transects from various studies redrawn from Scott and McCreary, 2001 (H – Hsiung, 1985; HG- Hastenrath and Greischar, 1993; LM – Lee and Marotzke, 1998; GS – Gasternicht and Schott, 1997; WP – Wacongne and Pacanowski, 1996; EC – Present study with ECMWF winds; QS- Present study with QuikSCAT wind stresses).

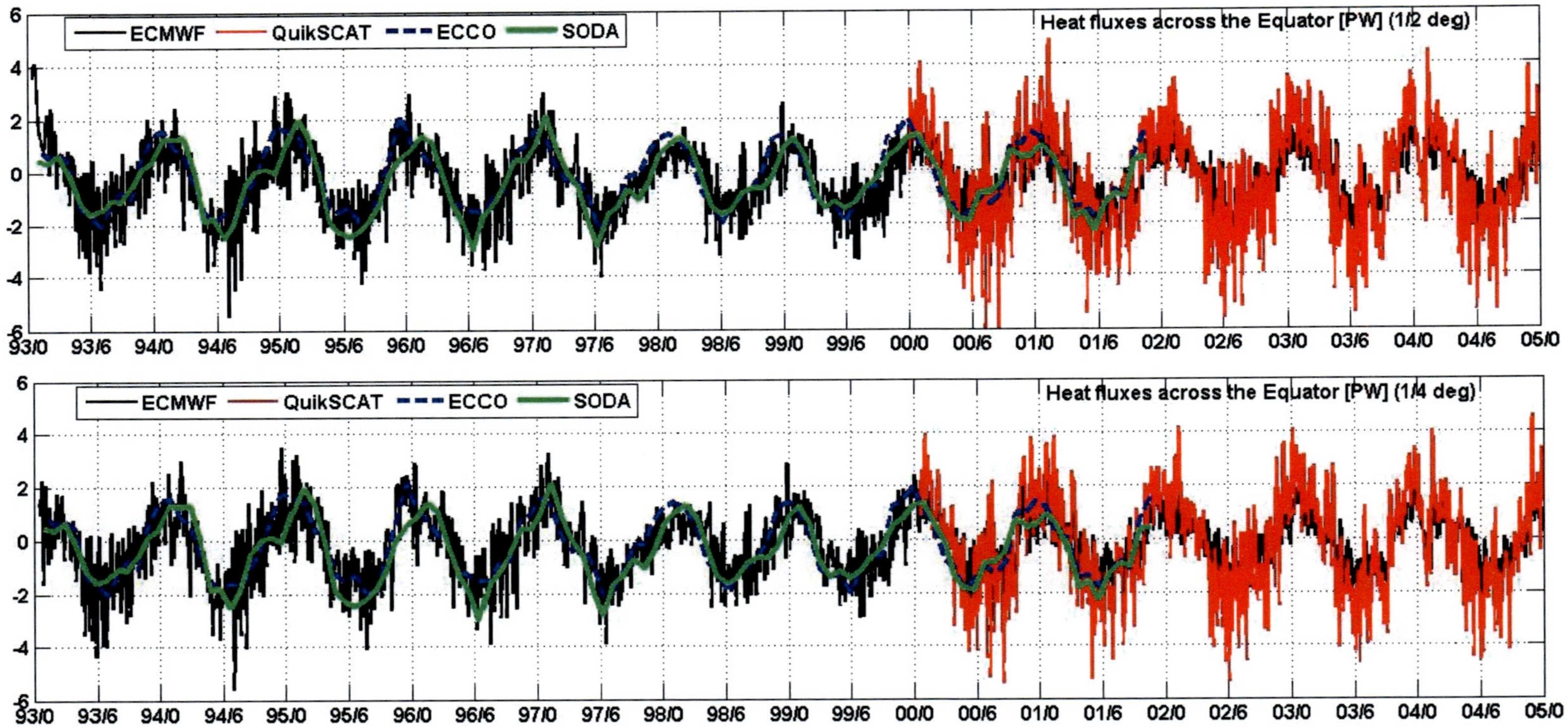

Figure 3. Meridional heat flux (in PW) across the equator from 1/2° (top panel) and 1/4° (bottom panel) models compared to that from ECCO (blue dotted line) and SODA (green) global models. Both the results from the model driven by ECMWF winds (black line) and QuikSCAT wind stresses (red line) show considerable seasonal and inter-annual variability, with the latter being significantly larger than the former. Note the anomalously weak fluxes between 1998 and 2000.

Figure 3 shows the 7-day running mean of the cross-equatorial heat flux for the years 1993 to 2005. While the seasonal signal is dominant, considerable intra-seasonal and inter-annual variability are present. The values from the University of Maryland SODA model, which uses the GFDL MOM2.b driven by monthly winds derived from COADS (1950-1991) and NCEP (1992-2001), are also shown (green line). SODA assimilates hydrographic data, SST and altimetric SSHA (Carton et al. 2000). The dotted blue line shows values from the ECCO project that uses the MIT non-hydrostatic GCM driven by NCEP products (Koehl 2002). Both ECCO and SODA have coarse 1°x1° horizontal resolutions, although SODA's resolution increases to 0.5^ox1^o in the tropics. There is general agreement between all three model results, although the overall mean values differ (1/2° EC: -0.42 PW, 1/2° QS: -0.52 PW, 1/4° EC: -0.26, 1/4° QS: -0.31, SODA: -0.41 PW, ECCO: -0.19 PW). Note the anomalously weak fluxes between 1998 and 2000, with a reduction of about 40% in peak-to-peak values compared to the normal years.

Figures 4a and b show the mean monthly meridional heat fluxes (averaged over the simulation years, 12 years for QC and 5 years for QS) compared with those from Hastenrath and Greischar (1993), Garternicht and Schott (1997) and Hsiung, Newell and Houghtby (1989). The pattern of variability is in good agreement. However, the heat fluxes are considerably larger, especially for the QS-driven hindcasts. This is not surprising because the QS wind stresses are observed values and EC winds are coarse resolution NWP model-generated values.

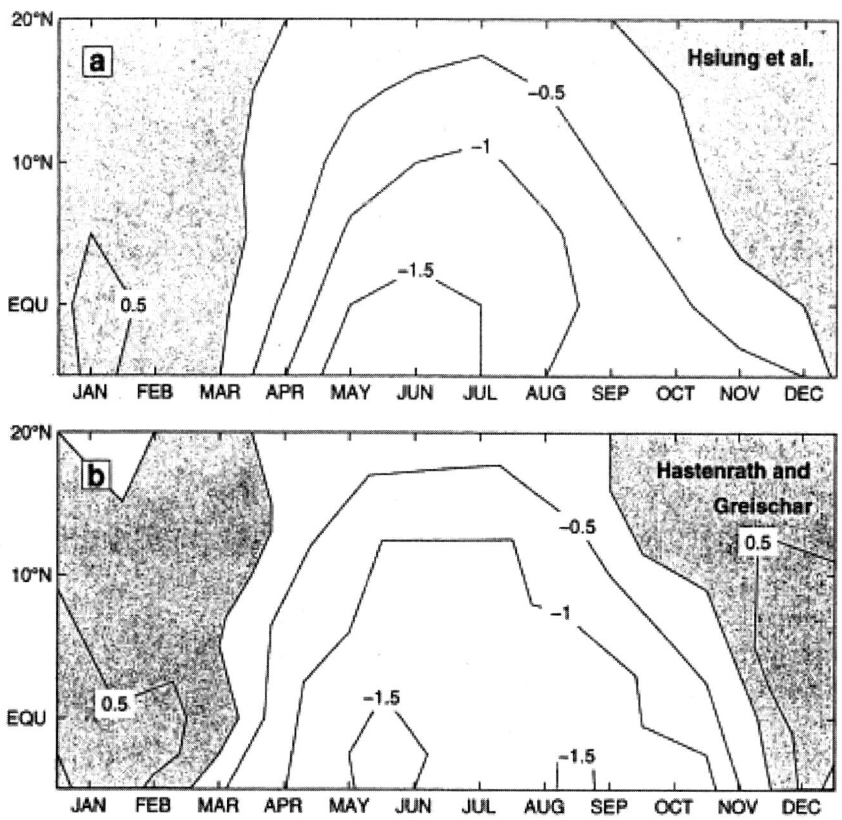

Figure 4a. (Continued).

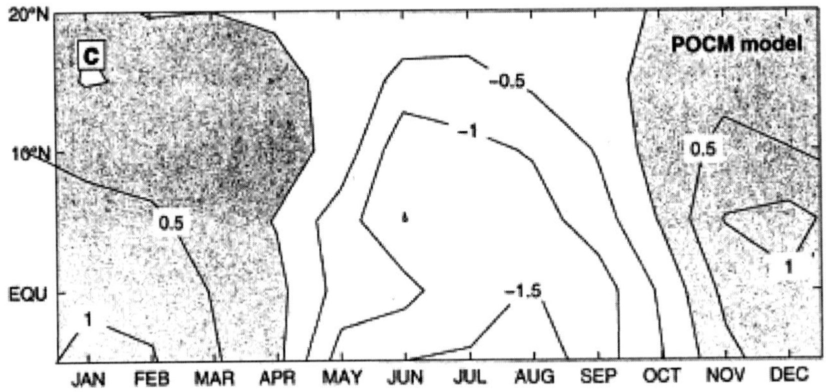

Figure 4a. Monthly meridional heat fluxes (in PW) from three different studies (adapted from Schott and McCreary 2001).

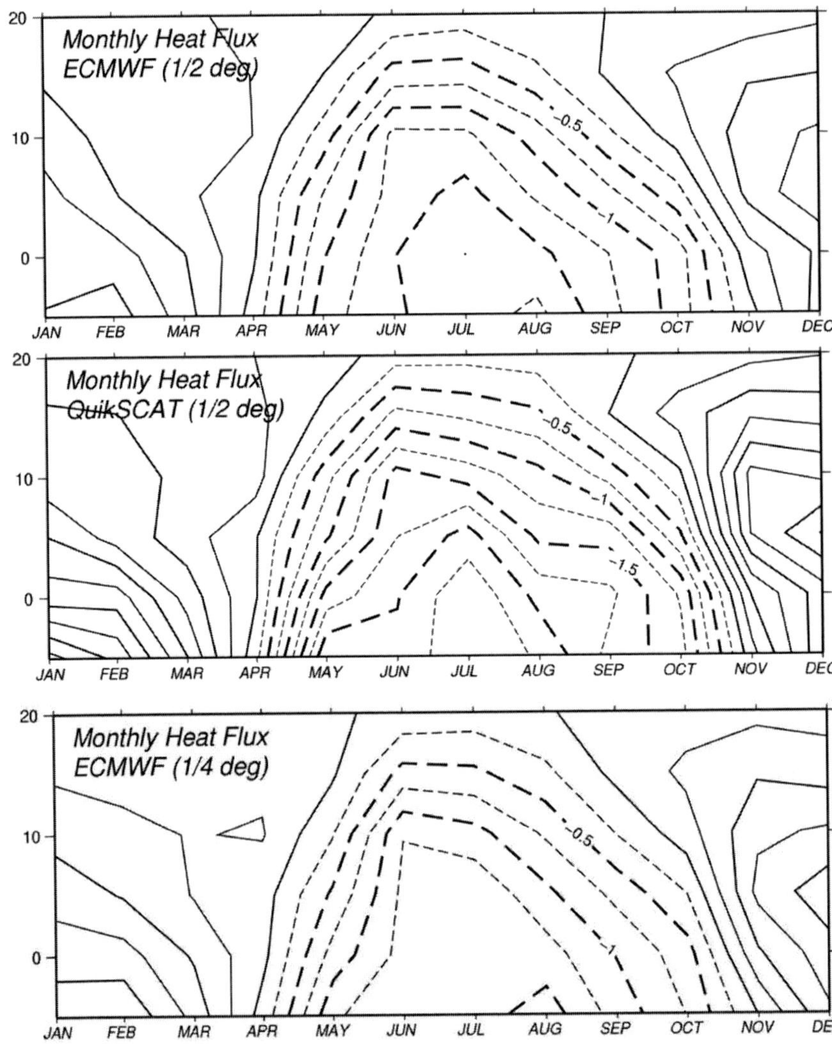

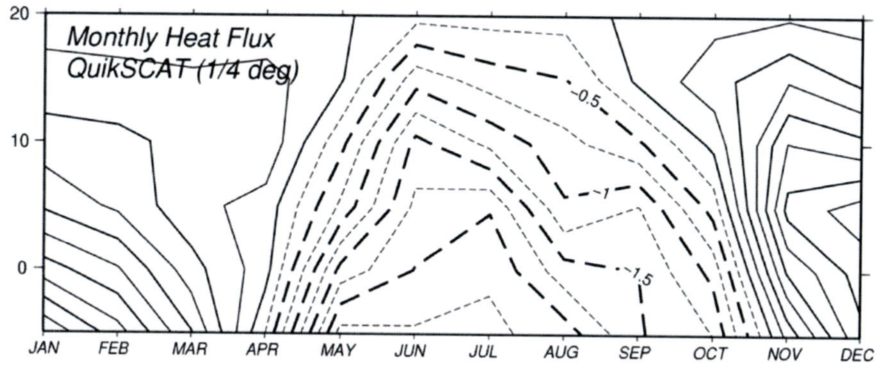

Figure 4b. Monthly meridional heat fluxes (in PW) from the hindcast.

Figure 5 shows meridional heat fluxes for the year 1996 (a normal year) and 1998 (an anomalous year), in comparison with the 12-year monthly means from the 1/4° model. The anomalous wind patterns during the anomalous year of 1998 give rise to stronger fluxes overall, compared to the normal year.

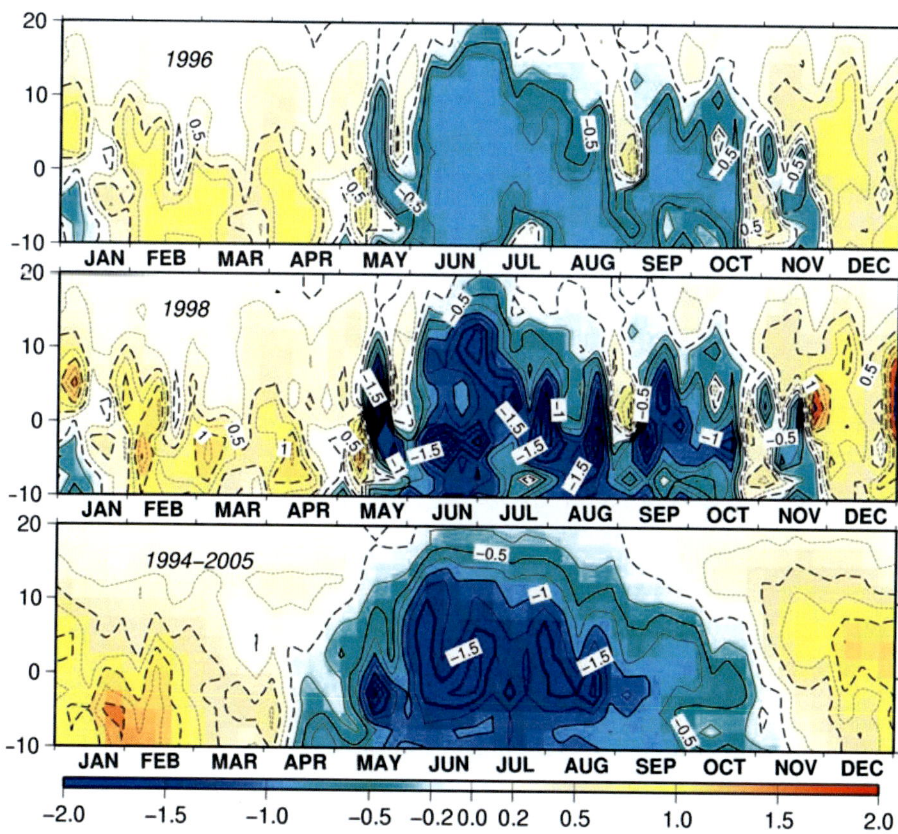

Figure 5. Monthly mean Meridional heat fluxes for the normal year 1996 and the anomalous year 1998 compared with the monthly mean over the 12-year hindcast period for the 1/4° model driven by ECMWF winds.

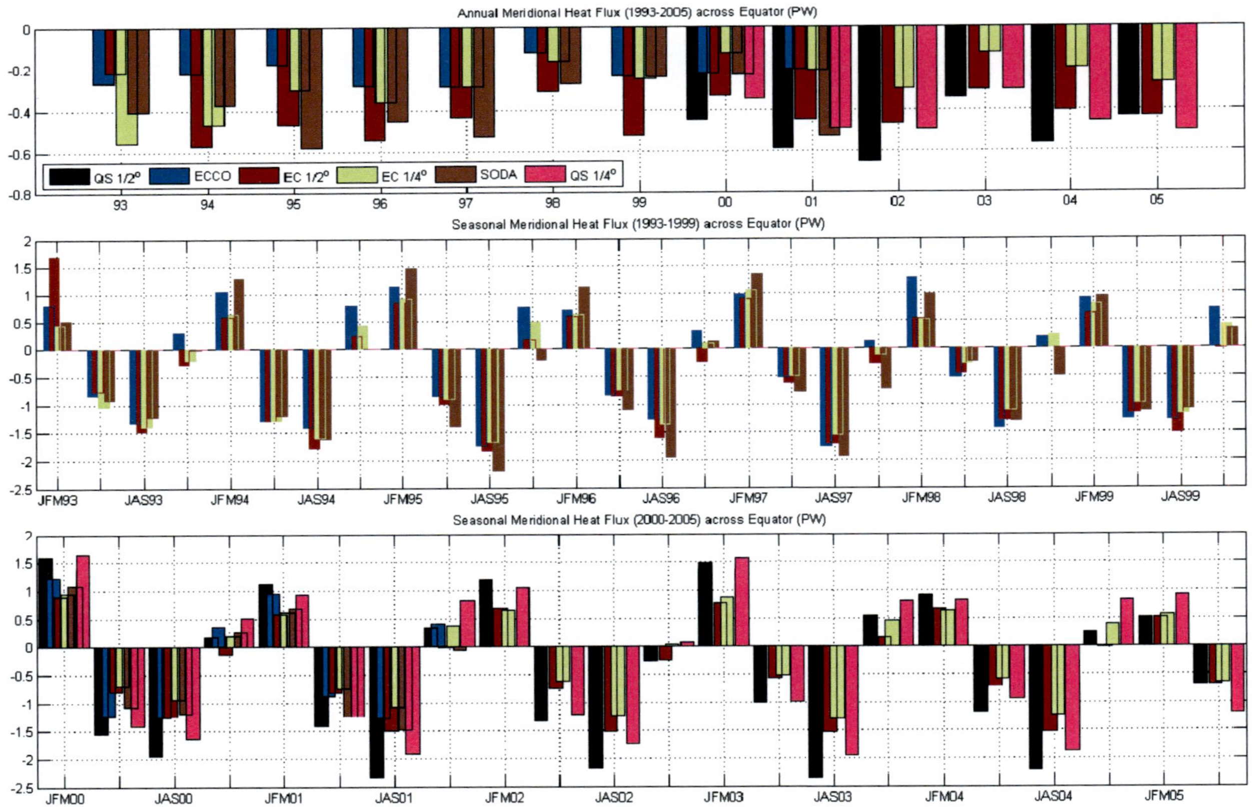

Figure 6. Annual (top panel) and seasonal (middle and bottom panels) means of the cross-equatorial heat flux (in PW) from 1/2° and 1/4° models compared to those from ECCO and SODA global models. The results from both the hindcasts driven by ECMWF winds and QuikSCAT wind stresses are shown.

Finally, Figure 6 shows annual and seasonal means of the cross-equatorial heat flux from various models for years 1993 to 2005. The annual mean shows considerable year-to-year variability, with values ranging from 0.1 PW to slightly above 0.6 PW southward. The 3-monthly means also exhibit considerable variability but with larger values of as large as 2.3 PW. Overall, all models indicate a northward transport during January-February-March and October-November-December, but southward during the rest of the year.

CONCLUSIONS

Using a simple 2½ layer model of *McCreary et al.* [1993], *Loschingg and Webster* [2000] found that the SST in the North Indian Ocean is regulated by strong advection across the equator, with a cross-equatorial heat flux of about 1.0 PW northward during the boreal winter and 0.5-1.0 PW southward during the boreal spring-summer, and a net transport southward of 0.2 PW. The strong seasonal variability (along with a small net southward transport) they infer is consistent with our findings here, reinforcing the notion that the cross-equatorial oceanic heat transport is essential to the seasonal cycle of the heat balance of the North Indian Ocean.

As can be seen from the results presented above, in addition to a strong seasonal variability, the cross-equatorial heat transport also exhibits considerable inter-annual variability. The anomalously weak fluxes are evident during 1998-2000, following the anomalous 1997-98 warming event. There are also strong, short period intra-seasonal fluctuations on time scales of 20-50 days. The meridional heat flux values are 20-30% higher when the model is driven by observed QuikSCAT wind stresses than when it is driven by ECMWF winds. Since most modeling studies use similar coarse resolution NWP winds, this point is of considerable interest in issues related to the oceanic heat transport in the global oceans.

ACKNOWLEDGEMENTS

LK acknowledges partial support by ONR through grant N00014-06-10287. Our sincere thanks to Dr. Rena Schoenefeldt for providing us with the ECCO and SODA cross-equatorial heat flux data.

REFERENCES

Carton, J. A., Chepurin, G., Cao X., and Giese, B. (2000). A Simple Ocean Data Assimilation analysis of the global upper ocean 1950-95. Part I: Methodology. *J. Phys. Oceanogr.*, 30, 294-309.

Garternicht, U., and Schott, F. (1997). Heat fluxes of the Indian Ocean from a global eddy-resolving model. *J. Geophys. Res.*, 102, 21147-21159.

Hastenrath, S. and Greischar, L. (1993). The monsoonal heat budget of the hydrosphere-atmosphere system in the Indian Ocean sector. *J. Geophys. Res.*, 98, 6869-6881.

Hsuing, J. (1985). Estimates of global oceanic meridional heat transport. *J. Phys. Oceanogr., 15,* 1405-1413.

Hsuing, J., Newell, R. E., and Houghtby, T. (1989).The annual cycle of oceanic heat storage and oceanic meridional heat transport. *Quart. J. Roy. Meteorol. Soc.,* 115, 1-28.

Kantha,L.H.,and C.A.Clayson (1994), An improved mixed layer model for geophysical applications, *J. Geophys. Res.*, 99, 25,235–25,266.

Kantha, L. H., and C. A. Clayson (2000), *Numerical Models of Oceans and Oceanic Processes*, 940 pp., Academic, SanDiego, Calif.

Kantha, L., T. Rojsiraphisal, and J. Lopez, 2008. The North Indian Ocean circulation and its variability as seen in a numerical hindcast of the years 1993 to 2004. *Prog. Oceanogr.*, 76, 111-147.

Koehl, A., Lu, Y., P. Heimbach, P., Cornuelle, B., Stammer, D., and Wunsch, C. (2002). The ECCO 1 degree global WOCE Synthesis, *ECCO Report No.20*, See also http://www.ecco–group.org/ reports.html.

Lee, T., and Marotzke, J. (1998). Seasonal cycles of meridional overturning and heat transport of the Indian Ocean. *J. Phys. Oceanogr.,* 28, 923-943.

Lopez, J. W. (1998). *A study of physical processes of the northern Indian Ocean using a comprehensive primitive equation numerical model.* Ph. D. dissertation, Aerospace Engineering Sciences, University of Colorado, Boulder, CO. pp.144.

Lopez, J. W., and Kantha, L. H. (2000a). Results from a numerical model of the northern Indian Ocean: Circulation in the South Arabian Sea. *J. Mar. Syst.*, *24*, 97-117.

Lopez, J. W., and Kantha, L. H. (2000b). A data-assimilative model of the North Indian Ocean. *J. Atmos. and Oceanic Tech.*, 17, 1525-1540.

Loschnigg, J., and P. J. Webster, (2000). A coupled ocean-atmosphere system of SST modulation for the Indian Ocean, *J. Climate*, 13, 3342-3360.

McCreary, J. P. Jr., Kundu, P. K., and Molinari, R. L. (1993). A numerical investigation of dynamics, thermodynamics and mixed layer processes in the Indian Ocean. *Prog. Oceanogr.*, 31, 181-244.

Schott, F., and McCreary, J. P. (2001). The monsoon circulation of the Indian Ocean. *Prog. Oceanogr.*, 51, 1-123.

Wacongne, S., and Pacanowski, R. (1996). Seasonal heat transport in a primitive equations model of the tropical Indian Ocean. *J. Phys. Oceanogr.,* 26, 2666-2699.

Submitted March 6, 2008

In: Atlantic and Indian Oceans
Editors: E. S. Askew and J. P. Bromley, pp. 179-199
ISBN 978-1-60692-475-4

Chapter 9

VARIABILITY OF CURRENTS AT 90°E IN THE EQUATORIAL INDIAN OCEAN

T. Rojsiraphisal[a], L. Kantha[b] and Y. Masumoto[c]
[a]Department of Mathematics, Burapha University, Thailand
[b]Department of Aerospace Engineering Sciences,
University of Colorado, Boulder, Colorado, 80309, USA
[c]Frontier Research Center for Global Change, JAMSTEC, Yokohama, Japan

ABSTRACT

Equatorial currents at 90° E from a numerical hindcast of the North Indian Ocean are analyzed for the years 2001 to 2004 along with ADCP currents from a mooring deployed there by the Japan Agency for Marine-Earth Science and Technology (JAMSTEC). The analysis indicates that the model captures a substantial portion of the low frequency variability in zonal, and to a lesser extent, meridional currents. Currents from the 1/4° model are more energetic than the 1/2° model, and QuikSCAT-derived wind stresses produce more energetic and more realistic currents than ECMWF-derived ones. The dominant variability is annual and semi-annual. Wavelet analysis highlights the significant events during this period. The 13-year hindcast of the years 1993-2005, driven by ECMWF winds, clearly highlights the anomalous 1997-98 period.

1. INTRODUCTION

The Indian Ocean plays an important role in Asiatic monsoons and yet it is the least explored of the three primary basins. A routine monitoring system of in-situ sensors such as the TOGA-TAO array in the equatorial Pacific is yet to be fully deployed. Consequently, important events, such as the 1997-98 anomalous warming, have gone mostly unobserved except by satellite-borne sensors such as the altimeter and the AVHRR. Therefore, it is hard to understand via observations alone, the state of the Indian Ocean during these interesting periods. A numerical hindcast using a reliable ocean model, assimilating altimetry and MCSST data, is useful in filling gaps in our knowledge of events and processes, and understanding the variability in circulation. With this in mind, we have carried out a 13-year hindcast from 1993 to 2005, with the model driven by ECMWF winds. We have also conducted hindcasts using QuikSCAT wind stress from 1999 onwards. Both these hindcasts have been performed at both 1/2° and 1/4° model resolutions. Details of the model and the hindcasts can be found in Kantha et al. (2008, see also Rojsiraphisal and Kantha 2008), as also the model skill assessment. Here we use the model hindcasts to analyze the variability of currents on the equator at 90°E during years 2001 to 2004, and compare the results with observed variability deduced from an ADCP mooring deployed there by the Japan Agency for Marine-Earth Science and Technology (JAMSTEC).

The circulation in the Indian Ocean is dominated by reversing seasonal monsoons. Most studies in the past have focused on the variability of Somali current and the western Indian Ocean, for example: development and evolution of circulation near the surface by Swallow et al., 1983; effect of winds on equatorial variability by Jensen (1993); and studies of seasonal variability in Somali Current by Schott et al. (1990). Very few studies of variability in the eastern Indian Ocean (EIO) have ever been done; notable exceptions are studies of: circulation in the Bay of Bengal using hydrographic data by Murthy et al., 1992(a) and 1992(b); variability along the southern coast of Sumatra (Iskandar, Masumoto and Yamagata, 2005); inter-seasonal current variability at 90°E, 0°N (Masumoto et al., 2005). Many of these studies are based upon hydrographic surveys. In this paper, we explore the variability of currents at 90°E, 0°N by analyzing and comparing currents obtained from numerical hindcasts and ADCP. The study uses power spectrum density and wavelet analyses of the time series of currents in the upper water column.

2. Model and Data

Model data used in this study are obtained from the University of Colorado version of the Princeton Ocean Model (CUPOM) applied to the Indian Ocean. The model covers the domain from 39°E to 120°E and from 20°S to 26°N. It is a primitive equation numerical circulation model using topographically conformal coordinates in the vertical and orthogonal curvilinear coordinates in the horizontal (Kantha and Clayson, 2000). Model simulations are done at two horizontal resolutions (1/2° and 1/4°). There are 38 sigma levels in the vertical. The model assimilates 3-day composite MCSST and altimetric SSHA from TOPEX/Poseidon, Jason1 and ERS-2 altimeters (Lopez and Kantha, 2000a, 2000b).

In this study, the model was forced by two sources of wind stresses; one is from the European Centre for Medium-Range Weather Forecasts (ECMWF) reanalysis winds with Smith drag coefficient [Smith, 1980], and the other is the QuikSCAT (QS) wind stress. Figure 1 shows the model domain and the bottom topography.

Figure 2 shows the SSH, SST and zonal currents along the equator from 1993 to 2004. The anomalous 1993-94 and even more anomalous 1997-98 years can be clearly seen, with SSH and SST anomalously depressed at the eastern end during the fall. The Wyrtki-Yoshida jet normally seen during the transition seasons disappeared in 1997 and the equatorial Indian Ocean behaved more like the equatorial Pacific. Comparatively speaking, the years 2001 to 2004 were near normal. The signature of the anomalous years can also be seen in the zonal currents at these stations (Figure 3). The eastward flowing EUC at a depth of 100 – 150 m can be seen during this anomalous period at all three stations.

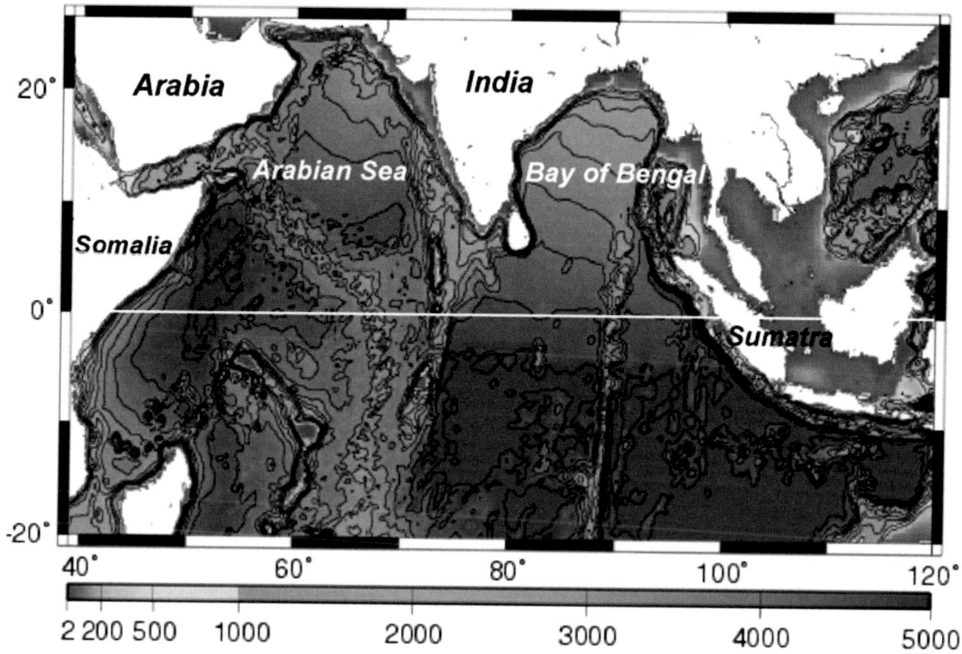

Figure 1. The model domain and bottom topography.

The model data were saved three-hourly at 90°E, 0°N at all vertical levels. The ADCP data is available hourly starting from November 14, 2000 to July12, 2004 with some but not very many short missing segments. The hourly ADCP data are simply averaged to obtain 3 hourly data for comparison with the model currents. In order to filter out the high frequency noise, we applied a 5-day running mean (Emery and Thomson, 2001) to ADCP and hence to all model data. The running mean also fills in short gaps in the ADCP data.

Vertical profiles of zonal and meridional currents at 90°E, 0°N from ADCP, as well as from the model are shown in Figures 4 and 5, respectively. The profiles extend from the surface to 400 m depth, except for ADCP currents, which are available only from 50 m down. Near the surface (50-100m), eastward currents are prominent most of the year and strong eastward currents are observed in April-May. This is due to the westerly winds along the equator which drive strong eastward surface jets known as Wyrtki-Yoshida jets (Wyrtki, 1973). *Luyten and Swallow* (1976) found an eastward upper layer flow and a westward flow underneath between May and June 1976. This event reoccurred between May and June during 2001 to 2004 as seen in this study. We also observe that the model zonal current forced by QS is stronger in magnitude than the one forced by ECMWF winds (Figure 1a and 1b -bottom two panels). Westward currents near the surface are found in January to March 2001to 2004. Deep layer currents reverse their directions periodically. The model currents correlate well with ADCP currents, although the agreement is by no means completely satisfactory.

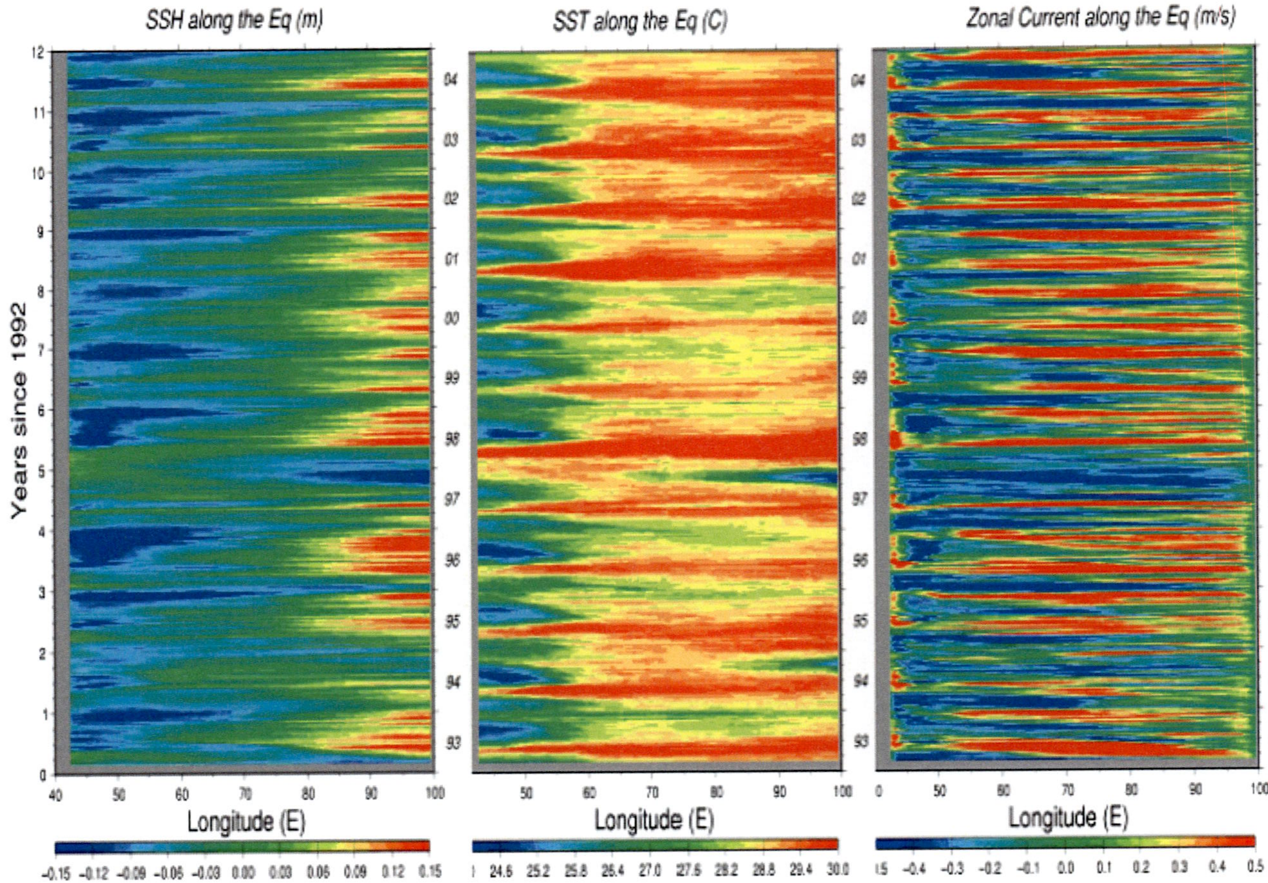

Figure 2. Hovmuller plots of SSH, SST and zonal currents along the equator over the 12 years of the hindcast run. Note the signature of the highly anomalous 1997-98 years, when SSH and SST were anomalously low at the eastern end of the waveguide and the zonal current was westward. Relatively weaker 1993-94 anomaly is also evident.

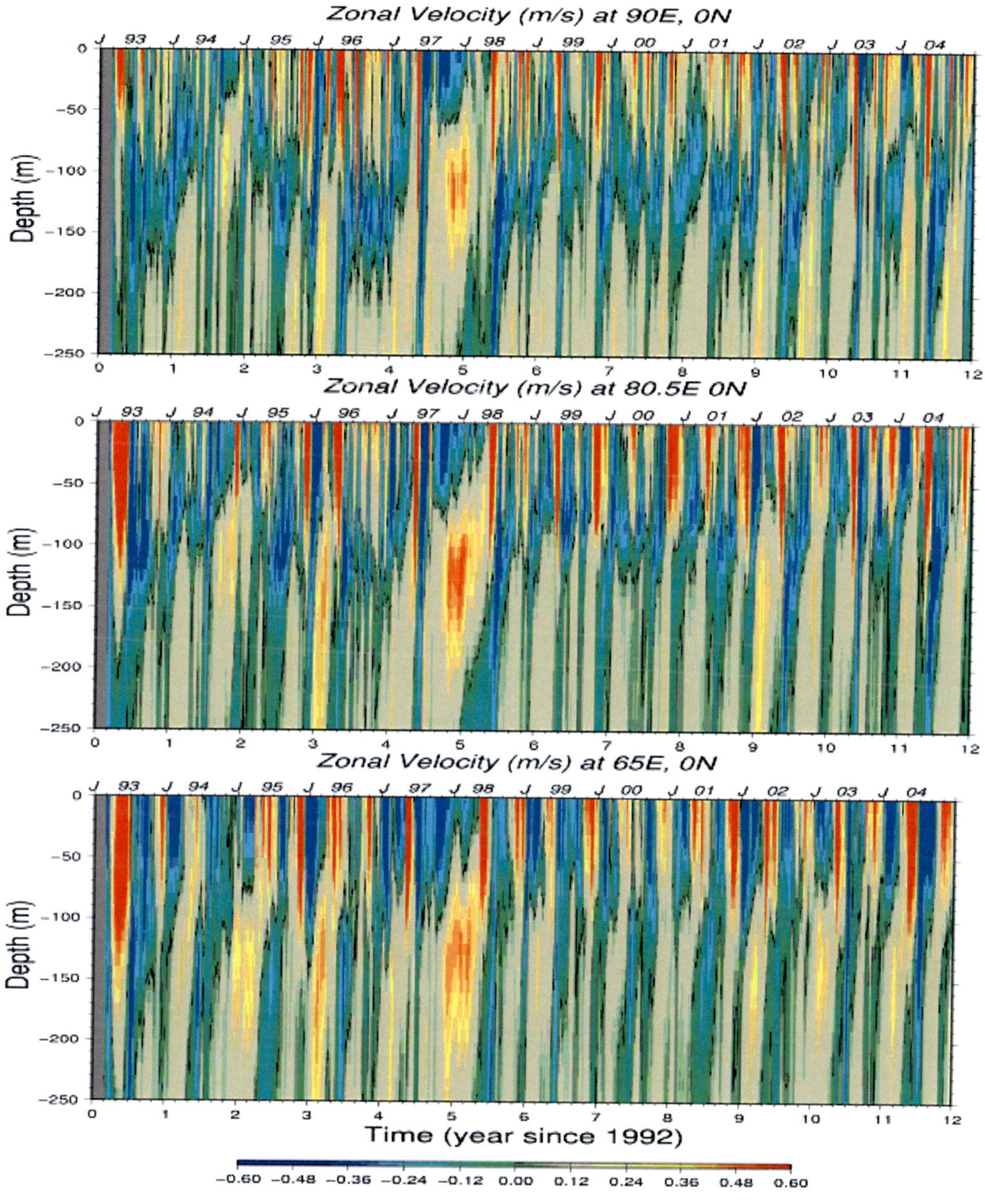

Figure 3. Zonal currents at the equator at 65 °E, 80.5 °E and 90 °E. The anomalous 1997-98 period with a eastward-flowing EUC at 100-150 m depth can be seen clearly at all 3 longitudes.

Meridional currents have a more complex pattern as shown in Figures 5a and 5b. Most of the time, currents are southward with fluid being advected out of the Bay of Bengal across the equator throughout the deep. Kantha et al. (2008) reported the annual heat flux and salt flux transport southward across the Equator. The observed currents with negative mean meridional velocities confirm their study. The variability in meridional currents is of higher frequency than the variability in the zonal currents. Once again, the 1/4° model currents are more energetic and variable than 1/2° model currents, and QS forcing yields currents that are closer in magnitude to the observed currents than ECMWF forcing.

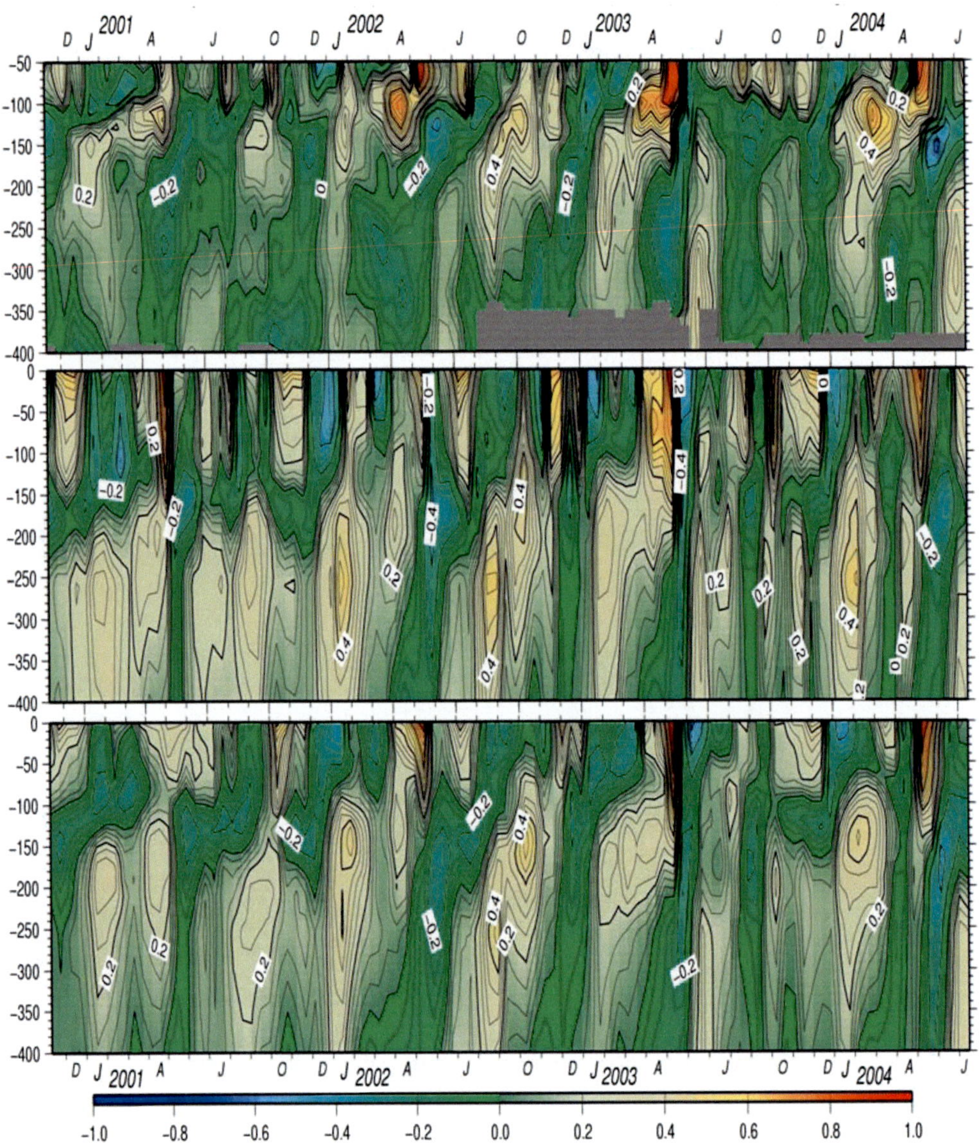

Figure 4a. Model (1/4°) zonal velocities compared with ADCP currents at 90°E, 0°N: ADCP currents (top), model currents with forcing by QuikSCAT wind stresses (middle), and model results with forcing by ECMWF winds (bottom). Contour interval is 0.05 m/s. Note that ADCP currents are not available above 50m depth.

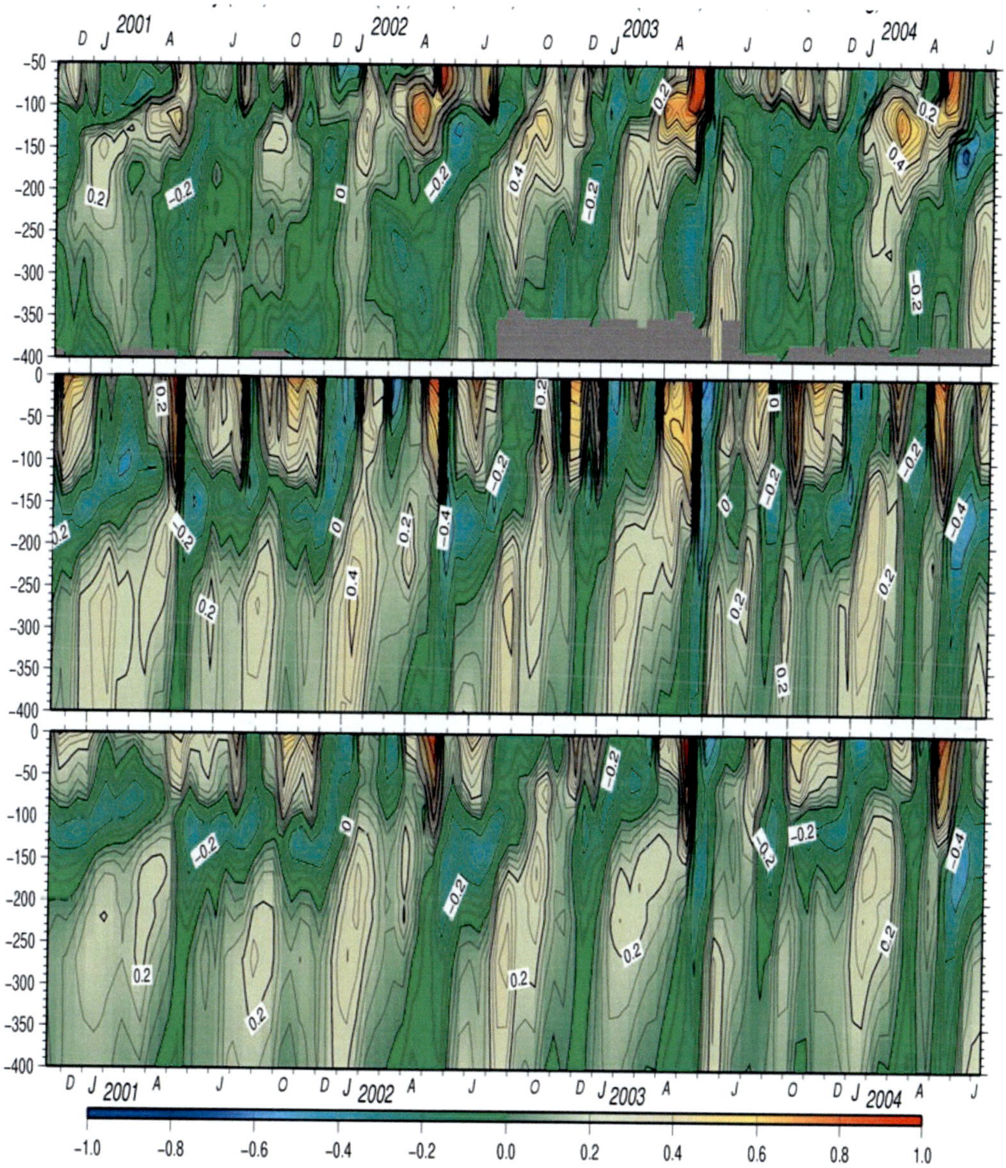

Figure 4b. Same as in Figure 4a but for 1/2° model.

This is to be expected because of the rather coarse resolution of the ECMWF grid (1.125°). The agreement between the model and ADCP meridional currents is poorer than in the zonal case.

The above qualitative findings will be quantified as discussed below. To further investigate the relationship between model currents and observations, we apply power spectral density (PSD) and wavelet spectrum (WS) analyses to investigate the variability of zonal and meridional currents at 90°E, 0°N.

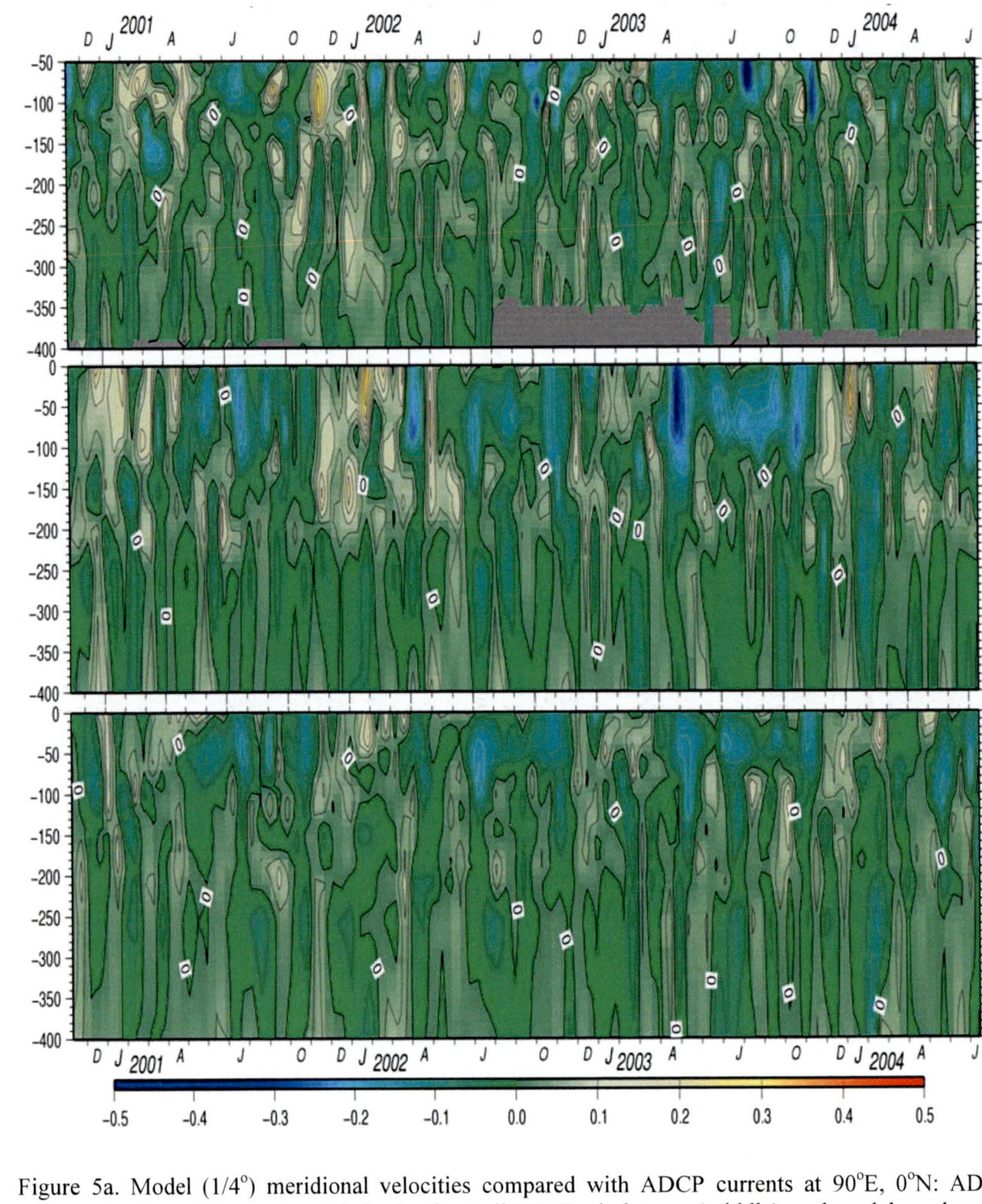

Figure 5a. Model (1/4°) meridional velocities compared with ADCP currents at 90°E, 0°N: ADCP currents (top), model currents with forcing by QuikSCAT wind stress (middle), and model results with forcing by ECMWF winds (bottom). Contour interval is 0.05 m/s. Note that ADCP currents are not available above 50m depth.

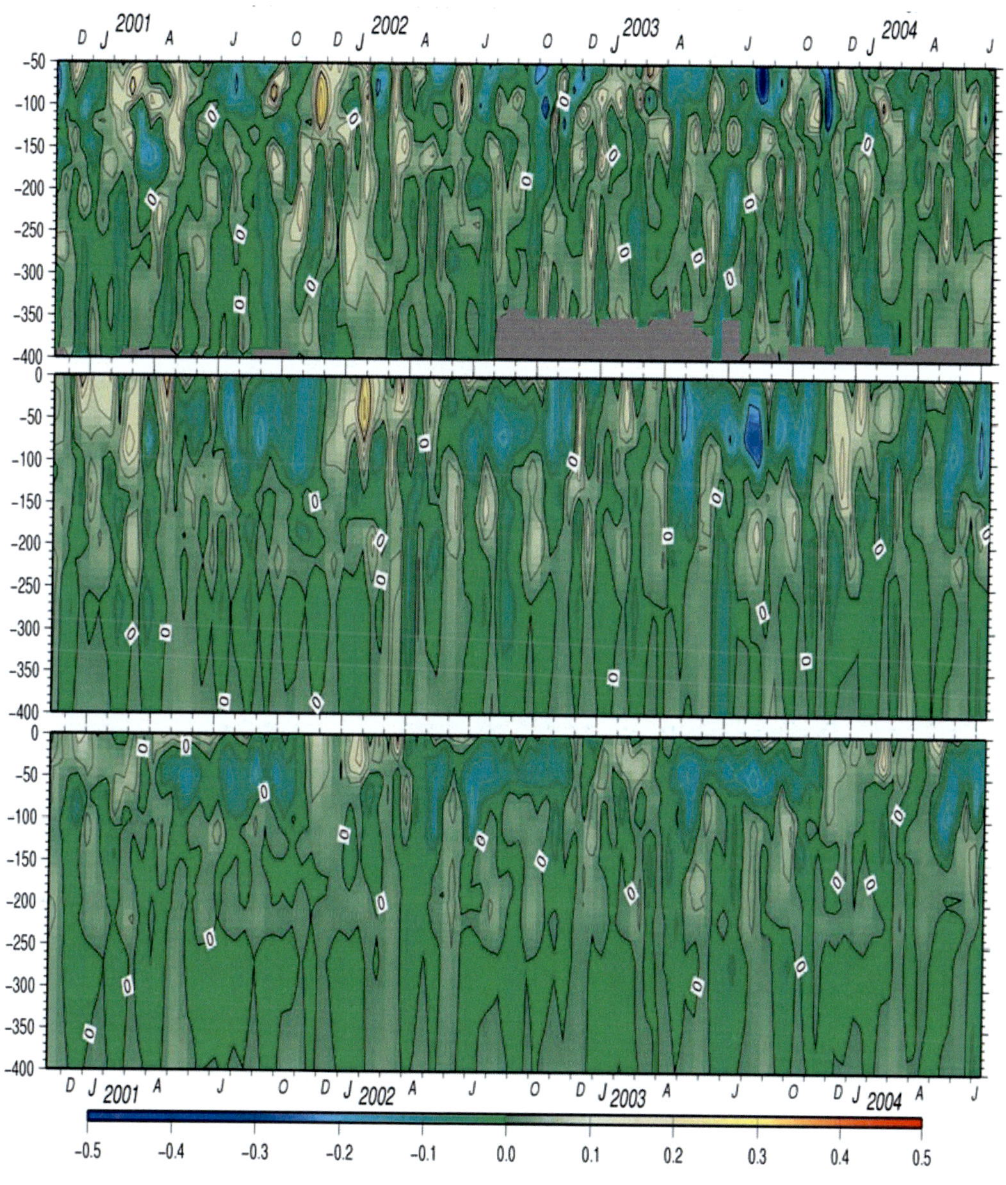

Figure 5b. As in Figure 5a but for 1/2° model.

3. RESULTS

3.1. Variability in Equatorial Currents during the 2001-2004 Observational Period

Figures 6 and 7 show the 15-day running mean time series of 1/4° model zonal and meridional velocities at 50, 100, 150 and 200 m depths: ADCP (green); and model results, QS (blue) and ECMWF (red). While it is difficult to see the correlation between model and observed currents in Figure 4, excellent correlation between the low frequency components of the two can be clearly in Figure 6.

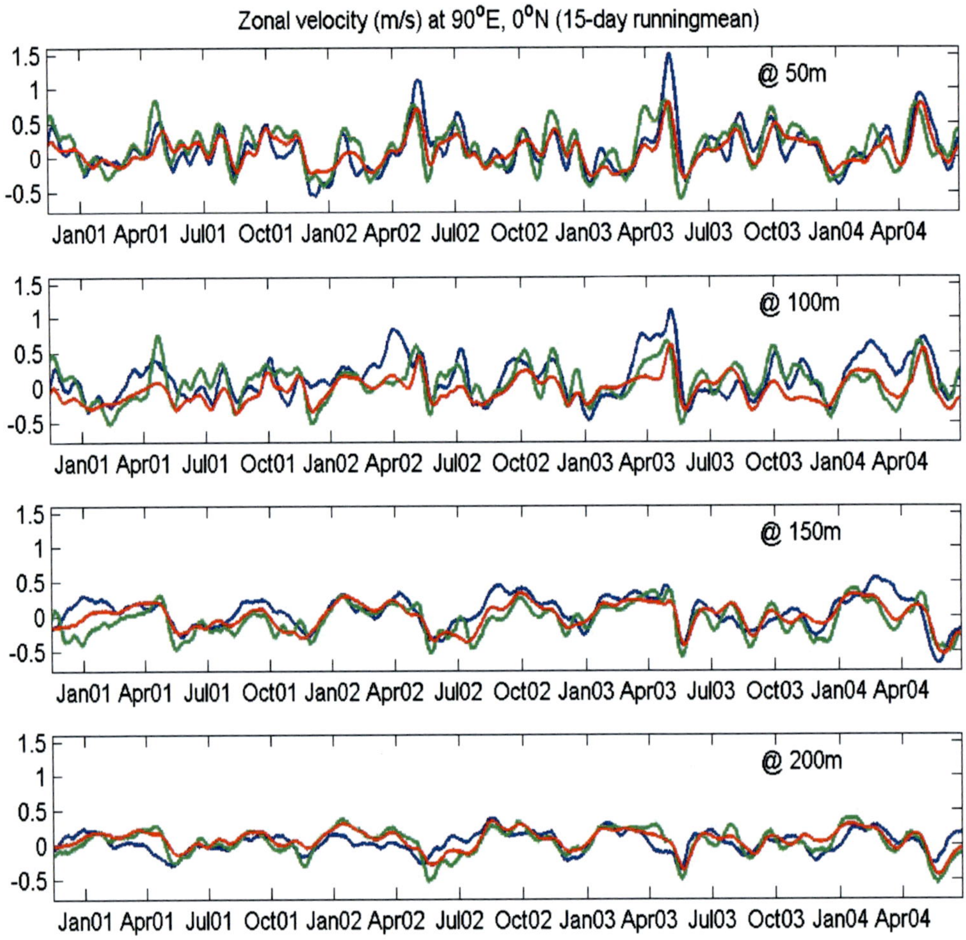

Figure 6. Time series of zonal velocities of 4 depths (50m, 100m, 150m, and 200m) at 90°E, 0°N from November 2000 to July 2004, with a 15-day running mean filter applied. ADCP currents (blue) and model currents forced with QuikSCAT (green) and ECMWF (red) are shown. There is generally a good agreement between the model and the observational data.

The correlation coefficient of the zonal velocity between ADCP and QS (ECMWF) are 0.84(0.66), 0.74(0.55), 0.79(0.66) and 0.61(0.56) at 50m, 100m, 150m and 200m, respectively. Strong transition around May/June from eastward to westward direction with the onset of the summer monsoons can be clearly seen in all the time series for all four years. The zonal currents are much stronger during the 2003 transition period than in other years. While the model underestimates the magnitude of the near-surface (50 and 100 m) currents, the deeper currents are not. The phases of the various events appear to be reasonably well produced by the model.

The meridional currents (Figure 7) are not as well reproduced by the model, especially for ECMWF forcing. The variability seen in meridional currents is of much higher frequency than that of the zonal currents. The magnitudes are underestimated and occasionally the phase too, as for example during November 2003.

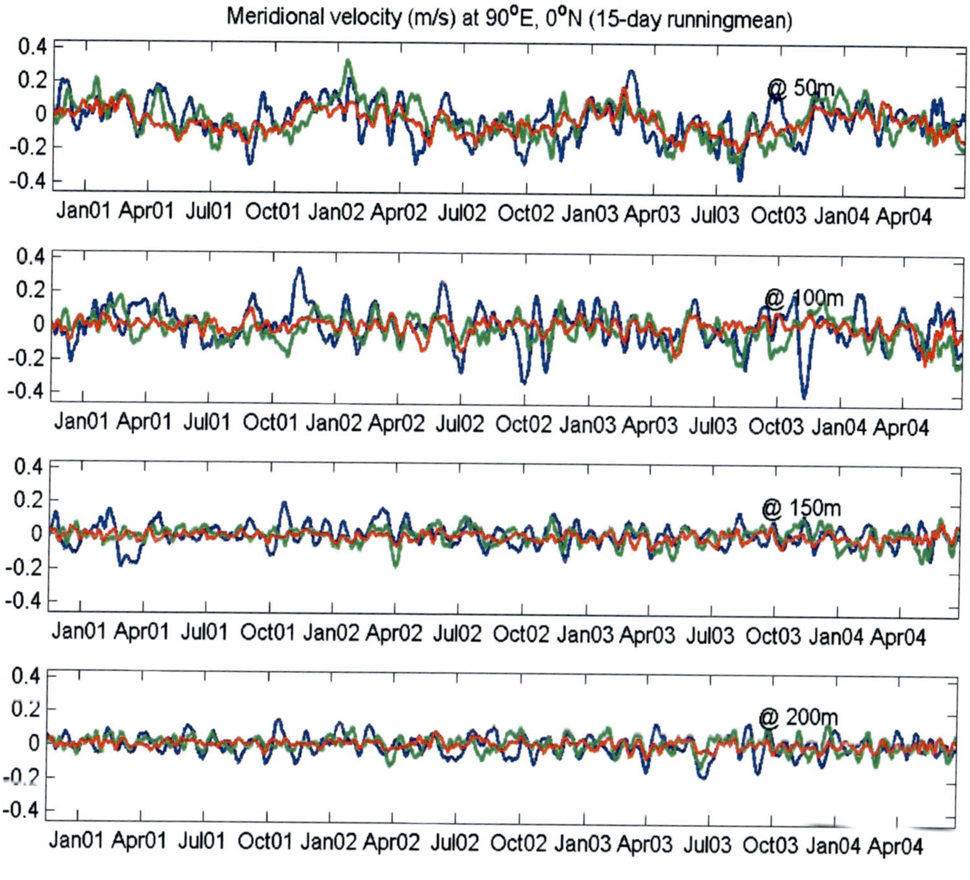

Figure 7. As in Figure 6 but for meridional currents. The agreement is not as good as for zonal currents with the model underestimating the magnitude significantly, especially during certain synoptic events.

The PSD is obtained by applying windowed Fourier transform (Tang et al., 2001, p 347-358). The WS as described by Torrence and Compo (1998) is used to investigate the dominant time-frequency events of the time series, with a focus on the cycles observed from PSD. The Degree of Gaussian (DOG) filter is used as the mother wavelet and a red noise background spectrum with autoregressive lag-0 is used to calculate the 95% confidence level.

Figures 8 and 9 show PSDs of the zonal and meridional velocities. PSDs of zonal velocities (Figure 8) at each depth show spectral peaks at roughly the same frequency. The spectra from the model forced by QS and the ADCP agree well with each other, except at 100 m depth. This is likely due to the undercurrent being located deeper in the model. Spectral peak at the low frequency suggests annual variability. The spectral magnitudes decrease with depth. Near the surface, the QS results produce better spectrum than ECMWF results, since QS wind stresses are more energetic (by around 30% in wind magnitude). PSDs of meridional currents (Figure 9) are relatively small compared to those from zonal currents. The most dominant peak at low frequency appears only at 50m, suggesting wind-related origin. Other spectral peaks are in high frequency at about 0.4 radians/day (15 days).

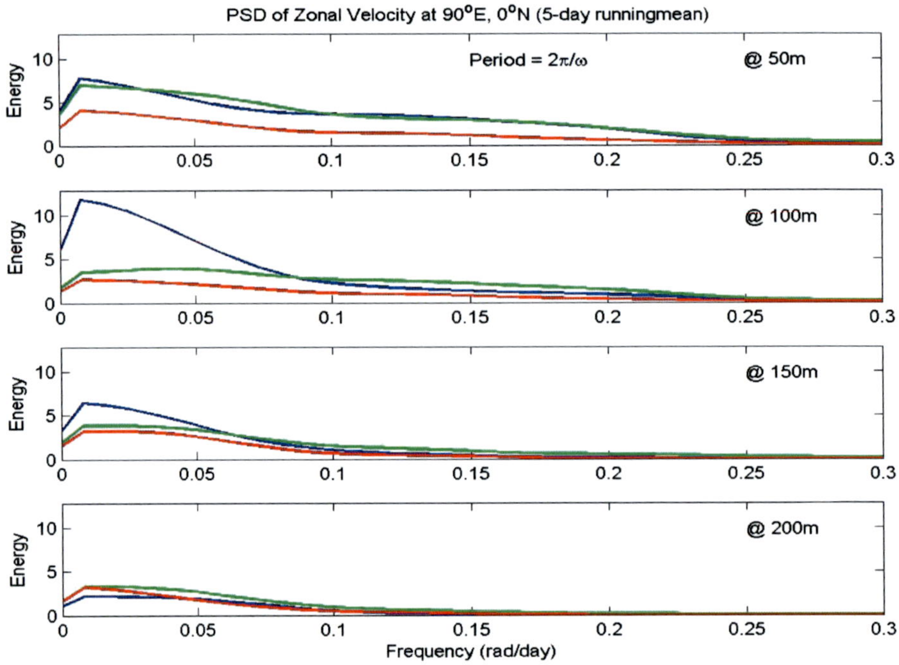

Figure 8. Power Spectrum Density (PSD) of zonal velocities at 4 depths (50m, 100m, 150m, and 200m) at 90°E, 0°N with 5-day running mean filter. ADCP currents (blue) and model currents resulting from forcing with QuikSCAT (green) and ECMWF (red) are shown.

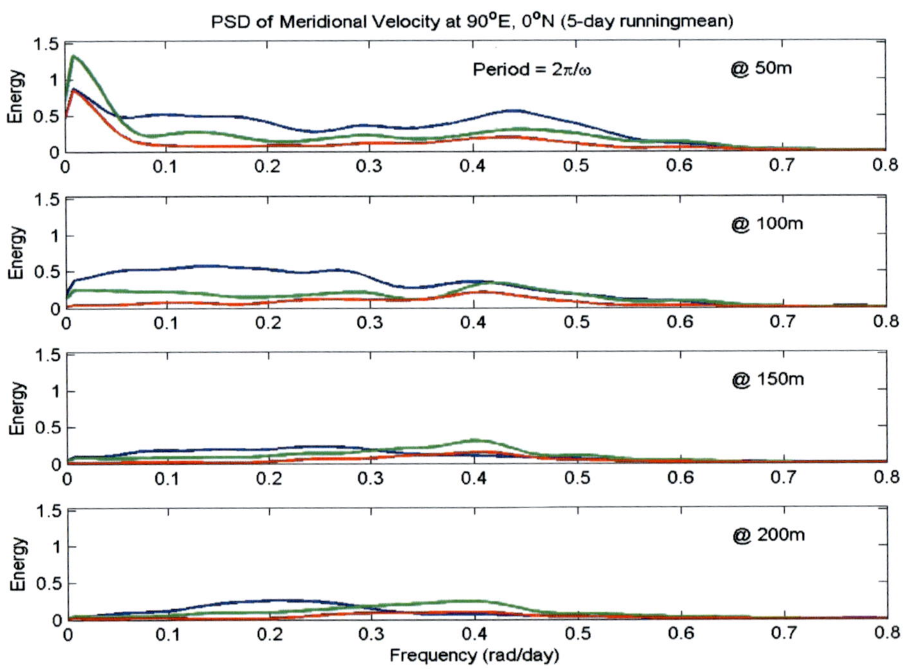

Figure 9. Same as Figure 8 except for meridional velocities.

These PSDs reveal that the meridional velocities have both *short-period* and *annual* variability, unlike zonal currents, which lack the high frequency variability of the meridional currents (compare Figures 4 and 5).

The PSDs reveal variability of currents at 90°E, 0°N in the frequency domain but they do not provide any time information on events causing this variability. We will therefore explore the time-frequency information of currents using WS next. Global wavelet spectrum (GWS) of zonal velocities at 50m and 150m from ADCP (top), QS (middle) and ECMWF (bottom) are shown in Figures 7 and 8. Pink contours indicate 95% significance and black line represents area of cone-of-influence [Torrence and Compo, 1998]. At all depths, GWS of ADCP is the strongest, although all three patterns are roughly similar. Dominant events in ADCP are found during Dec 2001-January 2002, May-June 2002, February, April and June 2003, and January and April 2004 time periods. Each of these events lasts a few months. In model results, these events are well reproduced at appropriate times but with slightly different periods as seen in the bottom two panels. Some events last for a few months, some last about one month. These figures reveal variability at various periods: a few months (interseasonal), and 6 to12 months (semiannual to annual). Notice that QS does not capture the May-June 2002 event but it does capture all other dominant events. As is to be expected, the ECMWF GWS are less energetic than the QS specra.

The variability of zonal currents at 150 m is very much similar to that at 50m depth, although the GWS is in general weaker. This is to be expected since the energetic surface winds are a predominant driver of the near-surface currents.

Figure 9 shows GWS of the meridional velocities at 50m depth. The model does reproduce all dominant events found every half-year *(semi-annually)* with period of 9-12 months. Note that GWS of QS is able to capture dominant events with period of a few months while GWS of ECMWF does not. At other depths, the model results have relatively small spectral magnitudes and do not reproduce any dominant event (not shown).

3.2. Variability in Equatorial Currents during the 1993-2005 Hindcast Period

Observed ADCP currents were not available prior to November 2000, and 2005 observations are not yet ready. Since QS is also not available prior to 1998, the only solution for a longer term hindcast starting from 1993, when altimetry data are available for assimilation into the model, is to use a NWP-generated surface forcing. If this solution were not adopted, it would be impossible to study the interesting 1993-94 and 1997-98 anomalous events. Consequently we carried out a 13-year hindcast from the beginning of 1993 to the end of 2005 using ECMWF surface forcing, with both 1/2° and 1/4° models. Only the results of 1/4° model are shown here. The 1/2° hindcast results can be found in Kantha et al. (2008).

The 13-year time series of currents at 90°E, 0°N have been subjected to a 5-day running mean and WS has been applied to this to explore the variability. Figures 13 and 14 show the WLs of zonal currents at 50 and 150 m depths, and Figure 15 shows the WS of the meridional currents at 50 m depth. WSs of meridional currents are relatively weak compared to those of zonal. The strongest spectrum is at 50m, showing annual variability. WS of zonal currents reveals variability at 6 months at all depths.

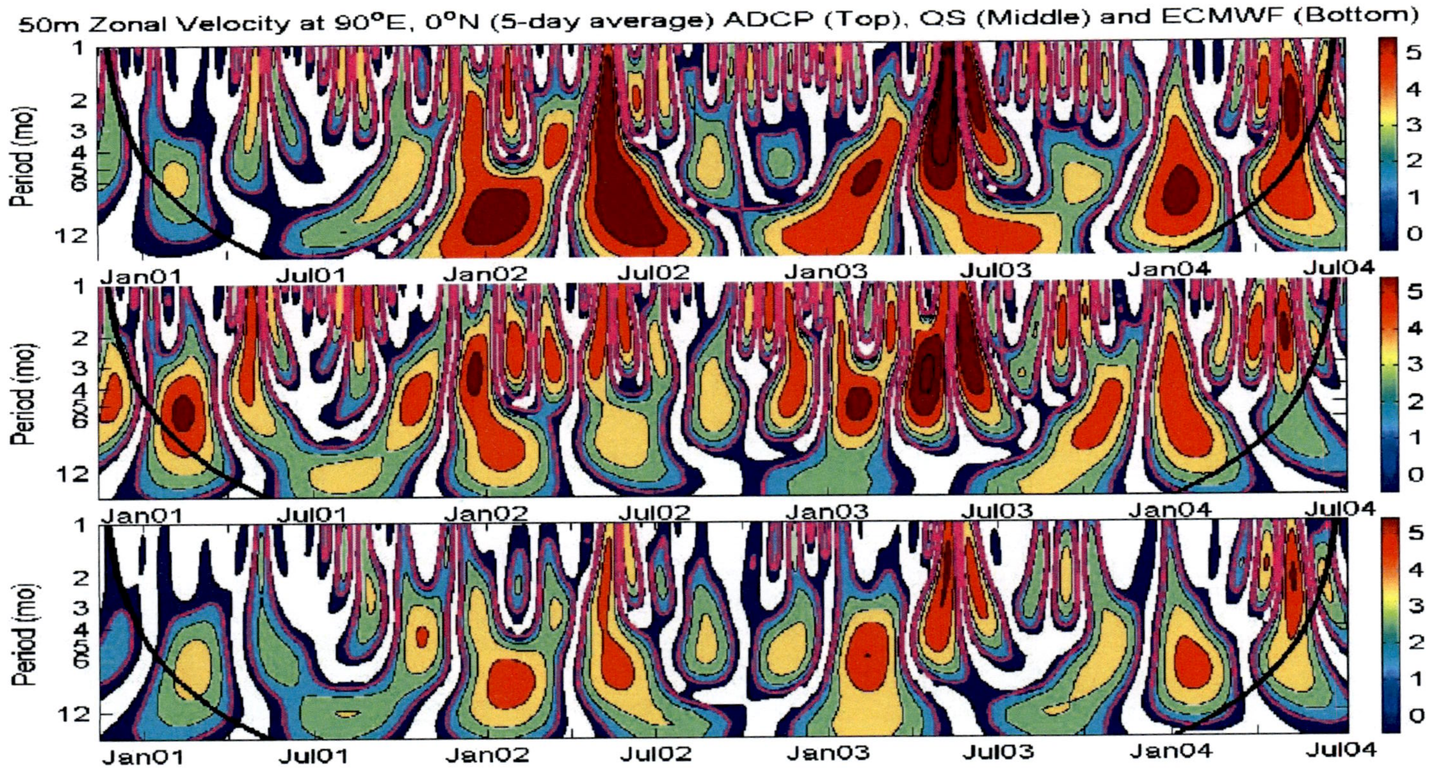

Figure 10. Global wavelet spectrum (with DOG basis function) of the zonal velocity at 50m depth at 90°E, 0°N: ADCP (top) and model currents, QuikSCAT (middle) and ECMWF (bottom). The pink thick contours corresponding to the 95% confidence level. Black line indicates area of cone-of-influence. The dominant events are shown in red. Two noticeable periods are between 3-6 months and between 6-12 periods. Notice that GWS of ADCP is strongest among all three as is to be expected.

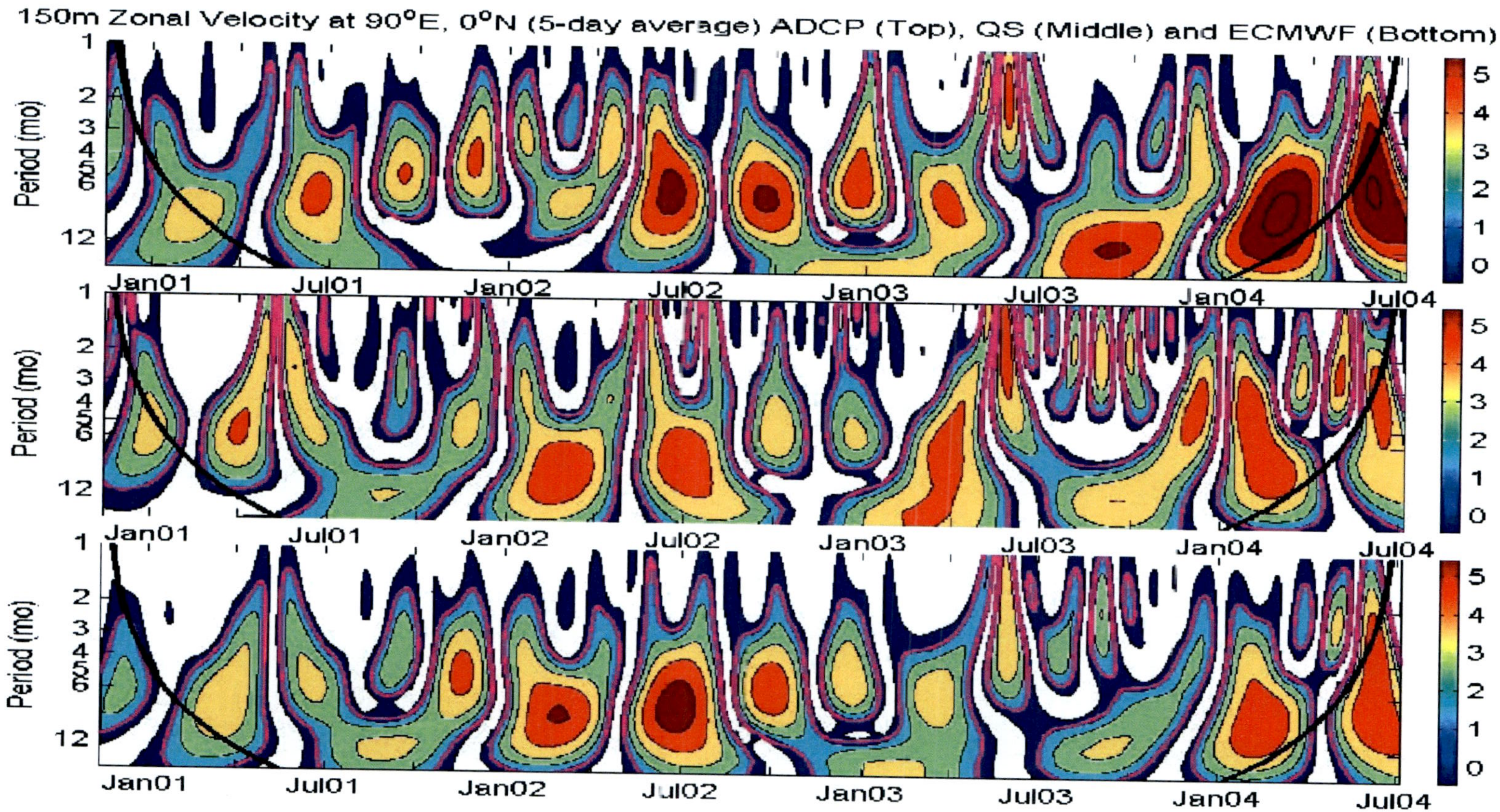

Figure 11. Same as Figure 10 except for 150m. Noticeable periods are between 6-12 months.

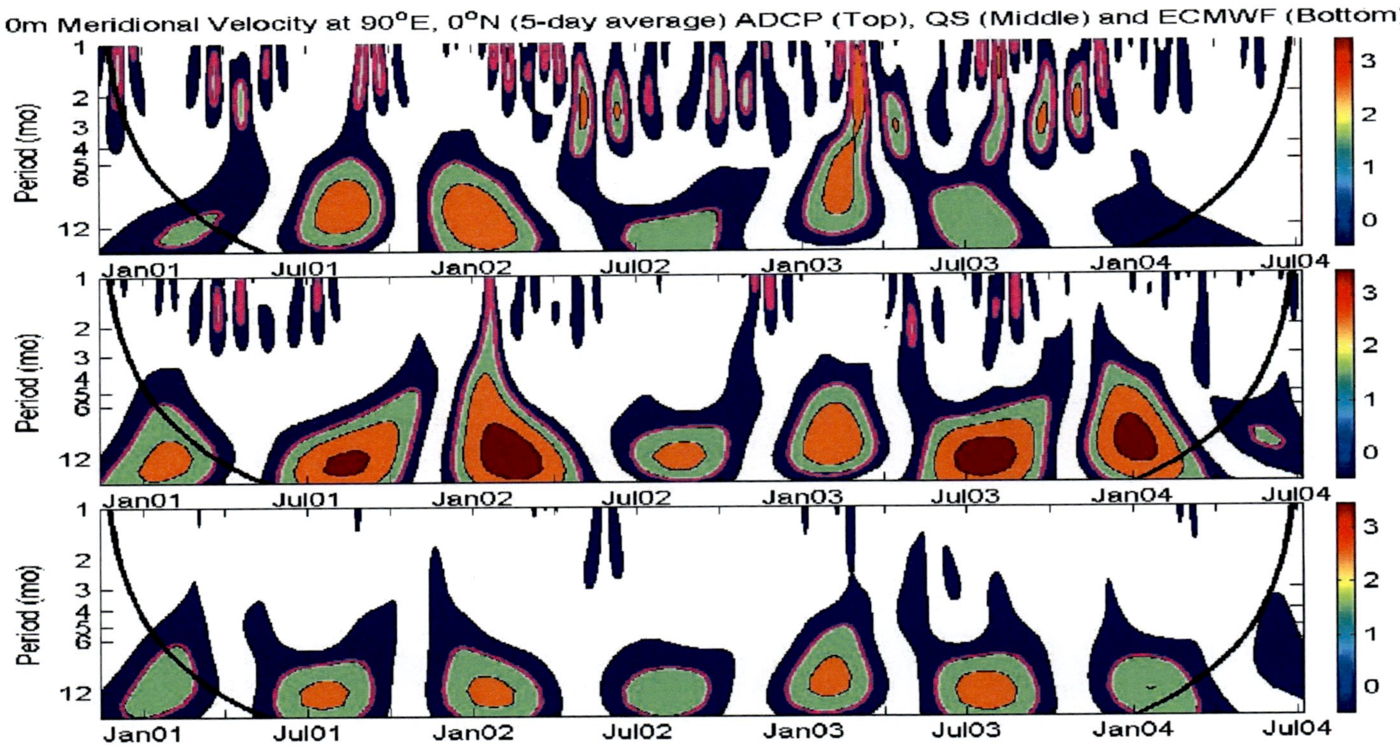

Figure 12. As in Figure 10 but for meridional velocity at 50m. Noticeable periods are in a few months which exist for a short period of time and between 6-12 months which occur longer.

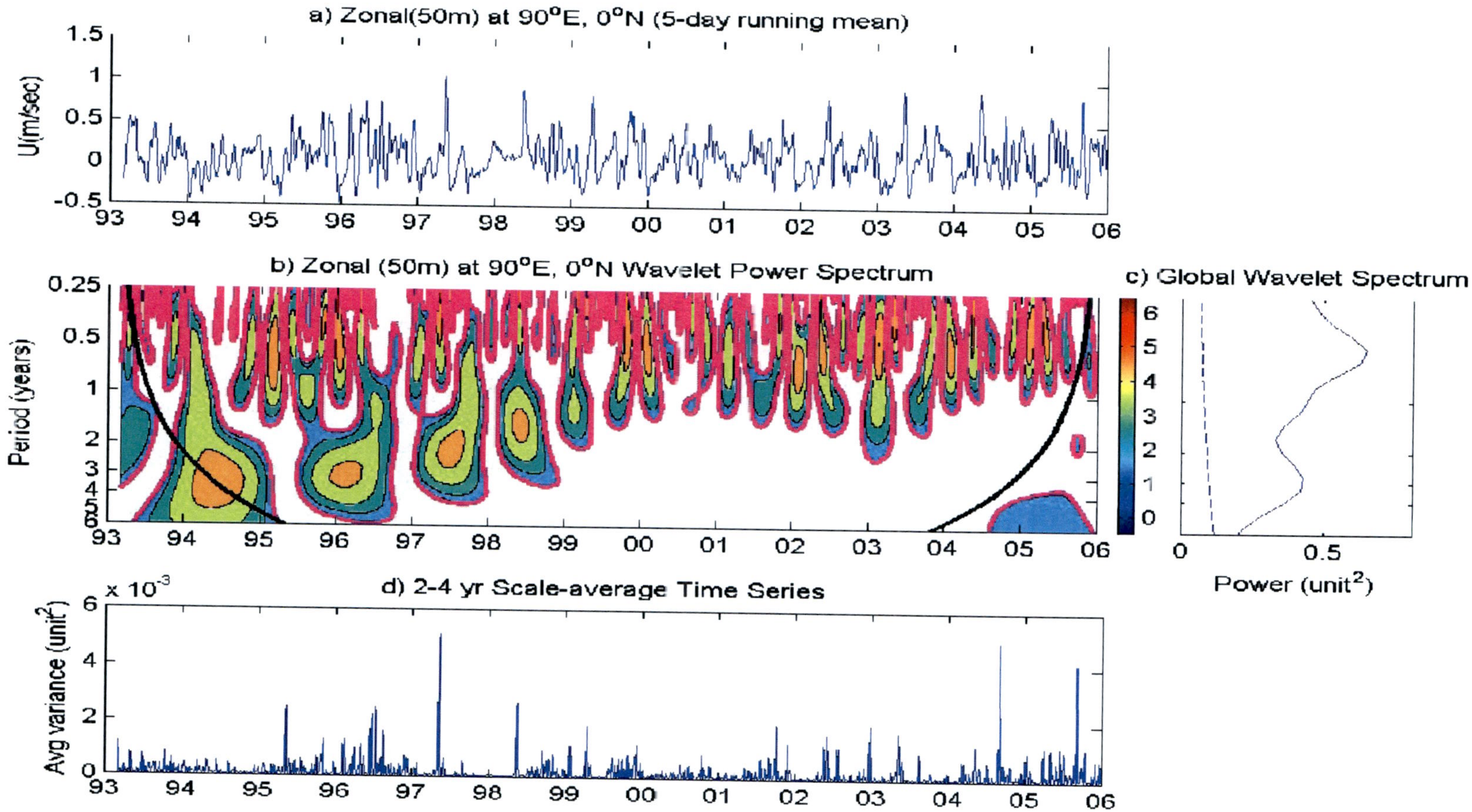

Figure 13. a) 13-year Time series of zonal velocity at 50m depth at 90°E 0°N with 5-day running mean. b) GWS associated with time series in a). Pink contour is 95% significant level and black line is area of cone-of-influence. c) Average GWS over the period. d) Scale-average with period of 2-4 years. Dashed line represents 95% significant level.

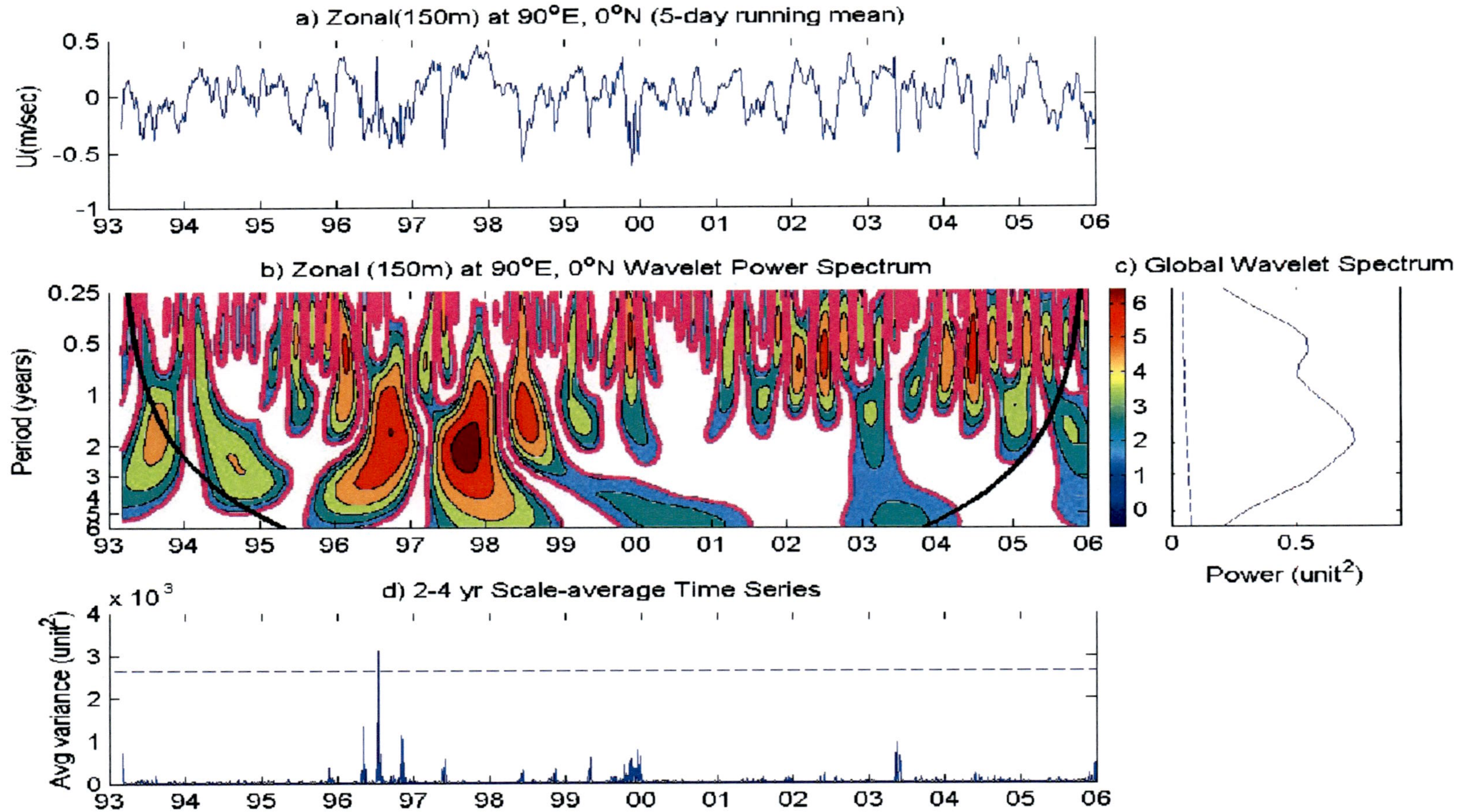

Figure 14. As in Figure 13 but at 150 m depth.

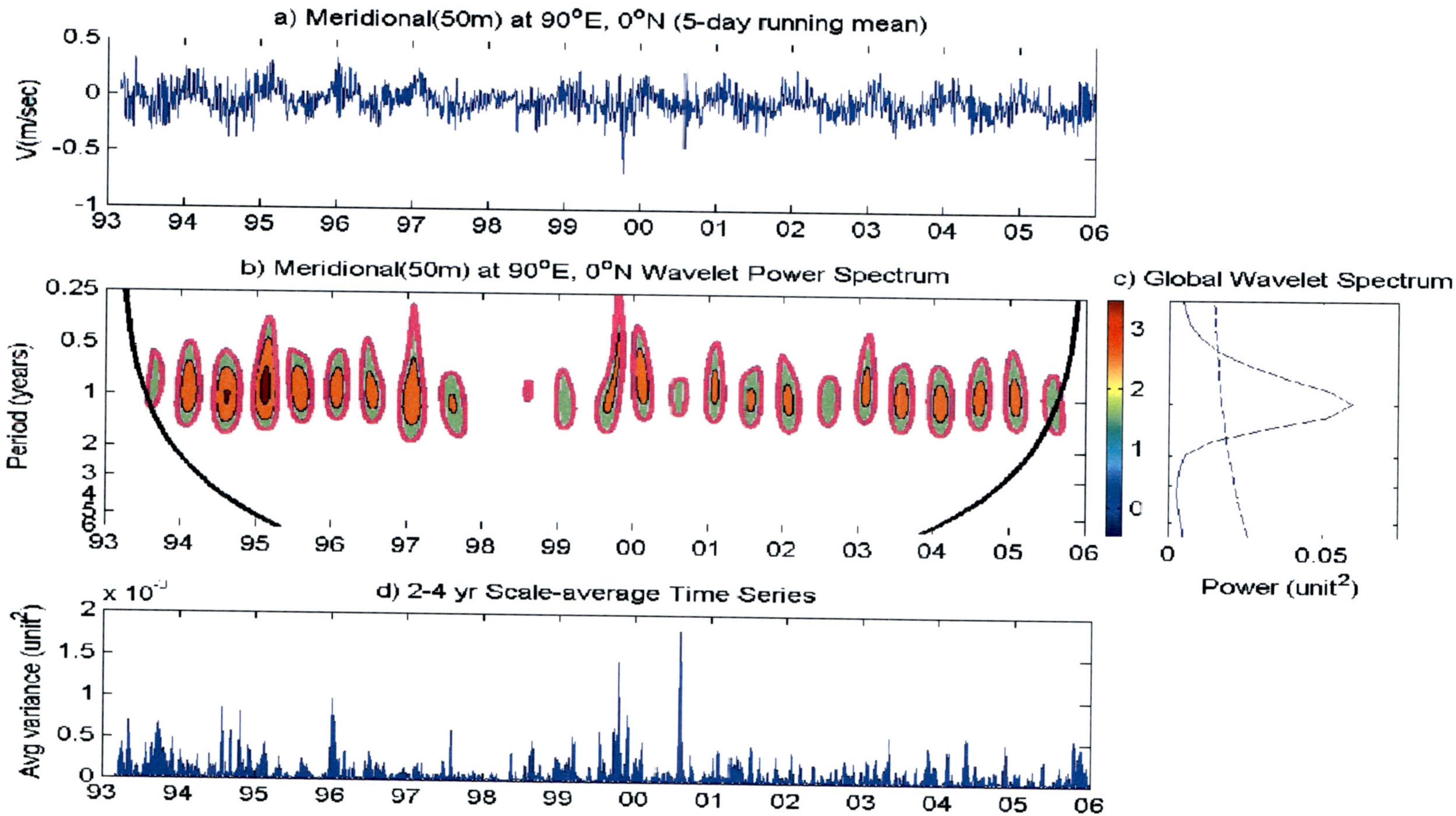

Figure 15. As in Figure 13 but for meridional currents at 50 m depth.

At all depths except 200m, WS also shows variability at a period of about 2-4 years. The anomalous 1993-94 and 1997-98 periods are clearly seen, especially at 50 m zonal currents. Kantha et al. (2008) report in detail on these anomalous periods. The meridional currents show variability at mostly annual frequency.

CONCLUSION

The PSD and WS of a time series are two useful tools to investigate variability. The PSD can identify regions of strong variability in frequency space, and the WS can isolate them in time as well. Unfortunately, no tool is perfect. PSD cannot provide information about time of the dominant events but WS can. WS cannot identify short period variability or when the spectrum is very small. Nevertheless, these methods have highlighted the variability of currents in the upper water column at 90° E and 0° N.The PSDs of modeled and measured currents at 90°E, 0°N indicate that the zonal velocity varies with a dominant period of 12 months, and the meridional velocity varies with periods as short as 15 days and as long as 12 months. WLs indicate that the variability of meridional velocity has periods of 1/2-1 month and 12 months. The variability of zonal velocity has periods of 3-4 months, 6-12 months as well as 2-4 years. While the results can be improved further by using higher resolutions and more accurate forcing, the modeled currents are in reasonably good agreement with observed currents.

ACKNOWLEDGEMENTS

LK thanks the Office of Naval Research for the partial support for this work through ONR grant N00014-06-10287.

REFERENCES

Iskandar, I., Mardiansyah W., Masumoto, Y., and Yamagata, T. (2005). Intraseasonal Kelvin waves along the southern coast of Sumatra and Java, *J. Geophys. Res.*, 110, C04013, doi:10.1029/2004JC002508.

Jensen, T. G. (1993). Equatorial variability and resonance in a wind-driven Indian Ocean model, *J. Geophys. Res.*, 98, 22 533-22 552.

Emery, W. J., and Thomson, R. E. (2001). *Data Analysis Methods in Physical Oceanography*, 2nd ed., 638 pp., Elsevier, New York.

Kantha, L. H., and C. A. Clayson, C. A. (2000), *Numerical Models of Oceans and Oceanic Processes,* 940 pp., Academic Press, San Diego, Calif.

Kantha, L., Rojsiraphisal, T., and Lopez, J. (2008). The north Indian Ocean circulation and its variability as seen in a numerical hindcast of the years 1993 to 2004, *Prog. Oceanogr.*, 76, 111-147.

Lopez, J. W., and Kantha, L. H. (2000a). Results from a numerical model of the northern Indian Ocean: circulation in the South Arabian Sea. *J. Mar. Syst.,* 24, 97-117.

Lopez, J. W., and Kantha, L. H. (2000b). A data-assimilative model of the north Indian Ocean, *J. Atmos. Oceanic Tech.*, 17, 1525-1540.

Luyten, J. R., and Swallow, J. C. (1976). Equatorial undercurrents, *Deep Sea Res.*, 23, 999-1001.

Murty V. S. N., Sarma Y. V. B., Rao D. P., and Murty, C. S. (1992a). Water characteristics, mixing and circulation in the Bay of Bengal during southwest monsoon. *J. Mar. Res.*, 40, 207-228.

Murty V. S. N., Sarma Y. V. B., Babu M. T., and Rao D. P. (1992b). Hydrography and circulation in the northwestern Bay of Bengal during the retreat of southwest monsoon. *Proc. Indian Academy Sci., Earth and Planetary Sciences*, 101, 67-75.

Masumoto, Y., Hase, H., Kuroda, Y., Matsuura, H., and Takeuchi, K. (2005). Intraseasonal variability in the upper layer currents observed in the equatorial Indian Ocean, *Geophys. Res. Let.*, 32, L02607, doi:10.1029/2004GL021896.

Rojsiraphisal, T., and Kantha, L. (2008). Meridional heat fluxes in the North Indian Ocean, *Chapter 8, this book.*

Schott, F, A., Swallow, J. C., and Fieux, M. (1990). The Somali current at the equator: Annual cycle of currents and transports in the upper 1000m, and connection to neighboring latitudes, *Deep Sea Res.*, 37, 1825-1848.

Smith, S. D. (1980). Wind stress and heat flux over the ocean in gale force winds. *J. Phys. Oceanogr.*, 10, 709-726.

Tang, Y. Y., Wickerhauser V., Yuen P. C., and Li, C. H. (2001). Wavelet Analysis and Its Applications, 2nd International Conference, WAA 2001, LNCS 2251, Springer-Verlag. Berlin

Torrence, C., and Compo, G. P. (1998). A practical guide to wavelet analysis, *Bull. Amer. Meteor. Soc.*, 79, 61-78.

Wyrtki, K. (1973), Equatorial jet in the Indian Ocean, *Science*, 181, 264-266.

Submitted March 10, 2008.

In: Atlantic and Indian Oceans
Editors: E. S. Askew and J. P. Bromley, pp. 201-210
ISBN 978-1-60692-475-4

Short Communication

REGENERATION OF NITROGEN IN THE SURFACE LAYER OF THE OCEAN: NEW CHALLENGES FOR THE NITROGEN CYCLE

Camila Fernández*
Center for Oceanographic Research in the Eastern South Pacific (COPAS)
Universidad de Concepción. Barrio Universitario S/N,
Casilla 160-C, Concepción, Chile

ABSTRACT

For the last 30 years, oceanographic research in the nitrogen cycle has stood firmly on two important concepts: new and regenerated production (Dugdale and Goering 1967). The former defines the part of primary production fueled by allochtonous nitrogen (mainly nitrate coming from vertical advection), while the latter represents the uptake of regenerated forms of N (usually ammonium). These concepts opened the way for estimating the strength of the "biological pump", i.e., the capacity of the biological compartment of the ocean to take carbon dioxide from the atmosphere and to export it to the deep ocean. Through the formulation of the *f* ratio (Eppley and Peterson 1979), the fraction of primary production sustained by allochtonous nitrate could be assimilated to the export of carbon to the deep ocean. In recent years, the assumption necessary for the use of *f*—that is, negligible nitrification (microbial oxidation of ammonium into nitrate) within the euphotic layer—has been challenged by increasing evidence of surface nitrogen regeneration in the open ocean. A few years ago, the largest data set with a seasonal frequency gathered in the NE Atlantic Ocean showed that in situ nitrification can account for up to 30% of nitrate assimilation within the euphotic zone (Fernández et al. 2005b). The consequences of this unaccounted nitrate supply into the euphotic zone (biasing our estimation of export production) are only beginning to be quantified at the global scale (Yool et al. 2007). In this chapter, I will focus on these recent developments in the nitrogen cycle and discuss the methodological improvements that allowed us to get to the present point in our understanding of nitrogen regeneration. Also, I will point out the new challenges that have arisen from these advancements, which should outline future research.

---

* fernandez@profc.udec.cl

Introduction

Nitrogen is one of the key elements for marine life. Often limiting, it shapes microbial community structure and nutrient availability for primary producers.

Research of the marine nitrogen cycle has a long history that has seen significant advances in the last 30 years. Phytoplankton is the main player modulating the concentrations of fixed nitrogen (NO_3, NH_4) in sea water. This is mainly done through biological assimilation and excretion, participating also in the formation of distinct features, such as the primary nitrite maximum (PNM).

Since the recognition of the role played by the microbial loop in the nitrogen and carbon cycles (Azam et al. 1983), bacteria have been considered essential vehicles for organic matter regeneration in the deep water column. Physics, through vertical transport and/or upwelling, complete the basic functioning of the "biological pump" by carrying inorganic nutrients from the deep ocean into the well-lit layers of the ocean.

During the last decade, however, a number of discoveries have taken place, which include reports of new metabolic pathways in marine environments (Dalsgaard et al. 2003; Kuypers et al. 2003; Wuchter et al. 2006) as well as crucial reevaluations of already well-known ones (Karl 2000; Capone et al. 2005; Yool et al. 2007). This has marked a new step in scientific research of the nitrogen cycle (Brandes et al. 2007; Francis et al. 2007; Ward et al. 2007).

2. Recent History of Nitrogen Biogeochemistry

New and Regenerated Production in Marine Environments

During marine photosynthesis (which can be quantified as primary production), carbon is fixed into organic matter and is accompanied by the assimilation of other inorganic elements (mainly nitrogen and phosphorous) following specific proportions determined by the Redfield ratio (Redfield 1934).

Through the nitrogen cycle, primary production can be assessed as a function of the origin of the assimilated nutrients. In doing so, Dugdale and Goering (1967) defined two important concepts, new and regenerated production. *New production* is defined as primary productivity fueled by nitrogen coming from allochtonous sources such as vertical transport (mainly NO_3^-) and (more recently) atmospheric fixation of N_2 gas (Karl 2000). In contrast, nitrogen resulting from remineralization of organic matter (mainly in the form of ammonium) is also able to sustain a fraction of primary production, termed *regenerated production* (Dugdale and Goering 1967). This distinction allowed characterizing different marine ecosystems, with regenerated production being predominant in nutrient-poor oligotrophic systems, while in nutrient-rich waters, primary production is mainly new-production–based.

Estimating Export Production

Because primary production based on advected nitrate could be assimilated with new production, it was also possible to quantify export production. The definition of the *f* ratio (Eppley and Peterson 1979), the balance between the assimilation of new forms of N and total primary production, allowed estimating the downward flux of particulate organic matter (POM, and hence carbon) to the deep ocean (Eppley and Peterson 1979; Lewis et al. 1986; Legendre and Gosselin 1989). In some cases, through independent estimations of *f*, it would be used to indirectly compute new production (Harrison et al. 1987). Since its definition, the *f* ratio has been of crucial importance for assessing the "strength" of the biological pump, globally or regionally. However, a sporadic debate concerning its significance and possible evolution has been ever-present (Legendre and Gosselin 1989).

In recent years, the use of this empirical tool for evaluating the export of carbon (also called carbon sequestration) in open ocean environments became increasingly complicated. First, the original assumption that all organic matter exported to the deep ocean is particulate had to face a significant reevaluation after the recognition of the importance of dissolved organic nitrogen (Glibert et al. 1985; Slawyk and Raimbault 1995; Slawyk et al. 2000).

Second, a fundamental assumption already stated in the seed work of Dugdale and Goering (1967) was that nitrification (the microbiologically-mediated oxidation of ammonium to nitrite and then to nitrate) only occurs below the euphotic zone. If the opposite turned out to be true, then the assimilation of regenerated nitrate should be considered as recycled production. As a consequence, total nitrate assimilation could no longer be equated to new production in surface waters. Although overlooked for more than 30 years, field evidence against this assumption slowly accumulated through local field measurements until the first large-scale seasonal nitrification study was achieved in the northeast Atlantic Ocean (Fernández 2003).

3. Nitrification within the Euphotic Zone

Nitrification is a key process in the global nitrogen cycle, as it is responsible for producing the entire stock of deep preformed nitrate. It is also acknowledged as an important component in the formation of the primary nitrite maximum (PNM) often observed at the base of the euphotic zone (Lomas and Lipschultz 2006).

In terms of the environmental regulations of this process, the most important control factors proposed are organic matter supply and light. Concerning the latter, studies using cultured and natural bacterial communities suggested a photoinhibition of ammonium-oxidizing bacteria at high light intensities (Horrigan et al. 1981; Olson 1981), which would make nitrification unlikely to occur in surface waters. Further laboratory evidence showed a different response (in terms inhibition threshold as well as recovery) between ammonium and nitrite-oxidizing bacterial communities (Guerrero and Jones 1996a; Guerrero and Jones 1996b). In recent years, however, laboratory studies concerning the relationship between nitrification and light have been scarce, and therefore this matter remains an open question.

Biogeochemical measurements of nitrification are not particularly abundant in open ocean environments. However, measurements within the euphotic layer have shown the occurrence of ammonium oxidation in well lit waters (Feliatra and Bianchi 1993; Bianchi et al. 1994; Dore and Karl 1996; Bianchi et al. 1997; Raimbault et al. 1999; Diaz and Raimbault 2000; Ward 2005; Fernández and Raimbault 2007; Clark et al. 2008). Already in 1996, Dore and karl reported for the North Pacific Ocean that as much as 142% of the nitrate assimilation rate could be explained by nitrification. They stated that "...we are beginning to see a convergence of estimates that indicates that the two processes [nitrate uptake and nitrification] have rates on the same order of magnitude and that suggests that a substantial fraction of nitrate assimilated is regenerated and not new". Previously, results pointing in the same direction were obtained in New Zealand, and the Southern California bight (Priscu and Downes 1985; Ward et al. 1989). Recently, Ward (2005) assessed the seasonality of this process in Monterrey Bay, stating that "the relationship between nitrification and light intensity is important, but light is obviously not the only determinant of the [ammonium oxidation] rate. Highest rates always occurred at the lowest light intensities, but the lowest rates occurred at all light intensities".

In 2000–2001, a series of oceanographic cruises were carried out in the mode water subduction zone of the northeast Atlantic (Mémery et al. 2005). This geographic area has attracted considerable attention as one of the greatest CO_2 sinks of the global ocean (Takahashi et al. 1993; Takahashi et al. 1997; Takahashi et al. 2002). It has been therefore intensively studied (Ducklow and Harris 1993; Savidge and Williams 2001; Mémery et al. 2005; Robinson et al. 2006). During a seasonal survey, $^{15}N/^{13}C$ tracer experiments were conducted, both in situ and on deck. The objective was to evaluate primary productivity as well as nitrogen biogeochemical cycling on a vaste mesoscale-influenced region (Fernández et al. 2005a). The data gathered during this program (POMME) still represents the largest nitrification data set to date, with more than 200 samples collected in over a 100 incubation experiments covering the winter, spring and late summer seasons (Fernández et al. 2005b; Fernández and Raimbault 2007).

The results showed the seasonal variability of the nitrogen cycle and revealed the role of ammonium oxidation during the year. Nutrient concentrations in surface waters were always above detection limits. Nitrate reached 4.3 ± 1.65 μM in winter and 3.3 ± 1.83 μM in spring while ammonium showed a subsurface maximum that was around 50m shallower than the primary nitrite maximum (PNM) within the euphotic zone (Fernández and Raimbault 2007). Nitrate regeneration through ammonium oxidation was observed in the first 80m of the water column (largely coinciding with the euphotic zone) at all seasons and particularly during winter and spring (Figure 1). Winter rates of nitrification reached 0.009 ± 0.015 $\mu M\ d^{-1}$ in surface waters (max. 0.017 $\mu M\ d^{-1}$) and increased slightly with depth to 0.015 ± 0.008 $\mu M\ d^{-1}$ (Table 1). Furthermore, two exceptionally high values during that season were obtained in the top 20 m, reaching up to 0.077 $\mu M\ d^{-1}$). The spring season exhibited surface nitrification rates close to winter values (0.009 ± 0.014 $\mu M\ d^{-1}$) with maximum rates approaching 0.02 to 0.06 $\mu M\ d^{-1}$). In deeper layers, nitrification rates increased to 0.016 ± 0.016 $\mu M\ d^{-1}$ (max. 0.06 $\mu M\ d^{-1}$ at 80 m, Table 1). As nitrification rates were always detected in surface waters and did not decrease between winter and spring, no clear evidence of light inhibition could be reported.

By correcting total nitrate assimilation, and by differentiating regenerated nitrate from new nitrate, it was possible to assess the effect of nitrification upon estimations of new production. This comprehensive study was able to reveal a dramatic overestimation of new production going from 10 to over 30% during the entire seasonal cycle (Fernández 2003).

Table 1. Rates of ammonium oxidation (nitrification) obtained during field experiments in the Northeast Atlantic Ocean (POMME program; Fernández and Raimbault, 2007)

	Winter	Spring
Average Nitrification in surface waters	0.009 ± 0.015 μM d^{-1}	0.009 ± 0.014 μM d^{-1}
Max. Nitrification rate in surface waters	0.17 μM d^{-1}	0.06 μM d^{-1}
Average Nitrification below euphotic zone	0.015 ± 0.008 μM d^{-1}	0.016 ± 0.016 μM d^{-1}
Average depth 1%light	60 m	50 m

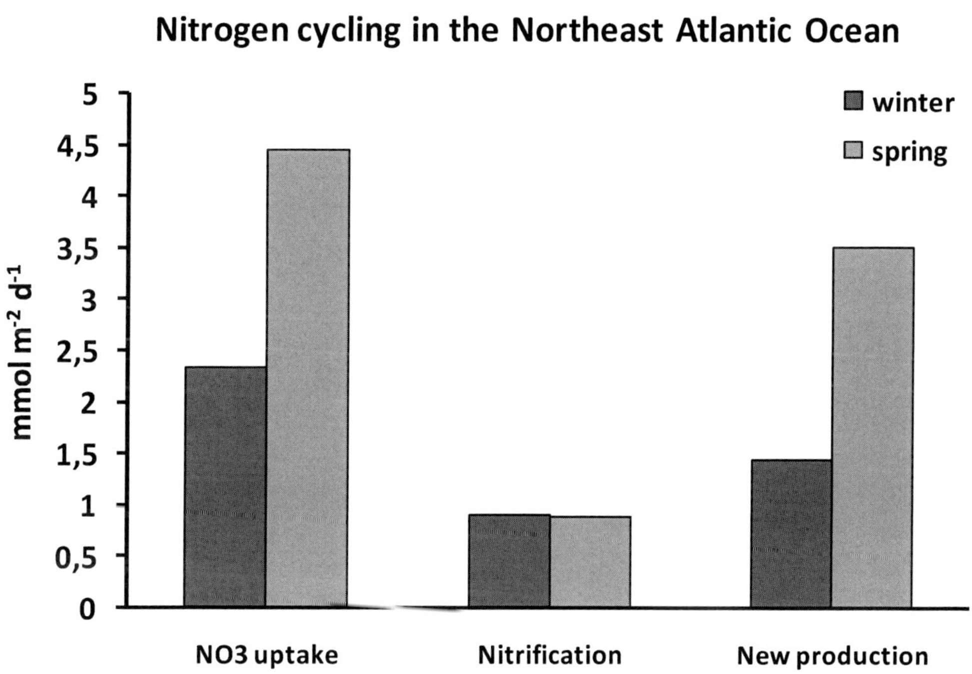

Figure 1. Integrated rates of total nitrate assimilation, nitrification and new production based on data collected during the POMME program in the Northeast Atlantic.

The Global Picture

In 2007, a modeling study was made for evaluating the global role of nitrification in sustaining primary production (Yool et al. 2007). In an unprecedented effort, all extant data sets from open ocean studies were gathered and evaluated (a total of nine studies and 427 measurements). The resulting global data set was used to provide, for the first time, a global estimation of the fraction of regenerated nitrate used by primary producers. The results showed that around 30% of nitrate available in surface waters is regenerated within the euphotic layer (see Figure 2 in Yool et al. 2007). Implications for the use of the *f* ratio as a proxy for export production were then clear. Total nitrate assimilation can no longer be assimilated as new production without simultaneous measurements of nitrification. More importantly, it seems now difficult to safely assimilate nitrate uptake to export production since not all nitrate assimilated in surface waters is "new". Additionally, a large scale systematic correction (30% for instance) cannot be easily applied.

This first global estimation of surface nitrification should be considered as a starting point for future research. More data is urgently needed if we are to better quantify this process.

4. Perspectives for Future Research

Nitrification: Are We Looking in the Right Place?

The need for increasing observations of ammonium and nitrite oxidation as well as the urgency for sampling at a high temporal frequency and spatial scale have been recognized for some time (e.g Lomas and Lipschultz 2006). However, measuring nitrification is not easy. During more than 30 years of nitrogen research, ammonium and nitrite oxidation have been approached through chemical and isotopic techniques, the later involving both carbon (^{14}C) and nitrogen (^{15}N). The ^{14}C technique relies on the assumption of nitrification being a carbon fixing process (bacteria assimilate carbon while reducing ammonium) and uses theoretically estimated C:N ratios (Feliatra and Bianchi 1993; Dore and Karl 1996; Bianchi et al. 1997; Rees et al. 2006). The use of ^{15}N stable isotopic solutions relies on measuring nitrate formed during incubation bearing the 15-N mark of the originally added substrate as $^{15}NH_4$ (Slawyk and Raimbault 1995; Raimbault et al. 1999; Diaz and Raimbault 2000; Fernández and Raimbault 2007) or $^{15}NO_2$ (Clark et al. 2007; Clark et al. 2008). Currently, high sensitivity GC-MS or IRMS instruments are available for assessing both steps of nitrification. Promising results have been reported in the north and south basins of the Atlantic Ocean (Clark et al. 2007; Clark et al. 2008) in the frame of the Atlantic Meridional Transect program (AMT), which is currently starting its third phase.

The list of organisms able to nitrify is no longer restricted to the originally described bacterial groups, Proteobacteria and Nitrospira. With the discovery of Chrenarchaeota ammonium oxidizers (Konneke et al. 2005; Wuchter et al. 2006), our vision of the nitrogen cycle have suffered a further twist. Not only Chenarchaota can surpass bacterial abundances in certain ecosystems (Santoro et al. 2008) but light inhibition does not seem to affect them. Furthermore, their ability for oxidizing ammonium seems to be close to rates known for ammonium oxidizing bacteria (Konneke et al. 2005).

In summary, biogeochemical measurements of nitrification have at least two major upcoming challenges: first, the relationship between nitrification and light needs to be determined precisely. Strong evidence in natural environments contradicts the photoinhibition hypothesis from previous laboratory experiments. This is probably because not all the components of the nitrifying community encountered in natural environments are sensitive to sunlight. The possible biogeochemical consequences of light exposure among nitrifying microorganisms is currently difficult to predict because of our lack of knowledge concerning the diversity of the nitrifying community. Lower nitrification rates in surface waters compared to deeper layers may be the result of substrate competition with phytoplankton, partial light inhibition, or both. As the nitrifying community expands (Coolen et al. 2007;

Beman et al. 2008), the effect of light on their metabolic performance will need to be studied in detail.

Second, it is crucial to precisely estimate the extent of the archean contribution to this process. Archaea diversity is high, and our knowledge of their control mechanisms, as well as the functional structure of their community, is still scarce. During ammonium oxidation by bacteria, energy is released and CO_2 is fixed via the Calvin-Benson cycle. In the case of archaea, the only cultured representative of AOA also fixes carbon, but presumably through the Krebs cycle. Furthermore, even if both communities provide the same final metabolic product (NO_3), little is known about the capacity of archaea to perform nitrite oxidation or to release nitrous oxide as an intermediate product as it is done by bacteria (Ward and Zafiriou 1988; Farías et al. 2007). Future research should achieve concrete results in these matters.

Conclusion

It is now clear that nitrification as a process needs to go back to our list of research priorities. The determination of the source of nitrogen species within the euphotic zone (such as nitrite and nitrate) has crucial implications for the definition of new and regenerated production and thus for our appreciation of the capacity of the ocean to sequester atmospheric CO_2 through the biological pump. The nitrogen cycle is intimately linked to climate change, and nitrogen of anthropogenic origin is now the focus of much scientific attention for elucidating its interactions with natural sources and sinks (Duce et al. 2008; Galloway et al. 2008). In order to reach a detailed understanding of the impact of anthropogenic nitrogen on natural cycles, we will need to revise some old issues that, until recently, were believed to be resolved.

References

Azam F, Fenchel T, Field JG, Gray JS, Meyer-Reil LA and Thingstad F (1983) The ecological role of water-column microbes in the sea. *Marine Ecology Progress Series* 10: 257-263.

Beman JM, Popp BN and Francis CA (2008) Molecular and biogeochemical evidence for ammonia oxidation by marine crenarchaeota in the Gulf of California. *The ISME Journal*: 1-13.

Bianchi M, Feliatra F, tréguer P, Vincendeau MA and morvan J (1997) Nitrification rates, ammonium and nitrate distribution in upper layers of the water column and in sediments of the indian sector of the southern ocean. *Deep Sea Research II* 44: 1017-1032.

Bianchi M, Morin P and Corre PL (1994) Nitrification rates, nitrite and nitrate distribution in the Almeria-Oran frontal systems (eastern Alboran Sea). *Journal of Marine Systems* 5: 327-342.

Brandes JA, Devol AH and Deutsch C (2007) New developments in the marine nitrogen cycle. *Chemical Reviews* 107: 577-589.

Capone DG, Burns JA, Montoya JP, Subramaniam A, Mahaffey C, Gunderson T, Michaels AF and Carpenter EJ (2005) Nitrogen fixation by trichodesmium spp.: An important source of new nitrogen to the tropical and subtropical North Atlantic Ocean. *Global Biogeochemical Cycles* 19: doi:10.1029/2004CB002331.

Clark DR, Rees AP and Joint I (2007) A method for the determination of nitrification rates in oligotrophic marine seawater by gas chromatography/mass spectrometry. *Marine chemistry* 103: 84-96.

Clark DR, Rees AP and Joint I (2008) Ammonium regeneration and nitrification rates in the oligotrophic Atlantic Ocean: implications for new production estimates. *Limnology and Oceanography* 53: 52-62.

Coolen MJL, Abbas B, Bleijswijk Jv, Hopmans EC, Kuypers MMM, Wakeham S and Damsté JSS (2007) Putative ammonia-oxidizing crenarchaeota in suboxic watrs of the Black Sea: a basin-wide ecological study using 16S ribosomal and functional genes and membrane lipids. *Environmental Microbiology* 9: 1001-1006.

Dalsgaard T, Canfield DE, Petersen J, Thamdrup B and Acuña-González J (2003) N_2 production by the anammox reaction in the anoxic water column of Golfo Dulce, Costa Rica. *Nature* 422: 606-608.

Diaz F and Raimbault P (2000) Nitrogen regeneration and dissolved organic nitrogen release during spring in a NW Mediterranean coastal zone (Gulf of Lions): implications for the estimation of new production. *Marine Ecology Progress Series* 197: 51-65.

Dore JE and Karl DM (1996) Nitrification in the euphotic zone as a source for nitrite, nitrate and nitrous oxide at station ALOHA. *Limnology and Oceanography* 41: 1619-1628.

Duce RA, LaRoche J, Altieri K, Arrigo KR, Baker AR, Capone DG, Cornell S, Dentener F, Galloway J, Ganesham RS, Geider RJ, Jikells T, Kuypers MM, Langlois R, Liss PS, Liu SM, Middelburg JJ, Moore SM, Nickovic S, Oschlies A, Pedersen T, Prospero J, Schlitzer R, Seitzinger S, Sorensen LL, Uematsu M, Ulloa O, M.Voss, Ward B and Zamora L (2008) Impact of atmospheric anthropogenic nitrogen on the open ocean. *Science* 320: 893-899.

Ducklow HW and Harris RP (1993) Introduction to the JGOFS North Atlantic bloom experiment. *Deep Sea Research II* 40: 1-8.

Dugdale RC and Goering JJ (1967) Uptake of new and regenerated forms of nitrogen in primary productivity. *Limnology and Oceanography* 12: 196-206.

Eppley R and Peterson BJ (1979) Particulate organic matter flux and planktonic new production in the deep ocean. *Nature* 282: 677-679.

Farías L, Paulmier A and Gallegos M (2007) Nitrous oxide and N-nutrient cycling in the oxygen minimum zone off northern Chile. *Deep Sea Research I* 54: 164-180.

Feliatra F and Bianchi M (1993) Rates of nitrification and Carbon uptake in the Rhône river plume (Northwestern Mediterranean Sea). *Microbial Ecology* 26: 21 – 28.

Fernández C (2003) Cycle de l'azote et production primaire dans l'Atlantique Nord-Est: suivi saisonnier et influence de la mésoéchelle. (p 300). Université de la Méditerranée, Marseille.

Fernández C and Raimbault P (2007) Nitrogen regeneration in the NE Atlantic Ocean and its impact on seasonal new, regenerated and export production. *Marine Ecology Progress Series* 337: 79-92.

Fernández C, Raimbault P, Caniaux G, Garcia N and Rimmelin P (2005a) Influence of mesoscale eddies on nitrate distribution during the POMME program in the northeast Atlantic Ocean. *Journal of Marine Systems* 55: 155-175.

Fernández C, Raimbault P, Garcia N and Rimmelin P (2005b) An estimation of annual new production and carbon fluxes in the northeast Atlantic Ocean during 2001. *Journal of Geophysical Research* 110: DOI:10.1029/2004jc002616.

Francis CA, Beman JM and Kuypers MM (2007) New processes and players in the nitrogen cycle: the microbial ecology of anaerobic and archaeal ammonia oxidation. *The ISME Journal* 1: 19-27.

Galloway JN, Townsend AR, Erisman JW, Bekunda M, Cai Z, Freney JR, Martinelli LA, Seitzinger SP and Sutton MA (2008) Transformation of the nitrogen cycle. recent trends, questions, and potential solutions. *Science* 320: 889-895.

Glibert PM, Lipschultz F, McCarthy JJ and Altabet MA (1985) Has the mystery of the vanishing 15N in isotope dilution experiments been solved? *Limnology and Oceanography* 30: 444-447.

Guerrero M and Jones R (1996a) Photoinhibition of marine nitrifying bacteria I: wavelength response. *Marine Ecology Progress Series* 141: 193-198.

Guerrero M and Jones R (1996b) Photoinhibition of marine nitrifying bacteria II: dark recovery after monochromatic or polychromatic irradiation. *Marine Ecology Progress Series* 141: 183-192.

Harrison WG, Platt T and Lewis MR (1987) f-ratio and its relationship to ambient nitrate concentration in coastal waters. *Journal of Plankton Research* 9: 235-248.

Horrigan S, Carlucci A and Williams P (1981) Light inhibition of nitrification in sea surface films. *Journal of Marine Research* 39: 557-565.

Karl D (2000) A new source of 'new' nitrogen in the sea. Trends in microbiology 8: 301.

Konneke M, Bernhard AE, de la Torre JR, Walker CB, Waterbury JB and Stahl DA (2005) Isolation of an autotrophic ammonia-oxidizing marine archaeon. *Nature* 437: 543-546.

Kuypers MM, Sliekers AO, Lavik G, Schmid M, Jorgensen BB, Kuenen JG, Damsté JSS, Strous M and Jetten MSM (2003) Anaerobic ammonium oxidation by anammox bacteria in the Black Sea. *Nature* 422: 608-611.

Legendre L and Gosselin M (1989) New production and export of organic matter to the deep ocean: Consequences of some recent discoveries. *Limnology and Oceanography* 34: 1374 – 1380.

Lewis MR, Hebert D, Harrison WG, Platt T and Oakey NS (1986) Vertical Nitrate Fluxes in the Oligotrophic Ocean. *Science* 234: 870-873.

Lomas MW and Lipschultz F (2006) Forming the primary nitrite maximum: Nitrifiers or phytoplankton? *Limnology and Oceanography* 51: 2453-2467.

Mémery L, Reverdin G, Paillet J and OSchlies A (2005) introduction to the POMME special section: Thermocline ventilation and biogeochemical tracer distribution in the northeast Atlantic Ocean and impact of mesoscale dynamics. *Journal of Geophysical Research* 110: doi: 10.1029/2005JC002976.

Olson RJ (1981) Differential photoinhibition of marine nitrifying bacteria: a possible mechanism for the formation of the primary nitrite maximum. *Journal of Marine Research* 39: 227-238.

Priscu JC and Downes MT (1985) Nitrogen uptake, ammonium oxidation and nitrous oxide (N2O) levels in the coastal waters of western Cook strait, New Zealand. *Estuarine Coastal Shelf Science* 20: 529-542.

Raimbault P, Slawyk G, Boudjellal B, Coatanoan C, Conan P, Coste B, Garcia N, Moutin T and Pujo-Pay M (1999) Carbon and Nitrogen uptake and export in the equatorial Pacific at 150°W: Evidence of an efficient regenerated production cycle. *Journal of Geophysical Research* 104: 3341-3356.

Redfield AC (1934) On the proportions of organic derivations in sea water and their relation to the composition of plankton. In: Daniel RJ (Ed) James Johnston Memorial Volume pp 176-192). Liverpool University Press.

Rees A, Law CS, Malcolm E and Woodward S (2006) High rates of nitrogen fixation during an in situ phosphate release experiment in the Eastern Mediterranean Sea. *Geophysical Research Letters* 33: doi10.1029/2006CL025791.

Robinson C, Poulton AJ, Holligan PM, Baker AR, Forster G, Gist N, Jickells TD, Malin G, Upstill-Goddard R, Williams RG, Woodward EMS and Zubkov MV (2006) The Atlantic Meridional Transect (AMT) Programme: A contextual view 1995-2005. *Deep Sea Research Part II* 53: 1485-1515.

Santoro AE, Francis CA, de Sieyes NR and Boehm AB (2008) Shifts in the relative abundance of ammonia-oxidizing bacteria and archaea across physicochemical gradients in a subterranean estuary. *Environmental Microbiology* 10: 1068-1079.

Savidge G and Williams PJlB (2001) The PRIME 1996 cruise: an overview. *Deep Sea Research part II* 48: 687-704.

Slawyk G and Raimbault P (1995) Simple procedure for simultaneous recovery of dissolved inorganic and organic nitrogen in 15N-tracer experiments and improving the isotopic mass balance. *Marine Ecology Progress Series* 124: 289-299.

Slawyk G, Raimbault P and Garcia N (2000) Use of 15N to measure dissolved organic nitrogen release by marine phytoplankton (reply to comment by Bronk and Ward). *Limnology and Oceanography* 45: 1884-1886.

Takahashi T, Feely RA, Weiss RF, Wanninkhof RH, Chipman DW, Sutherland SC and Takahashi TT (1997) Global air - sea flux of CO2: An estimate based on measurements of sea - air pCO2 difference. *PNAS* 94: 8292-8299.

Takahashi T, Olafsson J, Goddard JG, Chipman DW and Sutherland SC (1993) Seasonal variations of CO2 and nutrients in the high latitude surface oceans: a comparative study. *Global Biogeochemical Cycles* 7: 843-878.

Takahashi T, Sutherland SC, Sweeney C, Poisson A, Metzl N, Tillbrook B, Bates N, Wanninklof R, Feely RA, Sabine C, Olafson J and Nojiri Y (2002) Global sea-air CO2 flux based on climatological surface ocean pCO2 and seasonal biological and temperature effects. *Deep Sea Research II* 49: 1601-1622.

Ward B (2005) Temporal variability in nitrification rates and related biogeochemical factors in monterrey Bay, California, USA. Marine Ecology Progress Series 292: 97-109.

Ward BB, Capone DG and Zehr JP (2007) What's new in the nitrogen cycle? *Oceanography* 20: 101-109.

Ward BB, Kilkpatrick KA, Renger EH and Eppley RW (1989) Biological nitrogen cycling in the nitracline. *Limnology and Oceanography* 34: 493 – 512.

Ward BB and Zafiriou OC (1988) Nitrification and nitric oxide in the oxygen minimum of the eastern tropical north Pacific. *Deep Sea Research* 35 7: 1127-1142.

Wuchter C, Abbas B, Coolen MJL, Herfort L, Bleijswijk Jv, timmers P, Strous M, Teira E, Herndl GJ, Middelburg JL, Schouten S and Damstçe JSS (2006) Archaeal nitrification in the ocean. *PNAS* 103: 12317-12322.

Yool A, Martin A, Fernández C and Clark D (2007) The significance of nitrification for oceanic new production. *Nature* 447: 999-1002.

In: Atlantic and Indian Oceans ISBN 978-1-60692-475-4
Editors: E.S. Askew and J.P. Bromley, pp. 211-215

Expert Commentary

CHALLENGES IN THE TROPICAL ATLANTIC: UNDERSTANDING ITS INTERANNUAL TO DECADAL VARIABILITY

Annalisa Bracco[1,*] ***Fred Kucharski***[2] ***and Brad E. Rosenheim***[3]
[1]EAS, Georgia Institute of Technology, Atlanta, Georgia, USA
[2]The Abdus Salam International Centre for Theoretical Physics, Earth System Physics Section, Trieste, IT
[3]Dept. of Earth and Environmental Sciences, Tulane University, New Orleans, Louisiana, USA

Abstract

Atmospheric teleconnections between the tropical Atlantic and the African continent, extending to the Indian Ocean basin, modulate the interannual variability of precipitation in highly vulnerable and densely populated regions like the Sahel and the Indian peninsula. Additionally, increasing evidence supports the notion that in the Atlantic basin low-latitude, wind-driven convection cells contribute to the transport of heat and salt potentially interacting with the Atlantic Meridional Overturning Circulation (AMOC). Nevertheless, while the progress in our understanding of high latitude oceanography has been remarkable, the nature and causes of the variability of tropical Atlantic complex dynamics are poorly understood. Alone, observations and interpretations of Atlantic tropical variability have been controversial. Debate about interpretations of both trends and natural cycles within the observational data set is expected because the instrumental record is not long enough and too sparse. Furthermore, tropical Atlantic sea surface temperatures (SSTs) have been identified as a major source of errors in coupled general circulation models. It is therefore extremely difficult to discern the interannual variability signal, which is potentially linked both to the Atlantic Ocean internal dynamics and to the El Nino Southern Oscillation (ENSO), from longer period oscillations such as the Atlantic Multidecadal Oscillation (AMO) and secular trends perhaps related to anthropogenic activity. Here we explore recent research opportunities and challenges that the scientific community is facing in the

*E-mail address: abracco@gatech.edu. (Corresponding author address: School of Earth and Atmospheric Science, Georgia Institute of Technology, 311 Ferst Dr, Atlanta GA 30332, USA.

Tropical Atlantic and we suggest possible avenues of collaboration between atmospheric scientists and physical and paleo-oceanographers, linking modeling efforts and oceanographic observational campaigns with the goal of promoting a coordinated and multi-disciplinary research effort in this region.

1. Commentary

Recent studies using observational data-sets and coupled atmosphere-ocean models have demonstrated that sea surface temperatures (SSTs) in the tropical Atlantic modulates the interannual variability of precipitations over Africa and the Indian subcontinent. In particular Vizy and Cook (2001) and Giannini et al. (2003) found a direct correlation between tropical Atlantic ocean temperatures and precipitations along to the Gulf of Guinea coast, while, more recently, Kucharski et al. (2007) and Kucharski et al. (2008) have shown that tropical Atlantic SSTs influence the Indian monsoon interannual variability contributing to the decadal modulation of the El Nino Southern Oscillation (ENSO) - Indian monsoon relationship. As explained in Kucharski et al. (2009), the teleconnection over Africa and the Indian Ocean is, to a first approximation, a simple Gill-Matsuno-type response of the atmosphere to a heating anomaly in the equatorial Atlantic region (Matsuno 1966; Gill 1980). In simple terms, the heating anomaly excites two waves in the atmosphere: A tropically trapped, fast moving, Kelvin wave, which transports the signal from the heating source to the east, and a slower Rossby wave, responsible for the westward teleconnection. The resulting response projects largely onto the time-mean circulation, and a warm (cold) tropical Atlantic SST anomaly may weaken (strengthen) the time-mean Walker circulation. Obviously, the existence of such a teleconnection has great potential for improving our capabilities to predict variations of rainfall in poor and densely populated areas at seasonal iscales. Practically, though, a good understanding of the physical processes influencing the variability of Tropical Atlantic SSTs is required.

The circulation in the Tropical Atlantic is particularly complicated. North of the equator, the North Atlantic imports a large amount of heat (1.4 PW) from the tropical latitudes. The majority of it is lost sensibly to the atmosphere as warm saline waters are shuttled into the high latitudes by the Gulf Stream. The result is the formation of a cold, salty water mass that convects downward and drifts gradually southwestward to the Bermuda rise, mixing with Antarctic Bottom Water (Luyten et al., 1993) and Mediterranean Outflow to establish the characteristic deep water tongue known as the North Atlantic Deep Water (NADW). Water in the subtropical gyre that is not removed from the surface with the NADW is re-circulated diffusely southward along the eastern margins of the Atlantic via the North Equatorial Current and eventually re-enters the mid-latitudes after mixing with equatorial surface waters. Once under the evaporative, sinking leg of the atmospheric Hadley cell, the surface salinity of this water mass increases as a result of latent heat loss. Here, the surface ocean reaches its salinity maximum and this Salinity Maximum Water (SMW) is subducted into a shallow tongue of warm, salty water, known as the subtropical underwater. The formation of this dense water mass releases 0.1 - 0.4 PW of heat (Klinger and Marotzke 1999; Talley 2003). In the South Atlantic a complementary, and possibly more complicated, picture can be depicted. There the meridional overturning circulation (MOC) carries about 16 Sv of warm, intermediate water northward to the mid-latitutes via the South Equatorial Current. Part of this water mass then crosses into the northern hemisphere via the North Brazilian Current, in the alongshore currents along Guiana and in the North Brazilian Eddies (Frantantoni et al. 2000; Ganachaud and Wunsch, 2002; Schott et al., 1998), compensating the southward drift of NADW. The surface water in the subtropical cell contributes to the northward transfer of water with its surface layer, while water below the mixed layer is likely advected southward. A complex structure of zonal currents near the equator, together with mesoscale eddies along the western boundary, and air-sea interactions along the coast of Africa with intensification by local physical processes, including slow Rossby wave propagation, Ekman pumping and Bjerknes feedbacks (Hu and Huang, 2007), complete the circulation.

Based on the current knowledge of the Tropical Atlantic dynamics, several mechanisms can explain changes in the heat transport and heat content of the subtropical cells. If tradewinds strengthen and evaporation over the existing SMW pool becomes more intense, higher relative salinity will be accompanied by more vigorous Ekman pumping. Similarly, an increase in the size of the SMW pool due to a larger area of persistent evaporation can increase the amount of water unstable enough to subduct. In both cases more latent heat is exchanged to the atmosphere and several outcomes can be foreseen. If the subtropical cells simply become more vigorous the system must compensate the subduction with an increase of Ekman pumping. Otherwise, either density surfaces will heave to greater depths or the SMW will achieve greater stability -via less salinity gradient- allowing regional salinity anomalies to be carried around the subtropical gyres. Mesoscale feautures, eddies and Rossby waves, can propagate the anomalies in ways not yet fully understood.

From the above digression it is evident that variability in the tropical Atlantic circulation has to be studied in conjunction with high-latitude oceanography and that the low latitude, wind-driven convection cells contribute to the transport of heat and salt in interaction with the Atlantic MOC. Therefore the release of heat into the ocean-atmosphere system, accelerated by emissions of anthropogenic CO_2, may not only affect the MOC through accelerated melting of polar ice, but it may also modify the interplay between the deep MOC and the wind-driven poleward heat transport at lower latitudes, as shown by idealized and high resolution ocean models (eg. Hazelger and Drijghout, 2006; Pasquero and Tziperman, 2004). Recent observations confirm this scenario suggesting that the Atlantic shallow overturning subtropical cells (STCs) have increased complimentary to a decrease in high-latitude deep overturning (Bryden et al., 2005; Cunningham and Alderson, 2007; Curry et al., 2003). Interpretations of Atlantic tropical variability from observations, however, is controversial, due to spatially rich but temporally poor datasets.

The lack of oceanographic observations in the tropical Atlantic has not escaped the modeling literature. As noted by Brandt et al. (2006) "reliable transports and seasonal cycle analyses of the subsurface flows from the interior tropical Atlantic that could serve as a calibration base for model simulations have not become available." One exception is the few moored Acoustic Doppler current profilers (ADCPs) deployed since 1998 as part of the Pilot Research Moored Array in the Tropical Atlantic (PIRATA) project (Servain et al. 1998). But those time-series are too short to identify trends from natural variability, or to help calibrate General Circulation Models (GCMs). In this context Stockdale et al. (2006) investigated the capabilities of seven of the latest generation of coupled GCM seasonal forecast systems to predict tropical Atlantic SST anomalies over the period 1982-2001 as part of the DEMETER (Development of a European Multimodel Ensemble System for Seasonal to Interannual Prediction) project. Little to no predictive skill was found for all models, with worse performances south of the equator. The models did not even correctly simulate the mean seasonal cycle. In comparison, the same models performed better in both the tropical Pacific and Indian oceans. Analyzing the causes of model failure in the tropical Atlantic, the authors identify the poor quality of ocean initial conditions and the lack of good quality in-situ data, together with model imperfections in representing convection and winds over the Atlantic. Difficulties in modeling Tropical Atlantic variability are also evident in the intercomparison of third Coupled Model Intercomparison Project (CMIP3) dataset performed by Biasutti et al. (2008) for both XX and XXI century, under two different emission scenarios. Different models produce very different responses for the XXI century, and the authors suggest that aerosols may play a fundamental role in determining SST changes in the Tropical Atlantic in response to antrophogenic activities.

The scientific community has a strong need of supplementing instrumental records in the tropical Atlantic. This can be achieved in three ways: 1) by collecting data for a sufficiently long time period, 2) by looking to the past for proxy records capable of expanding the time domain of the instrumental record, or 3) by simulating natural systems with numerical models. Whereas it is likely that the former option will provide unparalleled quality of data, we do believe that the latter two options, in concert, will offer the best opportunity to understand the dynamics of tropical Atlantic variability, to improve our seasonal forecasting capabilities and, potentially, to adapt to and mitigate the effects of the anthropogenic climate change without sacrificing any sense of urgency. This will require a collaborative and synergistic effort between the physical-, paleo- oceanography and atmospheric science communities and we strongly encourage interdisciplinary projects to address some of the open questions outlined.

References

Biasutti, M.; Held, I. M.; Sobel, A. H., Giannini A. *J. Climate* 2008, vol. 21, 3471-2486.

Brandt, P.; Schott, F.A.; Provost, C.; Kartavtseff, A.; Hormann, V.; Bourles, B.; Fisher, *J. Geoph. Res. Letters* 2006, vol. 33, L07609, doi:10.1029/2005GL025498.

Bryden, H.L.; Longworth, H.R.; Cunningham, S.A., *Nature* 2005, vol. 438, 655-657.

Cunningham, S.A.; Alderson, *S. Geoph. Res. Letters* 2007, vol. 34, doi:10.1029/2007GL029821.

Curry, R.; Dickson, B.; Yashahaev, I. *Nature* 2003, vol. 426, p. 826-829.

Frantantoni, D.M.; Johns, W.E.; Townsend, T.L.; Hurlburt, H.E. *J. Physical Ocean* 2000, vol. 30, 1944-1966.

Ganachaud, A.; Wunsch, C. *J. Climate* 2002, vol. 16, 696-705.

Giannini, A.; Saravanan, R.; Chang, P. *Science* 2003, vol. 302, 1027-1030

Gill, A.E. *Quart. J. Roy. Meteor. Soc.* 1980, vol. 106, 447-462.

Hazeleger, W.; Drijfhout, S. *J. Geophys. Res.* 2006, vol. 111, doi:101029/2005JC002942.

Hu, Z.-Z.; Huang, B. *J. Climate*, 2007, vol. 20, 3366-3378.

Klinger, B.A.; Marotzke, J. *J. Phys. Ocean.*, vol. 29, 382-399.

Kucharski F.; Bracco A.; Yoo J. H.; Molteni F. *J. Climate* 2007, vol. 20, 4255-4266

Kucharski F.; Bracco A.; Yoo J. H.; Molteni F. *Geophys. Res. Lett.* 2008, vol. 33, L04706, doi:10.1029/2007GL033037

Kucharski F.; Bracco A.; Yoo J. H.; Tompkins A.; Feudale L.; Ruti P.; Dell'Aquila A. *Quart. J. Roy. Meteor. Soc.* 2009, In Press.

Matsuno, T. J. *Meteor. Soc. Japan* 1966, vol. 44, 25-43.

Luyten, J.; McCartney, M.S.; Stommel, H.; Dickson, R.; Gmitrowicz, E. *J. Phys. Ocean.* 1993, vol. 23, 1885-1892.

Pasquero, C.; Tziperman, E. *J. Phys. Ocean.* 2004, vol. 34, 805-816.

Schott, F.A.; Fisher, J.; Stramma, L. *J. Phys. Ocean.* 1998, vol. 28, 1904-1928.

Servain, J.; Bussalacchi, A.J.; McPhaden, M.J.; Moura, A.E.; Reverdin, G.; Vianna, M.; Zebiak, S.E. *Bull. Amer. Meteor. Soc.* 1998, vol. 79, 2019-2031.

Stockdale, T.N.; Balmaseda, M.A.; Vidard, A. *J. Climate,* vol. 19, 6047-6061.

Vizy E.K.; Cook K. H. *J. Climate* 2001, vol. 14, 795-821

INDEX

A

B

C

D

E

F

G

H

I

N

O

P

T